PROCEEDINGS OF THE INTERNATIONAL SYMPOSIUM HELD AT ASIAN
INSTITUTE OF TECHNOLOGY / BANGKOK / 1-15 DECEMBER 1980

# Geotechnical Problems and Practice of Dam Engineering

*Edited by*
A.S.BALASUBRAMANIAM / YUDHBIR / A.TOMIOLO
*Asian Institute of Technology, Bangkok*

J.S.YOUNGER
*Vallentine Laurie & Davies, Bangkok*

A.A.BALKEMA / ROTTERDAM / 1982

*The texts of the various papers in this volume were set individually by typists under the supervision of each of the authors concerned.*

ISBN 90 6191 265 2

© 1982 A.A.Balkema, P.O.Box 1675, 3000 BR Rotterdam, Netherlands
Distributed in USA & Canada by: MBS, 99 Main Street, Salem, NH 03079, USA
Printed in the Netherlands

*Symposium on Problems and Practice of Dam Engineering / Bangkok / 1-15 December 1980*

# Table of contents

| | |
|---|---|
| Preface | VII |

## Section A: *Principles, problems and practice of dam engineering*

| | |
|---|---|
| Practice, precedents, principles, problems, and prudence in embankment dam engineering<br>*Victor F.B.de Mello* | 3 |
| Lessons from earth dam failures<br>*P.Londe* | 19 |
| Recent trends in the design of earth-rock dams<br>*Piero Sembenelli* | 29 |
| Earthquake-resistant design of earth dams<br>*H.Bolton Seed* | 41 |
| Comparative behaviors of similar compacted earthrock dams in basalt geology in Brazil<br>*Victor F.B.de Mello* | 61 |
| Some problems and revisions regarding slope stability assessment in embankment dams<br>*Victor F.B.de Mello* | 81 |
| Some examples of design changes required during construction<br>*J.R.Hunter* | 99 |
| Dam engineering activities in Japan<br>*E.Mikuni* | 109 |
| The problems and practice of dam engineering in India<br>*Yudhbir* | 131 |
| The problems and practice of dam engineering in Hong Kong<br>*A.J.Vail & D.J.Eastaff* | 145 |

## Section B: *Investigations, instrumentation and foundation treatment*

| | |
|---|---|
| Geotechnical investigations for dams<br>*A.C.Meigh* | 163 |
| Instrumentation requirements for earth and rockfill dams<br>*A.D.M.Penman* | 183 |
| Grouting works in dam engineering<br>*S.Marchini* | 211 |
| Seepage through jointed rock – Sealing and drainage measures for earthfill and masonry dams<br>*K.H.Idel* | 225 |

## Section C: *Environmental considerations, geologic control and case histories of dam construction*

| | |
|---|---:|
| Environmental parameters in planning and operation of multi-purpose reservoir projects<br>*H.F.Ludwig* | 235 |
| Construction works and geology<br>*Hikoji Takahashi* | 241 |
| Systematic weak seams in dam foundations<br>*Ranji Casinader* | 253 |
| Use of Roller Compacted Concrete in dam construction<br>*Jack C.Jones & Gary R.Mass* | 265 |
| Geotechnical aspects of the Larona Hydroelectric Project in Sulawesi, Indonesia<br>*Richard L.Kulesza* | 281 |
| Core material for Lahor Dam<br>*Soerjono* | 295 |
| Stability of rock cavern for underground pumped-storage power stations<br>*Keiichi Fujita* | 301 |

## Section D: *Country reports*

| | |
|---|---:|
| Dam engineering in Pakistan<br>*Amjad Agha* | 311 |
| Dam engineering in Philippines<br>*Eduardo P.Abesamis* | 347 |
| Dam engineering in Taiwan<br>*David S.L.Chu* | 359 |
| Dam engineering in Indonesia<br>*Soerjono* | 371 |
| Dam engineering in Singapore and west Malaysia, 1967-1979<br>*Hooi Kah Hung* | 383 |
| Dam engineering in Korea<br>*S.K.Kim & W.T.Kim* | 387 |

*Symposium on Problems and Practice of Dam Engineering / Bangkok / 1-15 December 1980*

# Preface

A large dam is the most complex and hazardous of all civil engineering structures. Designing and building a dam successfully relies more on art than on science. There is still a continuing debate in profession as to the places of precedent, theory and judgement in the design of dams. In the fifth Laurits Bjerrum memorial lecture, Professor Peck asserted that "engineering judgement and earth dam-design go hand in hand", and "modern dams seldom if ever fail because of incorrect or inadequate analyses..., they fail because inadequate judgement is brought to bear on the problems...". Professor Peck would like to see that dams are "designed and constructed not to fail, even if a probability of failure is incorporated into the benefit-cost analysis". He has warned that the achievement of such an objective "does not depend on the acquisition of new knowledge,...... it depends on our ability to bring the best engineering judgement to bear on problems that are essentially non quantitative, having solutions that are non-numerical".

This memorial lecture in honour of the memory of Laurits Bjerrum was delivered on May 5, 1980 and it certainly appears more than appropriate that the International Symposium on Problems and Practice of dam engineering, held at AIT, Bangkok in December of the same year focussed attention on some of the issues raised by Professor Peck.

This volume presents contributions made at this symposium. Mr. Pierre Londe, President of the Commission on Large Dams, in his presentation on Lessons from Earth Dam Failures, focussed attention on the fact that failure of a dam is the result of a complex concourse of causes and mechanisms, which should be interpreted with extreme care. Professor Victor F.B. de Mello, currently President of the International Society of Soil Mechanics and Foundation Engineering, made stimulating contributions on "Practice, Precedents, Principles, Problems and Prudence in Embankment Dam Engineering"; "Comparative Behaviour of Similar Compacted Earth-Rock Dams in Basalt Geology in Brazil", and "Some Problems and Revisions Regarding Slope Stability Assessment in Embankment Dams". Professor Seed reviewed the state-of-the-art regarding "Earthquake-Resistant Design of Dams", and Dr. Sembenelli reviewed the "Recent Trends in the Design of Earth-Rock Dams". Dr. Meigh and Dr. Penman presented state-of-the-art reports on "Geotechnical Investigations for Dams", and "Instrumentation Requirements for Earth and Rockfill Dams" respectively. Dr. Marchini and Dr. Idel dealt with foundation treatment aspects, viz; "Grouting works in Dam Engineering" and "Seepage through Jointed Rock-sealing and Drainage Measures for Earthfill and Masonary of Dams" respectively. Hunter, Mikuni, Professor Yudhbir, Vail and Eastaff presented critical reviews of the practice of dam engineering under different geologic environments. Topics on environmental considerations, geologic control and case histories of dam construction were discussed by Ludwig, Takahashi, Casinader, Jones and Mass, Kulesza, Soerjono and Fujita. Country reports on problems and practice of dam engineering were presented by Agha, Abesamis, Chu, Soerjono, Hung and Kim for Pakistan, the Philippines, Republic of China, Indonesia, Singapore and West Malaysia, and Korea, respectively.

A total of twenty seven contributions are reported in this volume. For ease of reference these contributions are divided into four sections: Section A - Priciples, Problemes and Practice; Section B - Investigations, Instrumentation, and Foundation Treatment; Section C - Environmental considerations, Geologic Control and Case Histories of Dam Construction; Section D - Country Reports.

The editors wish to express their appreciation to several sponsoring agencies and individuals who contributed to the organization of this symposium. Special thanks are due to Dr.Chai Muktabhant, Mr.Nibon Rananand, Dr.Za-Chieh Moh, Dr.E.W.Brand, Dr.Tan Swan Beng and Prof.Chin Fung Kee and the General Committee Members of SEAGS. Thanks are also due to Professor R.B.Banks, President of AIT, Professor J.H.Jones, Dr.Peter Brenner, Dr.Hideki Ohta, Dr.Prinya Nutalaya, and Dr.F.Prinzl. To Mr.Sataporn Kuvijitjaru, Mrs.Vatinee Chern, Mrs.Uraivan Singchinsuk and Miss Suporn Arunyakanont, the editors are grateful for their help in the preparation of this volume.

Editors:

A.S.Balasubramaniam
Yudhbir
A.Tomiolo
J.S.Younger

# Section A:
# Principles, problems and practice of dam engineering

# Practice, precedents, principles, problems, and prudence in embankment dam engineering

VICTOR F.B.DE MELLO
*Consulting Engineer, Sao Paulo, Brazil*

SYNOPSIS. Practice and precedents are often quoted as supports for design decisions: this can be very wrong and dangerous unless case histories are analysed under tenable theoretical principles. However, even in the analysis of case histories we must be the devil's advocates in resisting the straightjacketing imposed by a given mental model against which an important unusual fact may be helpless. Examples are given of common fallacies. Some of the principal problems of design, materials selection, and construction specifications are shown to be formulated under intuitions that do not resist analysis. Examples of present most recognized problems are described. The place of statistics and determinism in design decisions are compared. The very fact that our theorization for quantification presupposes statistics of averages whereas catastrophic failures are events closer to extreme values, imposes a need for prudence in choice of design by physical model, such that a feared misbehavior be virtually excluded, in advance of any design computations.

## 1. INTRODUCTION

Practice and Precedents have been much lauded by the "experienced engineer" as the dictates of good design and construction in civil engineering. But it is herein emphasized that such exalting fails to recognize the true nature of Man as an animal, and of Society as an inexorably impelling corollary. Moreover, both Practice and Precedents always presuppose some hypotheses of Principles. The very desire to repeat a design embodies a principle, for instance, that what is, is good. If Design is Decision despite Doubts, it is inevitable that Decision is catalysed by Desire, and Desire is seldom (or never?) random.

Thereupon, Principles is what should dictate our approach to designs. Principles have been generated step-by-step, which is an inevitable burden: but at the other extreme lies the beckoning light that Principles really represent an abstraction of idealized knowledge and wisdom, applicable to a wide range of cases. So, hypothetically, at each instance we have the obligation, and the means, not merely to go from the particular to the general case, but also from the general to the particular. For the past decades we have been continually alerted that Principles are not deterministic, but statistical. I shall put forth my brief recommendations regarding the place of statistics in Principles of Design. Truly, despite the decades of warning, statistical thinking has not really begun to flavour our handling of theoretical Principles.

Thereupon, on the basis of present fairly well accepted Principles I shall discuss what seem to be the main Problems faced in embankment dam engineering today: many problems were illusions, some were even purposely cultivated, some were generated indirectly out of the best of intentions regarding other problems, some have not been recognized or honestly faced.

Finally my recommendations are of Prudence in the advance of notions, and of humility in recognizing what are personal errors, collective errors of the state-of-the-art, and what are situations inevitably beyond the reasonable duty of an Engineer, because optimization cannot possibly condone with the presumption of protecting against any and every possibility of problem.

Case histories represent an indispensable background to such a presentation. It is emphasized, however, that even in the analyses of case histories one must lean ever backwards in compensation for the strong interference of historical, geographical, contingency and subjective elements, in the very recording and transmission of would-be "facts".

At any given moment, if we are able to advance our Principles and Prudence far enough to accomodate the probable advance of Problems foreseeable, we should be treading a firm path in engineering Practice and staking out of Precedents.

## 2. PRACTICE

Practice plays a very fundamental role in any technology and/or engineering endeavour. It implies the distilling of experience into the so-called "common sense", and consequent prescriptions. Things have been done in a certain manner, and presumably would be most satisfactory if they continued being done in a similar manner. It is a very ponderous argument in the face of professionals of other branches (i.e. the lay in the specific specialty), obliged to judge, select, decide. Since most dam engineers (owners, designers, contractors, and consultants) have to act at the call, decision, and acceptance of other professionals (administrators, politicians, bankers, planners, and so forth) it is very important to emphasize some of the gross fallacies in the simple arguments in favour of Practice. Practice does imply a theory and Principle, and probably the most foolish of all: that what is, can continue to be, satisfactorily, without our facing the need to analyse and understand.

Practice is never static, but changes and grows continually. It is strongly influenced both by historical conditions, by temporary conditionings, and by the continued call of the nature of Man and Society to new challenges and to pushing forward the frontiers of impunity. Thus, whereas in principle Practice would embody the respectful constant repetition of things done, in practice it contradicts itself by always serving to push forward, hopefully by imperceptible increments. And sooner or later we are faced with the last straw that breaks the camel's back, a given Practice has been over-extended to the point where other parameters have become more conditioning.

Let us consider just two obvious examples.

### 2.1 Plasticity

When we imagined that it was of interest to employ earthcore materials of "high plasticity" we automatically fell into several fallacies historically comprehensible and pardonable, which must first be brought out by self-analysis and honest confession.

What was really desired was the ability of the material to deform to large strains without fissuring (First Approximation).

(a) Desired = high Plasticity ≈ (1) high deformability without fissuring.

Setting aside the criticisms on the (yet present) index tests on liquid and plastic limits (Plasticity Index Tests) as presumably applicable to compacted clay embankments, the first obvious association (merely due to the identity of the word Plasticity) was to assume that good plasticity was associated with a high Plasticity Index PI% (Fig. 1).

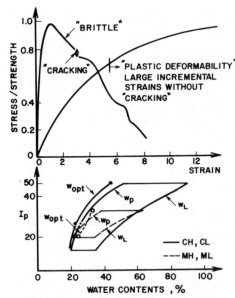

FIG.1 GOOD SOIL PLASTICITY VS. HIGH PI-THE FIRST FALLACY

(b) Automatic Association: high Plasticity ∴ high PI% (1) However, PI = $W_L - W_p$ = = range of water content $\Delta W$ over which the soil is "plastic".

And, core is compacted at some given water content
$W_c \approx W_{opt}$ (suppose)

(c) What logic is there in the association derived from reshuffling of words?
Soil "plastic" over big range $\Delta W \stackrel{?}{=}$ soil of big "plastic" stress-strain behavior at given $W_c$.

(d) At any rate, assuming that $W_c \approx W_{opt}$, a second approximation reasoning could be to compare $W_{opt}$ vs. $W_p$ (de Mello, 1973) since $W_p$ is an indicator of W at "similar" tendency to fissuring under major straining under atmospheric pressure, that is, simulat ing conditions near the crest or surface. And one can reason why in critical zones of core contact we gain by using $W_c > W_{opt}$ (e.g. $W_c \approx 1.1 W_{opt}$) is indirectly the desire to be at water contents above the plastic limit (Fig. 2).

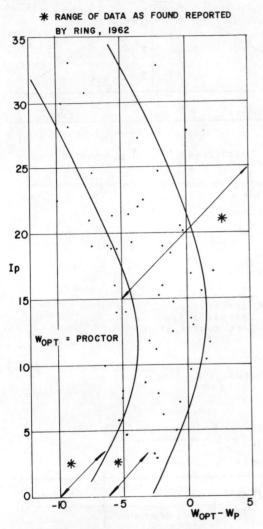

FIG.2 DATA FROM CLAYEY RESIDUAL SOILS. COMPARISON OF WATER CONTENTS AT OPTIMAM PROCTOR COMPACTION, WITH RESPECTIVE PLASTIC LIMIT.

(Second Approximation) Desired $W_c > W_p$ (2)

(e) Finally, we must still reason that what was and is really desired is not connected with any generalized fissuring under large strain, but principally protection against tensile cracking beyond some moderate strains.

Therefore, everything that causes tensile cracking upstream-downstream should matter. Of course, that includes the stress-strain curve, and thus directly includes (i) initial stresses built-in by compaction and partly retained (ii) the changes of stresses (iii) change of consequent strains (with time).

But, what compelling association is there with the stress-strain curve of a conventional "triaxial" (i.e. really biaxial) test? Even in such tests, has it not been repeatedly demonstrated that deformability moduli are quite different in extension vs. compression tests?

(f) And, if we are truly concerned with tensile cracking strains, is $\Delta$(total external stress) or $\Delta$(overburden stress) the only agent causing them? What about shrinkage, collapse, solution, colloid chemical action in void structure, and other volumetric strains generated quite independently of external (easily recognized) stress changes?

(g) As a first step, what is most effective is to resort to a dominant physical change of statistical universe: for instance to design so that only compressions and shears can occur, and/or to design for use of a material of such low shearing strength that movements and distortions are entirely taken up by shear. One cannot "crack" in tension a body of liquid (s = 0) or of pure cohesionless sand (s = 0 at $\sigma'$ = 0).

(h) To conclude the design discussion, let us summarily apply my DESIGN PRINCIPLE 5, DP5 (de Mello, 1977): "For every behaviour desired and assumed check what happens, of consequence, if it is not successful". In mentally checking what would happen if we accept that some tensile cracking might still occur, we would promptly recognize that what we really want is high erosion-resistance of the clay (coupled with moderate erodibility and selective clogging ability of the upstream cohesionless transition): that means high cohesive strength which depends on $\phi'$ and the compaction preconsolidation pressure $p_c$.

(i) In short, in revising the simple primitive Practice of requiring a high PI material, it appears that a heavy (high $p_c$) wet ($W_c > W_p$) compaction of a material of high PI and low $W_L$ (presumed higher $\phi'$) is a present approach to the desirable core material towards the top (where cracking can become tensile rather than shear). But consequent "rigidity" is highly undesirable if the top of the core be subject to delayed differential settlements. (Fig. 3)

And there are absolutely no test data supporting such intuitions on an all-important material detail. How to optimize

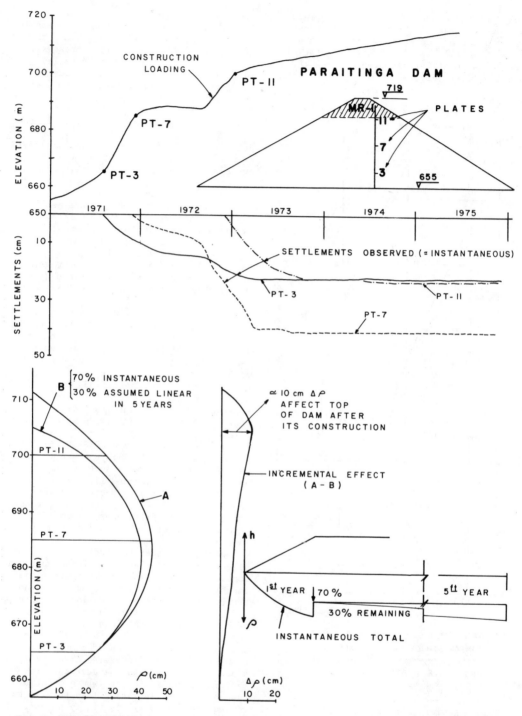

FIG. 3 IMPORTANCE OF DEFERRED SETTLEMENTS ON BEHAVIOR AT TOP OF DAM (EXAMPLE FROM PARAITINGA DAM DATA)

between frequently conflicting requirements? Engineering of dams must go on, while research institutions delay in furnishing the needed backup.

## 2.2 Dominance of visual-tactile culture

Practice is dominantly influenced by visual impressions, i.e. impressions at the time of building the dam, under visual-tactile observations that "are not more than skin-deep" (at $\sigma = 0$). Three obvious factors of such thinking have been mentioned (among others).

### 2.2.1 Homogeneity

Practice has automatically assumed that a material that is placed and constructed "homogeneous" will continue to behave as homogeneous (irrespective of being subjected inexorably to changes under different stress trajectories). However, it is obviously quite to the contrary, because any material that is constructed homogeneous, but before operation is subjected to different stresses and strains, will during operation behave as dutifully non-homogeneous. One first example to be noted concerns flownets. The idealized theories (e.g. flownet) required our assuming that a compacted clayey dam constructed as a "homogeneous section" would have a constant permeability across the section: inexorable fallacy (de Mello, 1977). Since the material compresses (settles) to different void ratios, it may indeed follow homogeneous laws of behaviour e vs. $\sigma$ and k vs. e, but ipso facto the body of the dam becomes law-abidingly "heterogeneous" in permeability. (Fig. 4)

Fig. 4 (Salto Santiago dam) presents data that are being noted more regularly in higher dams, indicating that a disproportionate amount of head loss takes place close to the inner end of the core. The fact that in upstream-inclined cores the compression, and therefore imperviousness, increases significantly towards the downstream face, would make such behavior reasonable. Of course, the seepage effective stresses will further affect the permeabilities and the flownet through a "secondary" effect.

A further very significant example of absolutely false intuitions of homogeneity concerns compacted sound rockfills, that have been mentally associated to a "big-size uniform sand". Fig. 5 gives data from the Salto Santiago dam confirming what true rockfill designers and builders know well, that each layer comprises two distinct sublayers, in grainsize and densities. A

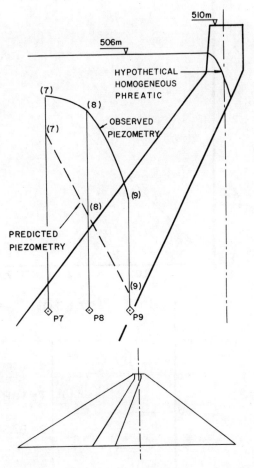

FIG. 4   SALTO SANTIAGO

better visualization of compacted rockfill is thus of a layered material. Figs. 6, 7, 8 concern the additional question of compression behavior of rockfills, and therefore, how far from homogeneous the mass will finally behave. Routine calculations on rockfill compressions (for moduli E) have assumed incremental stresses as directly the additional height of fill $\gamma h$ above the point: it is important to recognize the stress transmission influence factor I, and we may well use (Fig. 6) such factors from elastic solutions for the small increments. One immediate conclusion is that many a "delayed settlement"

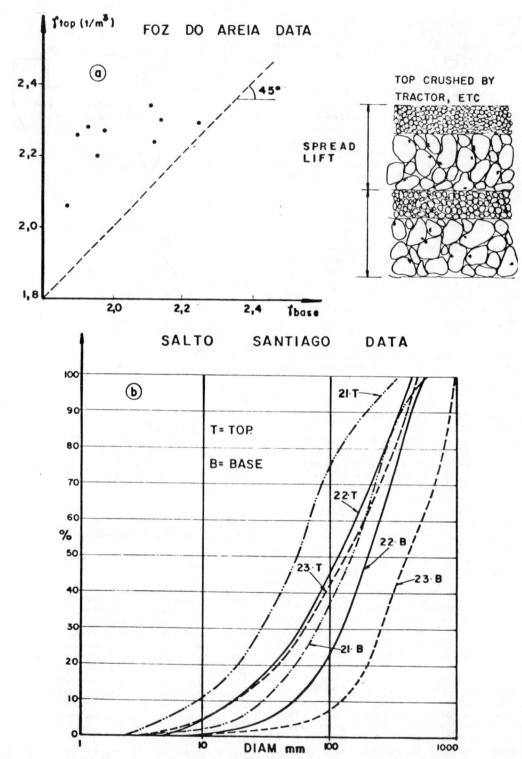

FIG.5 COMPARATIVE DENSITY TOP-BASE ON ROCKFILL LIFTS

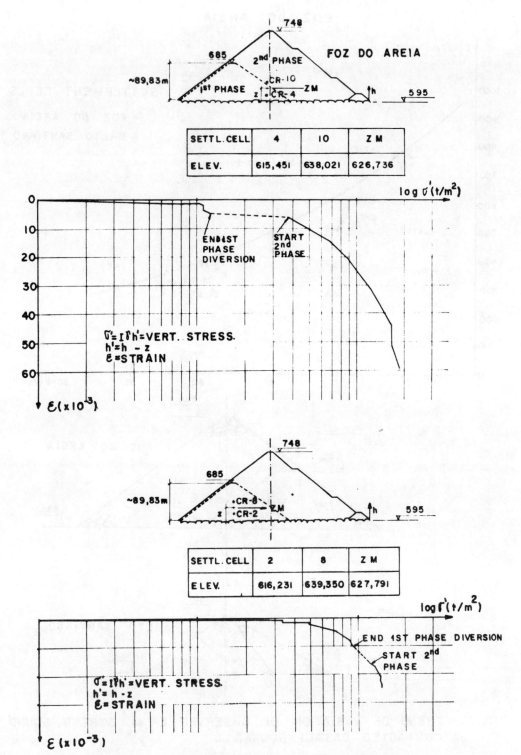

FIG. 6 DETAIL STRESS- STRAINS COMPUTED EXEMPLIFYING.

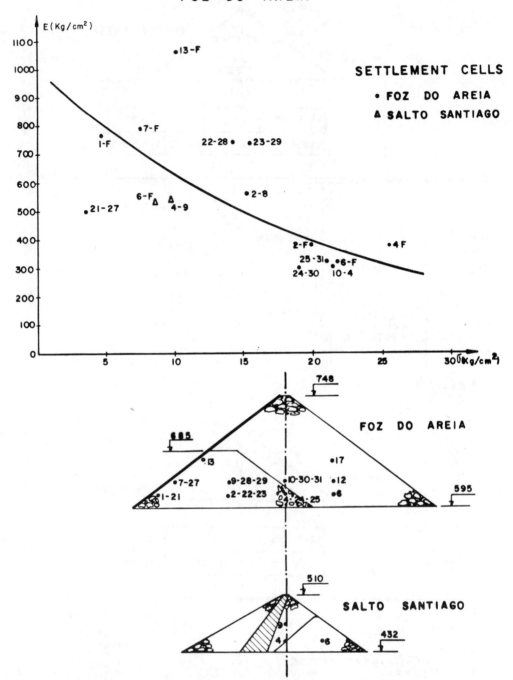

FIG. 7 TREND OF VARIATION OF OBSERVED E vs. STRESS, SOUND COMPACTED BASALT ROCKFILLS.

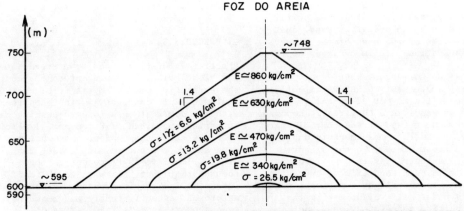

FIG. 8 FIRST CYCLE APPROXIMATION STRESSES FROM HOMOGENEOUS ELASTIC SOLUTION (POULOS + DAVIS) AND MODULI E VARYING "AS OBSERVED"

attributed to secondary compression, because fill directly above the point on the same vertical had ceased, is really found to be due to incremental stresses thrown by fill rising nearby. Another point, mentioned in an accompanying paper, is the apparent precompression pressure due to compaction. Finally Figs. 7 and 8 exemplify the very significant change of E with pressure, and therefore the dutifully non-homogeneous condition of the rockfill mass when the water load will come on the upstream concrete face. Incidentally, in a separate paper it is shown that the E moduli applicable for calculation of deformations under the hydrostatic load of reservoir filling, are obviously not the same E moduli as deduced from settlement observations of the rockfill under self-weight.

2.2.2 Geometric similitude

As I have repeatedly shown (de Mello 1972, 1977, etc.) in connection with slopes and crossections we find that our view of obeying Practice boils down to as simple and nonsensical a Principle as that of geometric similitude or even similarity. Satisfactory slopes are established as 1 on 2.0 or 1 on 2.5 etc. irrespective of heights of slopes, material properties, internal drainage details, etc. Dams varying in height from 20m to more than 200m have been and continue to be designed (i.e. drawn) as geometrically similar: if anything, with a disadvantage to the high dam because crest widths are not increased proportionately.

2.2.3 Symmetry

Of all the absurdities, the one most clearly atrocious is that of employing symmetry in a dam, built for a most unsymmetrical task. Symmetry is an inevitable corollary of gravity, and therefore comprehensibly dominates our visual culture. Moreover, during construction a dam grows against gravity: we can well see the kindergarten teacher requiring boys and girls (das kind in German, neuter gender) to skip-rope alike, dressed in like shorts and T-shirts. But the main function of US and DS zones is to complement each other in facing the reality of life which is with the reservoir full only on one side, (hopefully). How could anybody ever conceive of the temporary growing function as being the dominant one, and accept as reasonable a Practice of symmetrical sections?

Symmetry creeps in most imperceptibly as Practice in many other design endeavours. It belongs to the world of visual perceptions of laws of upright survival.

2.3 Summary conclusion

In short, Practice is an illusion, unless it is interpreted.

I recently read a very studious and well-documented paper on rockfills in dams. No distinction was made, either historical or behaviorwise between dumped and compacted rockfills. In that author's interpretation of Practice a rockfill was a rockfill, undistinguished between the very fundamental types, angular, rounded, dirty, dumped, compacted etc. A question of angle of vision and distance. To others a rockfill has seemed to be an overgrown sand. Just as beauty lies in the eyes of the viewer, so do observations lie in the eyes of the observer.

Our Russian colleagues have developed a remarkably successful technique of building hydraulic fill dams out of so-called homogeneous sands. Are they really homogeneous? How much of the satisfactory flownet behavior across the embankment is due to very wide crest widths, to the slight inevitable anisotropies in deposition of films of silt over each film of hydraulic-fill sand, to the slight additional compressions under self-weight making the central portion more impervious, and finally, to foundation conditions of pervious sands? Somebody designed and built a small homogeneous sand dam without any cutoff or drain: it failed, due to seepage exiting downstream (Florida, recent).

## 3. PRECEDENTS

Practice as distinguished from Precedents may be described as having predominated in the exaltation of the imperceptibly-moving status quo of conservatism, in comparison with the somewhat more forward-pushing case histories supported on Precedent. Implicit in Practice is the concept that what has been and is current in many cases, is proven. Implicit in Precedent is the thought that what "has been established" in one or few previous cases may be taken as proven, as sufficiently good: and may even support some (slight) extrapolation. The respect for Precedent is an anglo-saxon outgrowth of principles of jurisprudence and law: however respectable those might be, what possible connection might they

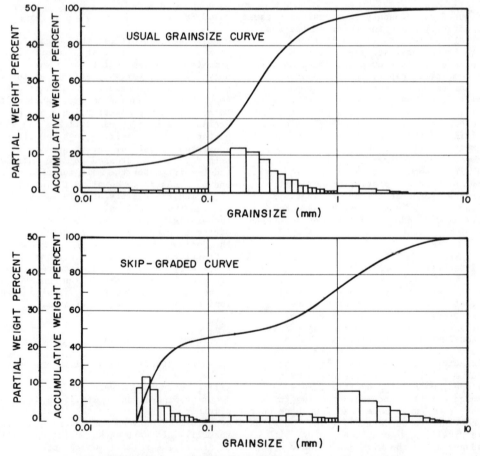

FIG.9 COMPARISON BETWEEN DIFFERENT GRAINSIZE CURVES PRESENTED AS HISTOGRMS

have with technological and statistical laws and behavior?

Inevitably, therefore, both are based on the Principle that what is, or has been, has been observed, and is satisfactory. They further imply the Prediction that what is, will be, will continue to be. Fundamental fallacies have already been discussed above:

(a) What is. Our cognizance of what is depends on the cultural tinge of the eyeglasses we happen to be wearing.

For instance, the presentation of grainsize curves in the traditional form of cumulative percentages of weights vs. log. diameters, was obviously directed towards facilitating observation of the fines. Concern centered on the fines. We easily see, however, (Fig. 9) that for the sake of impact visualization of gap-grading (de Mello, 1975 a) the preferable presentation would be by histograms of grain diameters (the latter in log scale for appropriate handling of the wide range).

How many problems and failures have been literally due to the practice of poor visual presentation of skip-graded grainsize compositions?

(b) "Has been proven". The very statement implies a deterministic causeeffect relationship that is quite fallacious. How many an action has been spared by the grace of God, or of statistics, despite its being inherently unsatisfactory? How often will we continue to test the recurrences of statistics, or the patience of God, in presuming to repeat as proven and good, what we have failed to interpret as really faulty but lucky? If Practice, applied repeatedly, fails to establish a proven Principle, Precedent (as above quoted) applied in pushing ahead as supported on few cases, can comprise an even more fallacious and dangerous concept.

(c) "Can be extended or extrapolated". Any extrapolation is always supported on the Principle of faith: and is exercised with greater confidence, and consequent ultimate danger, the greater the faith. It is of interest to compare (Hynes and Vanmarcke, 1977) the faith of few elaborate sophisticated solutions for predictions on Prof. Lambe's embankment failure problem, as compared with the audience's histogram of 26 estimates by "adjusted gut feeling". Since a significant Ingenious Engineering development frequently tends to would-be problems far beyond immediate needs (de Mello, 1975 b:118), quite often Precedent can be slowly extended for quite a stretch before finally being caught over some frontier of impunity. It is of interest to recall Terzaghi's words to Coyne regarding the Malpasset dam failure, to the effect that it could only be to a distinguished pioneer that the mishap could occur, of serving as the instrument to reveal a problem not yet brought out to the fore, in Man's gradual advance to greater needs and solutions.

Note, for instance, that when compaction of rockfill shells was developed as being good, the Swedish central-core wetcompacted earthrock dams that had behaved satisfactorily with dumped rockfills, ended up giving problems of silo effect and piping near the top.

Once again, therefore, in Precedents we recognize the intervenience of Principles -- of cognizance, of determinism, of faith, and so on.

4. PRINCIPLES

In the very discussion of Practice and Precedents we have been employing Principles, and recognizing their innate Principles. Of course, they were wrong Principles. And we now have Principles of dam design and construction; that are right: they are ours. We have Finite Element Analyses and computers.

Can we be so sure? Could it be that failures have not been mostly statistical (random), but rather very repeatedly deterministic, the main cause-effect parameter having been excessive faith in our own Principles? Was not Fontenelle Dam a dress rehearsal of Teton Dam, and both on a design crosssection faithfully employed most repeatedly? Is not each dam failure principally due to our faith that all factors have been rightly taken into account, so that "by accident" some additional factor shows up? Was Baldwin Hills a "calculated risk" or a calculated provocation?

We are, unfortunately, imbued in our likeness-of-God syndrome, and our exact--science syndrome. And faith is not scalar, but a vector: education is not scalar, but a vector. If we teach that overburden total stress is deterministically $\gamma z$, armed with our deterministic faith in the Effective Stress Principle $\gamma'z = \gamma z - u$, we leap forward into solving so many earlier problems with great success that we inevitably advance confidently towards many an unrecognized engineering solution before we are shocked out of our faith, by a failure. We learn much from shocking failures, but truly what do we learn for quantification, for developing statistical laws? (de Mello, 1977).

Principles are also adjustable, and our views of Problems depend on our Principles, and our Principles depend on the Problems that did beset us.

One fundamental Principle of the engineer is that any behaviour $X = f(a, b, c, d, ...z)$ is always a function of infinite number of parameters, and we have to synthesize immediate solutions, Prescribed as satisfactory, for our view of a finite number of Problems rated according to Priority. Scientific investigation and analysis proceeds in a diametrically opposite trend, picking out the knowledge of the behavior of X with regard to each separate parameter, all others maintained constant. Because of our finite capacity to recognize and face problems it is always dangerous to divert attention to non-problems (the classic scapegoat technique) and it will generally happen that the next accident will be due to a different problem. Churchill said that the trouble with Chiefs of Staff of armies is that they always prepared well how to fight the last war.

In my Rankine Lecture I tried to distinguish between problems associated with Extreme Value statistics, and those belonging to statistics of averages, repetitive, permitting formulation of laws, amenable to Bayesian adjustment, quantifiable within degrees of confidence. And I postulated that in Civil Engineering remarkable or catastrophic failures are Extreme Value cases. I have heard that statement questioned.

Let us first set aside some failures as Acts of God. Mount Saint Helen's volcanic explosion should be classified as an Act of God: we cannot propose to design our dams for such eventualities.

Civil Engineering always designs for conditions far from failure, and therefore when significant failures occur they are always an "accident", something beyond existing theory, something observed, analysed, and adjusted a posteriori. Hypothesis and theses may derive prematurely from intuitions, but theory and "laws" of behavior can only be formulated by repetitions of facts. Fortunately we can derive intuitions from assumed facts: but upon closer analysis such facts will turn out to have statistical dispersions. Our quantification and adjustment must insist on seeking highly repetitive conditions: therefore Failures are excluded.

Engineering really implies a sequence:

(a) Visualization of a physical model. Observations of Extreme Value conditions, failures, constitute a great support for such visualization. Create structures that avert the feared extreme value conditions: that is, in the face of possible extreme value failures in a given physical Universe (of statistics), use a change of Universe for a solution. Design Principles DP1 and DP2, Rankine Lecture.

(b) Employ nominal design-analysis procedures and observed great number of cases for "Satisfaction Indices". This is the quantification in statistics of averages.

(c) Refine steps (b) and consequences by repeated iterative adjustments. As far as possible employ the principle of Pre-testing so as to achieve Factors of Guarantee rather than nominal Factors of Safety.

(d) Thus hopefully move forward from knowledge of computations and behaviors to the wisdom of choosing a physical model (statistical universe) that literally dispenses analysis. Presumably it is guaranteed against failure, or at least against distressing failure. We have used Design Principle DP5 in mentally checking what can happen if our hypotheses and desired behavior do not fall within the presumed range.

5. PROBLEMS

Besides the overall Problem of time-lag in our redirecting the vectors of our deep faith in our Principles of cause-effect zero-dispersion determinism, what may be some of the specific technical problems I visualize being faced in dam engineering presently? Here go some examples.

5.1 Corrective measure vs. design solution. A good localized corrective measure to an extreme value problem is not necessarily a good overall design solution on average conditions. Exemplified by the case of filter-drainage at local seepage exits down-stream, in comparison with toe drainage (Fig. 10) (de Mello, 1977).

5.2 Variability of overburden stresses around $\gamma z$.

The average value $\gamma z$ is inexorable. However, the simple computation is based on the hypothesis of homogeneity and no shear stresses on the sides. In the cases of the dam superstructure we well recognize the silo effect. How can we be blind to significant variations in foundations, when heterogeneous. The more rigid elements carry most of the pressure (de Mello, 1972). In a silt lens beside big boulders it is not merely a statistical dispersion, but quite deterministic: the silt was deposited due to the protection from the boulder; and receives a small share of overburden for the same reason (Fig. 11).

5.3 Strong faith in flownets, highly aver-

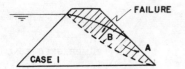

| | OBSERVATION | AVERAGE CONCLUSION |
|---|---|---|
| CASE 1 | REPEATED FAILURES EXIT A (EXTREME) MASS B (AVERAGE) | 1) FLATTER DS SLOPE..., LITTLE IMPROVEMENT. 2) TOE AND BLANKET FILTER-DRAIN OF CASE 2. |

| | OBSERVATION | AVERAGE CONCLUSION |
|---|---|---|
| KOZENY CASE 2 | VERY INFREQUENT FAILURES ; EXITS A (EXTREME) | FAILURES STILL BLAMED ON "UNUSUAL" CONDITIONS. COMPARE CONCEPTUALLY WITH SPILLWAY DESIGN FOR AVERAGE FLOOD. |

FIG.10  A TOE FILTER-DRAIN RAPIDLY APPLIED AT A POINT OF SEEPAGE EMERGENCE IS AN EXCELLENT CORRECTIVE MEASURE. AS GENERALIZED DESIGN IT IS INSUFFICIENT, CANNOT PRESUPPOSE POINTS OF EMERGENCE.

aged gradients i.

In fractured rocks how valid are "triangular diagrams" of uplift pressures, drainage tunnels etc... (Fig. 12)

In heterogeneous gravel-sand alluvia, how can we use reasonings based on limiting average $i = H/L$? What difference can it make, to local piping conditions, to change blankets from 10H to (15 or 20)H? Even the concept of $i_{crit} \approx 1.0$ grossly ignores directions of vectors $\sigma'g$ to be composed with $\sigma'i$, and assumes $\gamma'z$ as average overburden.

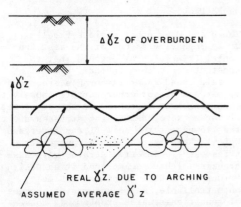

VERY DANGEROUS FOR PIPING BECAUSE THE FINE COMPRESSIBLE SILTS ARE EXACTLY (DETERMINISTICALLY) THE ONES ON WHICH THE $\gamma'z$ NECESSARY FOR S FOR EROSION RESISTANCE, DOES NOT ACT AS ASSUMED

FIG. 11

5.4 US impervious blanket as a badly conceived structure.

Represents an oversimplified attempt to solve a very idealized partial problem, seepage. Already inefficient if one considers tridimensional deposition of gravels-sands-silts. No thought to problems of loading due to reservoir, especially if there is time-lag in establishing underlying flownet (de Mello, 1977). Requires attention to $k_o$ (see 5.7).

5.5 Grouting and fixed-width diaphragm walls.

In my Rankine Lecture I discussed the error of the mental model of grout curtain as a fixed-width discontinuity. The inherent benefits of grouting are as a pretest treatment, more effective where most needed, and helping to exclude extreme conditions of perviousness. The diaphragm walls as presently executed are dangerous inasfar as they limit themselves to fixed width. The inherent error can be easily corrected by techniques long since developed in the grouting of alluvia.

5.6 Uniform filters, flat well-graded filter.

Design of filters for stereometric hindrance was considered a problem solved, but has turned up as a vexing problem. With well-graded non-uniform "highly desirable" materials the risks of segregation set in, depending on inexorable selectivity of construction operations. As an extreme value problem it must be solved by ap-

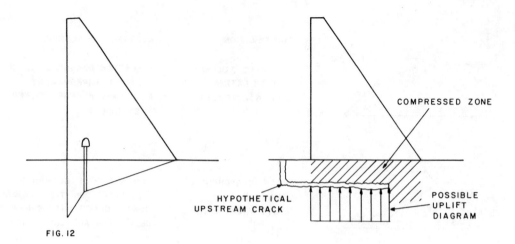

FIG. 12

propriate physical model: besides the appropriate grainsize for stereometric hindrance, promote compressive stresses (not exaggerated - cf. Prudence) in the material being filtered, so as to have increased arching, compression, resistance, around any start of washing-through.

5.7 Importance of $\sigma'_2$

Comprehensibly, while attention was directed to slope failures, the pair of stresses of interest $\sigma'_1$ and $\sigma'_3$ were in the plane US-DS. The function of the dam however, is to retain water and the principal risk is of transverse planes of low $\sigma'_2$. What do we know about this all-important item? Do Finite Element Analyses as presently available shed the necessary light? What do we know about the difference between the $K_o$ due to external (membrane) loading, as compared with body-stress effective stress loading? Is it valid, as regards volumetric strains to consider the classic simplification (Taylor, 1948) of soil mechanics that we can consider as equivalent

  Total stresses -
  - Boundary Neutral stresses $\stackrel{?}{=}$
  $\stackrel{?}{=}$ Gravity effective stresses $\sigma'_g$ coupled with seepage effective stresses $\sigma'_i$?

Do the finite element analyses presently conducted consider the hysteresis effects?

5.8 Influence of differential settlements on tensile cracking.
  The problem has been discussed for thirty years, but the all-important factor of time has not been discussed or considered. Obviously what matters is the delayed settlements that will affect the upper part of the dam (where tensile cracking can develop) after it has been built. Settlement occurring before a layer exists cannot possibly affect it.

5.9 Instrumentation for alerting on failure

A most dangerous fallacy to be guarded against is that of relying on instrumentation for indications of impending failure (de Mello, 1977). It is a most dangerous faith. Instrumentation can, and does indeed, furnish excellent information on average conditions, Satisfaction Indices.

6. PRUDENCE

These and many other serious Problems lead me to emphasize as one of the most fundamental Principles of engineering, the constant watch for Prudence. Even if a certain trend seems favourable, too much of it may not be so: other factors appear, to condition. For instance, we should prefer promoting some compressive stresses due to the flownet at the soil-filter interface: but if there is too high a compressive stress, the volumetric strains may cause cracking. I have summarized it as Design Principle 4, DP4: "Minimize untimely, uncontrollable, major and rapid, changes of condition towards problems of consequence". Indeed, from the solutions of one generation frequently arise the plagues of the next, because one of the greatest of all Problems is one placing too rabid a faith on one's Principles. We must lean over backwards to take ourselves with a pinch of salt.

An old Arab saying goes:

He who knows not, and knows not that he knows not; he is a fool:shun him.
He who knows not, and knows that he knows not; he is simple:teach him.
He who knows, and knows not that he knows; he is asleep:awake him.
He who knows, and knows that he knows; he is wise:follow him.

Upon analysis I would find that the last line would be quite comprehensible for a culture of yore. Unacceptable today. Moreover, there is one more combination that makes sense. So offer a revision:

He who knows and knows that he knows; he is useful:use him.
He who knows and knows that he knows not; he is wise:follow him.

## 7. REFERENCES

de Mello, V.F.B. (1972), "Thoughts on Soil Engineering Applicable to Residual Soils", Proceedings of The Third Southeast Asian Conference on Soil Engineering, Hong Kong, pp. 5-34.

de Mello, V.F.B. (1973), "Eleventh International Congress on Large Dams", Madrid, Spain, Vol. 5, pp. 394-406.

de Mello, V.F.B. (1975 a), "Some Lessons From Unsuspected, Real and Fictitious Problems in Earth Dam Engineering in Brazil", Proceedings of The Sixth Regional Conference for Africa on Soil Mechanics and Foundation Engineering, Durban, South Africa, Vol. 2, pp. 285-304.

de Mello, V.F.B. (1975 b), "The Philosophy of Statistics and Probability Applied in Soil Engineering" Proceedings of the $2^{nd}$ International Conference 'Applications of Statistics and Probability in Soil and Structural Engineering', Aachen, F.R.G., Vol. III, pp. 65-138.

de Mello, V.F.B. (1977), "Seventeenth Rankine Lecture: Reflections on Design Decisions of Practical Significance to Embankment Dams", Geotechnique, London, ICE, 27 (3), pp. 281-354.

Hynes, M.E. and Vanmarke, E.H. (1977), "Reliability of Embankment Performance Predictions", Mechanics in Engineering, University of Waterloo Press.

Ring, G-W. et al (1962), "Correlation of Compaction and Classification Test Data", Highway Research Board, Bulletin 325, pp. 55-75.

Taylor, D.W. (1948), Fundamentals of Soil Mechanics, New York, John Wiley and Sons, 1948.

*Symposium on Problems and Practice of Dam Engineering / Bangkok / 1-15 December 1980*

# Lessons from earth dam failures

P.LONDE
*International Commission on Large Dams & Coyne et Bellier, Paris, France*

Summary : The history of dam building, since the dawn of civilisation, is a long series of failures. Man learns little from success but a lot from his mistakes. The failure of a dam is however the result of a complex concourse of causes and mechanisms, which should be interpreted with extreme care.

Several case histories of earth dam failures are reported, and arranged in three groups : failure by sliding, failure by overspilling, failure by seeping water. Each group leads to specific lessons. Learning from our errors is vital for improving our knowledge and promoting safer designs, as designing and building a successful earth dam is still nowadays more an art than a science.

In the light of these lessons, ten recommendations are put forward for improving the overall safety in the design, construction and operation of earth dams.

## 1. INTRODUCTION

Designing and building a successful earth dam is still nowadays more an art than a science. Of course soil mechanics is a powerful tool which has all the characteristics of an experimental science. Its proper use in design, however, implies the use of simplified models and the selection of numerical parameters which are far from representing the actual complexity of nature. In addition, the floods in the river are ill-known and remain a threat over the period of construction and afterwards during operation of the dam, although hydrology also is an experimental science.

The best proof that designers are not yet able to control their works in a scientific manner is given by the recent spectacular failures of earthdams. These failures are less and less frequent but they still occur even in dams where conventional good engineering practice was followed.

The lessons to the profession from failures are vital for improving our knowlege and promoting safer designs.

The case histories reported in the following will be arranged in three groups, each group leading to specific lessons :

1. failure by *sliding* - concept of residual strength of overconsolidated clay, and of liquefaction potential of saturated sand,

2. failure by *overspilling* - need for hydrological studies and proper spillway design and operation,

3. failure by *seeping water* - basic function of filters and drains, instrumentation as a vital part of design, proper design of conduits.

## 1. FAILURE BY SLIDING

### WACO dam (Texas, USA) 1961

This homogeneous embankment, 43 m high and 5,500 m long, was almost completed when cracks appeared on the downstream face. From 4 October to 20 October 1961 the movement was slow but it accelerated then to come to a standstill early in November. The top of the embankment had dropped by 5 m and the downstream face had moved horizontally by 6 m over a total length of 230 m (Fig. 1). The width of the failed zone was 250 m. Fortunately the reservoir was not yet filled and no catastrophy resulted from the slide.

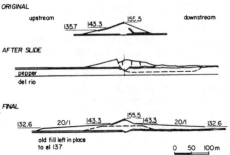

Fig. 1. WACO dam : cross sections

After the failure it was realized that the foundation shale was locally different from what it was for the rest of the dam. A formation, called the Pepper shale, was higher between two shear faults which were located just at the limits of the slide.

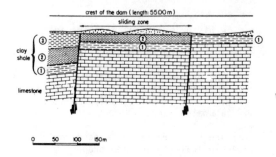

Fig. 2. WACO dam : geology

This feature (fig. 2) covered by alluvium, and not detected by the reconnaissance boreholes systematically drilled along the 5,500 long dam axis, was not known by the designers. The back-analysis and laboratory tests showed that the Pepper formation had a very low residual strength. In addition, its horizontal stratification allowed high excess pore pressure to spread over large areas. This type of behaviour has been recognized in many sites since that time. It is typical of plastic clays and shales, overconsolidated by tectonic forces or by drying. The mechanism leading to complete sliding involves a progressive shear failure along the slip surface. We know now that for such formations the *residual strength* has to be taken into account in the analysis.

Another lesson from the Waco dam accident is that the probability of foundation failure of a *long dam* is higher than that of a short dam, owing to the increased risk of a local detrimental geological feature remaining indetected by an apparently adequate number of boreholes.

The Waco dam was repaired using flatter slopes and wide berms (Fig. 1). The shear strength of the foundation was taken as low as $\phi_r = 8°$ and $c_r = 0$. The excess pore pressure in the shale was taken at $0.7\ \gamma h$, a pressure which was actually measured in the 127 piezometers installed after the accident, some zones giving up to $\gamma h$. Deep draining wells were also installed. With these severe assumptions the factor of safety used was only 1.20.

The danger of overconsolidated clays was gradually realized by engineers, following the studies made in London by Prof. A.W. Skempton. In the early 60's the Roseires dam which was under design in the Sudan was modified in the light of the new findings, using residual strength parameters $\phi_r = 15°$ and $c_r = 0$, whereas the unconfined compressive strength of the clay was as high as Rc = 2 MPa.

It was later recognized that the failures of Seven Sisters dam and North-Ridge dam in the USA were due to the presence of overconsolidated plastic clay of low residual strength.

Failures were necessary for disclosing the mechanics of progressive reduction of strength, a thing which is now well known but not always adequately studied during investigations for dams. The recommended guideline is to make residual strength measurements in all clayey soils, and to base the stability analysis on this low strength parameter, with a factor of safety not much above unity.

Mention should be made at this point of another possible cause of sliding : the *liquefaction* of the loose saturated sand. This sand may be part of the dam material (hydraulic fill) or a foundation layer.

Liquefaction could result from a sudden pore pressure increase which might be induced by an earthquake, or from a strain building up induced by a slide. Although no large dam has actually failed during an earthquake, we know of several near-failures (San Fernando dams, California) and the complete failure of a small dam (Sheffield dam, California).

As a result of the San Fernando earthquake, on February 9, 1971, the Upper-San Fernando dam developed large downwards displacements of the crest (about 1.5 m horizontally and 0.8 m vertically) and of the downstream slope (about 2 m horizontally), and a major slide occured in the upstream shell of the Lower San Fernando dam (Fig. 3), resulting from the liquefaction of the sandy hydraulic fill. Fortunately, there was no overtopping of either dam, although the remaining freeboard was extremely small.

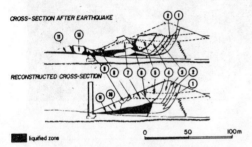

Fig. 3. LOWER SAN FERNANDO dam : cross section

The Sheffield dam, with a maximum height of about 7.5 m (Fig. 4) was built of sandy silt to silty sand taken from the reservoir and compacted by routing the construction equipment over the fill. The upstream slope was faced by a concrete facing overlaying a clay blanket, which extended as a cut-off wall into the sandy to silty foundation layer, down to the sandstone bedrock. As a result of the Santa Barbara earthquake, on June 29, 1925, the dam failed by sliding of the downstream slope on a nearly horizontal plane, near the base of the dam, initiated by liquefaction of the foundation layer.

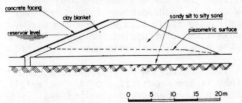

Fig. 4. SHEFFIELD dam : cross section

These two case histories led to extensive studies and back-analyses, which improved the earthquake resistant design procedures for earth dams. Good engineering has to consider liquefaction potential of sandy formations as a real danger in seismic areas.

## 2. FAILURE BY OVERSPILLING

### OROS dam (Brazil) 1960

North-East Brazil is a semi-arid region, with irregular rainfall, varying greatly from one year to another. In order to reduce the disastrous effects of dry spells on agricultural lands a large reservoir was built at Oros. The catchment area is 25,000 km2 with bare granitic ground. The runoff coefficient is high and therefore floods very sudden, with the river dry 6 months in the year. The gorge at the dam site created an erosion pit immediately downstream, leading to an unusual curved dam axis for the embankment (Fig. 5). The height of the dam is 54 m.

Construction started in 1958. Foundations were ready before the rainy season of 1959 and the embankment was due to be completed before the rainy season of 1969. Unfortunately administrative difficulties within the government offices led to a slowing down of the construction, the contractor not being paid. And that particular year, floods were early and high.

When the first flood started to fill the reservoir the embankment was as shown by Fig. 6. Then a dramatic fight took place between construction and rain. Workers did not stop day and night, hoping to keep up with the rising waters. But more than 500 million cubic meters in 5 days filled up the limited storage provided by the unfinished dam in spite of the emergency dike thrown up on the top of the embankment, as shown in Fig. 6.

On 26 March, in the morning, the water began to spill over the whole length of the embankment, in a sheet 35 cm thick. Then a breach opened in the central part (Fig. 7). The erosion was soon very rapid, cutting the embankment down to the foundation level. In 30 hours the total volume of water discharged downstream is estimated at 1 billion cubic meters, i.e a mean flow of 10,000 m3/s. Nearly 1 million cubic metres of embankment were washed away. One million people had to be evacuated before the rupture with the help of air force and army.

Fig. 5. OROS dam : view of the dam before failure

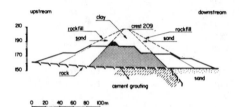

Fig. 6. OROS dam : cross section before failure

Fig. 7. OROS dam : view of the dam after opening of the breach

1) Had the delay from bureaucratic inertia been one month less, the dam would have suffered no damage.

2) The accident is a clear illustration of the *sensivity of an earth dam to overtopping*. It shows the large margins of safety that should be used in the planning of construction, especially when the hydrology of the river is uncertain.

1) I know too many earth dams which are a real danger to people living downstream at the time of the end of construction, when a slight delay could result in a catastrophic release of a large volume of water. This situation is very serious on big rivers, where diversion of the flood discharge through tunnels is not practicable owing to the capacity required and the spillway sill too high for being operative before the dam is completed.

## DA CUNHA and OLIVEIRA dams (Brazil) 1977

In 1977 two homogeneous earth dams failed in Brazil by overtopping. They were on the same river and the flood released by the failure of da Cunha dam induced the failure of Oliveira dam, 10 km downstream. Both were completed in the late 50's and their behaviour was entirely satisfactory for more than 15 years.

Their main purpose was power and the operating rules were such that the reservoir levels had to be kept as high as possible so as to use the maximum possible head on the turbines. These rules, underestimating the possibility of overtopping by high flood, were responsible for the failure of these dams (Fig. 8).

Fig. 8. DA CUNHA dam : view of the dam after failure

In addition to these *inadequate operating procedures,* technicians needed to open the spillway gates had left the powerhouse for lunch, and were unable to return that afternoon when fast-rising flood waters cut off the access road.

As reported in the Preliminary Report on the failure "the overtopping had been going for over a 7 hours time period at da Cunha dam, approximately between 20.00 hrs of 19 January and 03.30 hrs of 20 January 1977, before failure started. Half an hour thereafter the Oliveira dam breached. From the water marks it was concluded that the maximum depth of overflow over da Cunha dam was 1.26 m and over Oliveira dam about 1.30 m". One dam was able to stand overtopping for about 7 hours and the other 2 hours before breaching.

3) This time depends on many parameters. One of them is the *erodibility* of the material placed in the downstream shell of the embankment, particularly at the face. The worst material is a sandy soil with low cohesion.

I know of a recent failure in Greece where a low re-regulating dam was overtopped and nearly failed in less than one hour. It did not fail, as a matter of fact, because the flash flood was extremely short and the reservoir level receded faster than the time required for deepening the breach, which stopped at 2 m depth.

Close examination of failures by overtopping is necessary for the *proper design of fuse plug spillways*. It is my conviction that among the fuse plugs built all over the world, as an emergency device for avoiding overtopping the main enbankment, many would not be washed away in the short time required. Arrangements such as shown by Fig. 9 would give a proper and reliable functioning. On the other hand, overtopping of an embankment dam is generally a major threat.

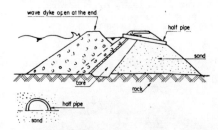

Fig. 9. Fuse plug (after Barry Cooke)

The lesson from this type of failure is that *ample spilling capacity* is required for earth dams. Many rivers in the world are not well known hydrologically. The determination of design flood is always uncertain, and any underestimation may lead to the complete failure of an earth dam. Among the possible causes of exceeded spillway capacity, one must mention the occurence of seiche in the reservoir, triggered by an earthquake, and of large slides or rockfalls into the reservoir. The better ability of concrete dams to withstand an overtopping has been shown by spectacular case histories (Vajont dam, Italy, in 1962 ; Gibson dam, USA, in 1964). The earth dam may require a much higher spillway capacity than the concrete dam for the same level of safety.

Another lesson is the reduction of safety of a spillway when gated. There are several reasons why gates are not operated properly : power failure, lack of maintenance, earthquake, human misoperation, sabotage. A *gated spillway,* at an earth dam, requires most careful study and organization of the operating rules, including provisions for emergency action in case of unlikely circumstances (failure of automatic control, absence of responsible staff, war, etc.). The failure of da Cunha dam in Brazil is typical of misoperation of the gates.

The recent improvements in hydrology, particularly the use of the Probable Maximum Flood concept, have resulted in a substantial increase in the safety of earth dams as far as overtopping is concerned. However, they make it clear that many earth dams designed and built in the past are not safe, and their spillways should be enlarged. In addition the probability of occurence of a catastrophic flood increases with time, at a given site.

## 3. FAILURE BY SEEPING WATER

### HELL HOLE dam (California, USA) 1964

The failure of Hell Hole earth and rockfill dam happened during construction. It could have been catastrophic without the presence of a very large reservoir downstream, which stored the released water. Fig. 10 shows the breach in the dam on December 1964 as a result of the exceptional flood of the river.

Fig. 10. HELL HOLE : view of the dam just before breaking (9.25 a.m)

For a period of 24 hours the water level in the reservoir stabilized at 22 m above the core surface. The flood passed through the diversion tunnel and partly through the rockfill shell. The flow through the shell is estimated at 350 m3/s. No erosion was observed (Fig. 11).

Then, the reservoir level rose within a couple of hours to 30 m above core surface. The flow through the rock shell is estimated at 500 m3/s. A hole started forming at point A on the downstream face just at the level of the saturation line.

This hole enlarged rapidly and *retrogressive erosion* developed, following the saturation line, as shown in Fig. 11. During this process both reservoir level and seeping discharge remained constant.

However as soon as erosion reached point B, the mechanics were different, since overtopping started. Then the discharges increased very rapidly to 7 000 m3/s, a value which remained about constant for one hour, until the breach had cut down the core level. The core was practically not eroded, nor the upstream shell and filters.

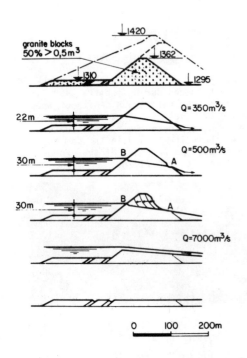

Fig. 11. HELL dam : sequence of sevents

This flood was far in excess of the maximum flood on record over a period of 40 years. The probability of experiencing such a high flood during construction was therefore very low, and it was reasonable not to design the diversion works for such an unlikely event.

The lesson however it that an homogeneous rockfill is in danger of *piping* like all soils if it is not protected by proper filter zones.

The proof of this statement is given by Nebaana dam (Tunisia). In October of the same year a flood occured when the dam was exactly in the same state as Hell Hole dam in December. Nothing happened to Nebaana dam. The main difference between the dams however was that at Nebaana the downstream shell was made of two zones, the upstream one of smaller rocks compacted in 0.50 m thick layers and the downtream one of bigger rocks compacted in 1.0 m thick layers. It was organized as a huge filter, preventing the seepage flow from daylighting high on the embankment face.

BALDWIN HILLS dam (California, USA) 1963

Baldwing Hills dam, creating a reservoir of domestic water right in the City of Los Angeles, failed in the afternoon of 14 December 1963. (Fig. 12). Fortunatly the outstanding *efficiency of the warning* and evacuation operations limited the casualties to five in a highly populated area.

Fig. 12. BALDWIN HILLS dam : view of the breach

At 7.45 in the morning the caretaker started his daily inspection. Nothing different from usual was observed. At 11.15, while walking near the spillway, he noticed a slight noise of running water in a culvert. Looking closer at the drainage system he realized that the discharge was about 5 times higher than usual and the water was muddy. He then walked to the chamber collecting all drains. Those coming from under the reservoir were discharging "like fire hoses", and the water was muddy.

He immediately went to his house and telephoned the engineer in charge, who arrived on the dam at 12.00. At that time it was no longer possible to enter the drain chamber, full of water. Twenty minutes later the decision was taken to empty the reservoir.

At 13.00 a muddy spring was noticed at the toe of the dam near the right bank. At 13.15 a crack appeared on the crest of the dam, just above the spring. At 13.30 the decision was taken to evacuate the threatened people downstream of the dam. When the Chief Engineer arrived at the site, at 13.45, he informed the Police Department that the dam would fail within two hours. Half an hour later an alert was broadcast by all radio and television channels in Los Angeles. Sixty police patrols were immediately sent to the area for evacuating everybody.

The crack at the top of the dam kept widening and the leak at the toe was steadily increasing. Sand bags were lowered along the crack on the upstream side of the embankment. They were swallowed by the water flowing through the erosion hole still under the reservoir level. This level was receding but not fast enough to empty the reservoir before the piping had entirely breached the embankment. This happened at 15.38, releasing a catastrophic flood in the streets downstream. The dam had failed only 7 minutes short of the time estimated by the Chief Engineer. In spite of the thorough evacuation of the flooded area, five people were drowned in a car which entered a street where no guards had been placed.

It took several years of investigations by different enquiry panels to arrive at a reliable explanation of the failure. Its causes are complex and even today there remains some room for controversial opinions on the mechanisms.

The dam, a homogeneous embankment, was well designed. The famous engineer R.R. Proctor was head of the design office of the Owner. The behaviour of the structure was entirely satisfactory for 15 years.

The reservoir was placed over a system of minor faults, a fact which was recognized by the designers. The foundation soils was a silty sand of low density, therefore quite erodible. In order to avoid any detrimental seepage in this foundation the whole surface of the reservoir was lined with a compacted earth layer, up to 3 m thick at the bottom. This impervious layer was covered by a continuous asphaltic paving, and underlain by a drainage layer made of pea-gravel placed on an asphaltic membrane (Fig. 13), associated at the bottom of the reservoir with a network of 10 cm clay tile drains.

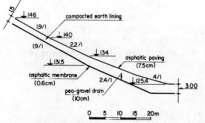

Fig. 13. BALDWIN HILLS dam : section on reservoir slope

These unusual precautions seemed perfectly safe, all the more so in that all drains were connected to a control chamber where discharges were monitored. *Instrumentation by clinometers and settlement devices was extensive.* The dam was inspected daily by the caretaker and monthly by the engineers. This exceptional surveillance set-up prevented the failure from being a catastrophy and saved the lives of several hundred people.

The main conclusions of the inquiry boards for explaining the accident are as follows :

1. the extraction of oil and gas in the close vicinity of the reservoir induced a general ground subsidence of 3m between 1917 and 1963 resulting in local differential settlements of faults.

2. A minor fault under the reservoir sheared with a total offset varying from 5 cm to 18 cm.

3. This movement broke the impervious lining at the bottom of the reservoir.

4. The seeping water penetrated the highly erodible foundation and a piping process developed rapidly along the fault, daylighting at the toe of the dam.

Several conditions are therefore required to explain the failure : subsidence of the area, presence of fault under reservoir, erodability of ground. Should only one of these conditions be not present the failure would not have occurred. This is a general rule in dam accidents, which always makes finding the causes difficult and too simple explanations usually unreliable and even misleading. Another lesson from the Baldwin Hills failure is the invaluable effect on public safety of good surveillance.

25

In populated areas *twenty-four hour surveillance* should be provided for all reservoirs. This can be arranged with suitable combinations of inspection personnel and automatic recording instrumentation" (Ref. Recommendation from City of Los Angeles - Baldwin Hills Board of Inquiry Report).

TETON dam (Idaho, USA) 1976

The failure of Teton dam is very recent and is the best illustration of the fact that modern dams still have a probability of failure which is not nil.

It is the highest dam which ever failed, with 93 m above river bed. Piping occurred through the core and the foundation rock during the *first filling* of the reservoir. No instrumentation whatsoever was installed and there was no inspection gallery under the core.

On 3 June 1976, a small spring appeared some distance downstream of the dam at the toe of the right bank. Its discharge of 1.3 l/s was steady for two days.

At 8.30 on 5 June, two large leaks appeared at the contact between the dam and its right abutment. One was at the bottom of the valley, with 1,400 l/s, the other 50 m higher up with 57 l/s. Both were slightly muddy.

At 9.45 the Owner (Bureau of Reclamation) asked the authorities to evacuate the area below the dam.

At 10.00 the high level leak developed into a heavier one (420 l/s) located in the embankment 5 m from the right bank. It increased rapidly in size and formed a crater in the embankment. The contractor used bulldozers to try and fill this crater with large blocks, but was not successful.

At 11.00 a vortex formed at the surface of the reservoir, just opposite the leak downstream.

At 11.30 all attempts to plug the piping hole were abandoned, the bulldozers being carried away by the flow.

At 11.57 the embankment was breached to the crest. This breach deepened rapidly. The erosion washed away 3 million cubic meters of earthfill, releasing a peak flow of 70,000 m3/s (Fig. 14).

Fig. 14.   TETON dam : view of the breach

Alarm was given to all living downstream in a very efficient way, so that only 11 people were drowned although the flood reached over more than 200 km of river plain.

Six months later, i.e in a very short time, the first Panel of Inquiry published their conclusions. They were definite in condemning *"faulty design"*.

The embankment itself, with a thick central core of silt and shells of sand and gravel, was conventional. However the fissured rhyolite foundation was highly pervious. A deep grout curtain made of three rows of holes was not deemed efficient in the upper part of the rock formation, where cracks were wide open, up to 10 cm width. That is why the central part of the core was compacted in a narrow deep cut-off trench excavated in the rock (Fig. 15). Unfortunately no treatment of the open cracks and no filters were provided at the contact between the rock and the core. This was the fatal error in design.

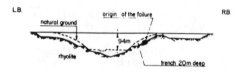

Fig. 15. TETON dam : profile

Several mechanisms can be used separately or simultaneously for explaining the piping. One of them is the erosive action of water flowing within cracks and through the grout curtain at the contact with the core (Fig. 16). Another one is erosion through cracks formed in the cut-off trench by hydraulic fracturing. A third is the formation of horizontal cracks in the narrow cut-off trench by arching.

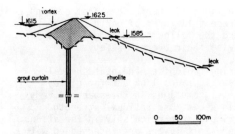

Fig. 16. TETON dam : cross section near the failure zone

A recent controversy arose at the ICOLD Congress in New Delhi, when Dr. Jack Hilf claimed that the failure was originated by wet seams in the silty core, probably created by frozen soil at the time of construction.

Whatever the details of the penetration of water through the core, it is certain that the silty material required protection by a filter, an arrangement which is conventionally used and was surprisingly omitted in this high dam. It seems that engineers responsible for the construction were well aware of this, but, owing to administrative rules in use in USBR, they had no contacts whatsoever with the engineers responsible for the design documents and specifications. It is remarkable that the erosion took place at the level where plugging of open cracks in the cut-off trench walls was discontinued. This plugging with concrete and mortar was done by the contractor although it was not specified. Nobody knows why it was discontinued.

This context led Dr. J.L. Sherard to declare in a recent public lecture that Teton did not offer any technical lessons to the designers, as it was "a bureaucratic failure".

What must be remembered is that piping is probably the most detrimental factor of failure in earthdams nowadays. The discovery, during the last fifteen years, of the *dispersivity* of some clays has emphasized our concern regarding the danger of piping through dam cores. Piping along the contact with concrete works, such as culverts or retaining walls, is also a threat. Fortunately, provided the designer is well aware of these dangers, he can remedy them in complete safety by proper *use of filters and drains*.

## CONCLUSION

The history of dam building, since the dawn of civilisation, is a long series of failures.

Man learns little from success but a lot from failure. Not to publish the facts about a failure, and we all know many reasons why not to do so, is a severe breach of our duties as engineers. Every failure, severe or minor, should be described by competent engineers and all relevant data made available to the profession. In this respect we are all most grateful to the American engineers, who have so carefully published their failures, as you have noticed in this lecture.

A serious difficulty however would still remain. The failure of a dam is the result of a complex concourse of causes and mechanisms. Very often several if not all of these mechanisms are not brought to light with certainty. It is exceptional for reliable witnesses to an accident to be able to report on the sequence of events, and after the accident most of the weak elements involved in the failure mechanisms have disappeared. In addition, genuinely or intentionally wrong accounts are often given by responsible men for psychological or legal reasons.

That is why all known failures should be interpreted with extreme care, and why statistics of accidents should be used cautiously.

Statistics are nevertheless invaluable for the profession as they point to the more common causes of damage and therefore where particular care has to be taken in *design, construction and operation* of the dams.

For earth dams, statistics were established in 1953 by an American engineer T.A. Middlebrooks, using 200 case histories of failures. Not all were catastrophic but all were serious enough for disrupting the operation of the dam. Fig. 17 gives the result in terms of percentage of each main category of cause of failure. This statistical result covers a period of more than 100 years, during which time spectacular improvements were brought in by dam engineers. It is interesting to note however that more recent studies, particularly the famous "Lessons from Dam Incidents" published by the International Commission on Large Dams, give very nearly the same percentages.

The most significant one is the low value of relative number of failures by sliding. Only 15 % of the total number of accidents is due to sliding, that is, could be evaluated by the usual concept of factor of safety. The remaining 85 % come from causes which are not covered by the conventional stability analysis but are relevant to design, construction and operation procedures which can hardly be computed.

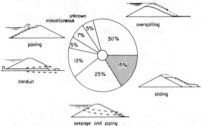

Fig. 17. Statistics on earth dam failures (after Middlebrooks, 1953)

This means that the so-called factor of safety can be dangerously misleading. Unfortunately it is much easier to carry out very sophisticated stability analyses, including by finite element method, than to know how to evaluate the risks of piping, of cracking, of uplift and to know how to design for minimizing these risks. That is why too many earth dams are designed and approved on the basis of stability analysis only, notwithstanding the fact that 6 dams out of 7 have failed for other reasons than an inadequate "factor of safety". Although improvements in our technology in past years has been spectacular, reducing the probability of failure of any new dam by a ratio of 10 to 1 between 1900 and the present time, (Fig. 18) there remains a number of catastrophic failures which show that engineers have still to learn and should still improve their practice. In addition, existing old dams, built to lower standards, require more and more surveillance and maintenance. Dams more than 50 years old represent 20 % of the existing dams.

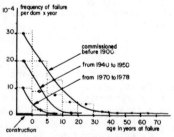

Fig. 18. Frequency of failure vs. age of dam

In the light of the lessons from old and more recent failures of earth dams, I put forward the following recommendations for improving their overall safety.

1. During geological and geotechnical investigations look out for clayey soils, sometimes in the shape of very thin seams. Use residual strength parameters, particularly in formations where soft layers are interbedded with stiffer layers. Overconsolidated clays are particularly relevant for use of residual strength parameters.

2. During geological and geotechnical investigations look out for loose silty or sandy soils, and study their liquefaction potential.

3. For design of sequence of construction stages and of diversion of river, make a thorough analysis of the floods in terms of probability of occurence and corresponding probability of damage downstream.

4. For design of spillway and outlets make the best use of most recent hydrological methods, with a clear appraisal of the catastrophic consequences of overtopping the main embankment. Fuse plug spillway generally is a good safety measure, provided it is correctly designed.

5. For a safe gated spillway, very detailed and strict operating rules are required. Any human or mechanical failure has to be envisaged and corrected by emergency arrangements.

6. Ample and well graded filters and drains are vital for preventing piping through dam and foundation.

7. All fine clayey soils should be tested for dispersion potential of clay particles.

8. Instrumentation is the only means of monitoring safety during operation of the dam. It is a vital part of the design of a new dam and must be incorporated in old dams where it is missing.

9. Inspection galleries, either at the base of the core or in the foundation rock below, are invaluable for placing instrumentation for direct observation and for quick remedial action if required.

10. Thorough and careful surveillance of dams safeguards safety of people living downstream. It should be done on a 24 hours basis, together with automatic recording instrumentation.

# Recent trends in the design of earth-rock dams

PIERO SEMBENELLI
*Electro Consult, Milano, Italy*

## 1 INTRODUCTION

Earth-rock dams, intended here as water retaining structures built with natural particulate materials, ranging from clay crystals up to gravel and quarry blocks, are the oldest such structures built by man. Although during the early years of the industrial revolution, dams were mostly built with masonry, concrete or other man-made materials, there is now a definite shift back toward earth and rock as construction materials. Progress made by earth moving equipment is partly responsible for this but, undoubtedly, one main reason is the reliability of concepts and methods developed by geotechnics. By far the majority of the world's dams is built today through the combined use of a variety of natural materials and the highest dams in the world are earth-rock dams.

If progress of design concepts and construction technologies for structures of such importance has been slower than their growth in number and size, we are now witnessing a fast and remarkable evolution of geotechnics related to dams. Further developments are needed and will come soon.

Large volumes of rock as construction material were already in use at the beginning of the century but investigations on rockfill properties started in the late fifties only, prompted by Terzaghi, Casagrande, Carillo and Marsal and spearheaded by the large scale testing facilities of the Infiernillo dam in Mexico.

Comprehensive measurements and behavioural observations of large dams did not start much before. Akosombo, Muddy Run and again Infiernillo were among the earliest such structures extensively monitored and controlled.

Only in the last decade, combining the analysis of material characteristics and the observations of prototype behaviour through new mathematical procedures, has produced a wealth of information that has allowed a global reinstatement of the problem and substantial developments of the Art.

Introducing new methods and criteria in the design and construction of earth-rock dams, like the deformational analysis or the assessment of the dynamic response is already producing a beneficial fall-out on the profession, owners and society. However, this is not all: there are other and equally meaningful developments taking place in the way we design and build the earth-rock dams of tomorrow. This paper is intended to focus on some of these developments for their methodological relevance and for the fact that they are often blurred by the glare of other components of the design process.

It may be unnecessary to say that this paper just reflects the writer's experience and practice, matured designing and building earth rock dams in different parts of the world, besides South East Asia. Contributions from many colleagues and consultants are closely knitted into it and duly acknowledged here.

## 2 CONSTRUCT TO DESIGN

Except for small and straightforward projects that happen to enjoy particularly simple conditions, conventional investigations are not suited to designing a dam making full use of modern methods and techniques. Today, a thorough and proper design almost invariably requires ample use of in situ investigations and special tests.

Topo mapping and geologic surveys are no longer enough to represent the physical reality of a damsite. Nor can drill holes, few or many, be considered a sufficient source of subsurface information and of specimens for exhaustive laboratory tests.

Designing a large earth-rock dam requires, at least, a sufficient knowledge of: lithology and structure of the geologic building in the surroundings of the dam, mass and joint permeability, average and minimum shear strengths, critical failure mechanisms and the definition of bulk strain characteristics, scale effects and zones of differentiated deformability. The assessment of how to treat the foundation soils and/or rocks in order to reduce or increase their permeability, to control failures and to improve deformations are equally essential.

Equal attention must be paid to natural materials locally available for construction because every natural material, if properly used, can be suitable for building a dam. Each source of practical significance should be thoroughly characterized and defined as far as deformability, strength and permeability are concerned. By no means should investigations on potential construction materials stop short of proving their suitability of being handled with conventional technologies or of assessing alternative processes.

Without this very type of information, the design is bound to be a blind alley that can be either overconservative without any specific reason or dangerously inadequate. Teton dam is a reminder on how far inadequate design can go.

Once the vital importance and the vastness of this introductory activity has been acknowledged, it is not difficult to realise that this is a matter that cannot possibly be dealt with by a small crew and simple equipment. Too often, the Owner imposes on the designer his own mapping, surveying and drilling teams and facilities as if they were the golden key to any possible knowledge and adequate enough to answer all possible requests. This type of attitude, often accepted and even indulged in by designers is, in the writer's opinion, beneficial to no one and often definitely unacceptable. It should, and often is at least by the more perspicacious Organizations, replaced by what I like to call a construct-to-design concept.

CONSTRUCT TO DESIGN means carrying out those activities that are a prerequisite for the proper design of an earth-rock dam, as the very first step of the construction process i.e. preparing a proper and comprehensive plan of work, producing programs, drawings, instructions and implementing them with the contribution of specialized contractors hired through fully fledged bids and with proper financing. Non-standard activities, usually performed when adopting this concept can be grouped, for our purposes, under five headings as follows:

- LARGE SCALE TESTING OF FOUNDATION MATERIALS
- GEOMECHANICAL MODELS
- LARGE SCALE TESTING OF CONSTRUCTION MATERIALS
- PRECONSTRUCTION INSTRUMENTATION
- PRECONSTRUCTION STRUCTURES.

To briefly recall some examples is probably the best way of illustrating a Construct to Design concept. In no way does what follows cover all the aspects of an approach still applied in isolated cases, if not erratically, and which in any event needs to be adapted to each particular project and site.

## 2.1 Large scale testing of foundation materials

The site of a large dam in South East Asia consisted of an orderly sequence of very competent crystalline rock and weak to very weak and compressible shale to clay layers. In addition, there were several indications of unisotropy within individual rock layers. The parameter of relevance to the dam structure was foundation deformability. Soon after engineering appraisal, it appeared that bulk deformability could not be inferred from tests on the individual materials of the geologic sequence. The need for large scale in situ tests was clear.

In view of the size and number of tests and because the time available was short, the testing program was transferred to a set of contract documents and bidded on an international basis. The work was awarded to a specialized firm and was completed in six months, an incredibly short time for most investigation campaigns.

At several locations, along adits excavated on purpose, radial press tests shown in Fig. 1 were carried out. In short, the tests consisted of loading the entire periphery of

Fig. 1 - A large scale loading test on foundation rock performed with radial press

a stretch of adit with several sets of flat jacks placed against the adit wall and reacted by a steel frame. More conventional plate loading tests and laboratory tests on intact cores were also carried out. The results shown in Table 1 show a striking scale effect. Clearly, the larger the rock mass involved, the lower the modulus. However, increasing the size of the testing device by one order of magnitude meant halving Young's modulus values and the lowest moduli could not have been measured without resorting to specially designed, large scale in situ tests.

Table 1. Young's moduli of a slate-quartzite formation

| | | |
|---|---|---|
| Sonic | $-$ | 25 000 MPa |
| Unconfined Compression | (5 10 mm) | 15 000 MPa |
| Plate loading | (5 $10^2$ mm) | 5 000 MPa |
| Radial press | (5 $10^3$ mm) | 2 000 MPa |

Without going any further, it is worth mentioning that under the same special investigation contract, large compression and shear tests on graphytic seams, as well as rock permeability and treatment tests, were performed.

## 2.2 Geomechanical Models

A good example of the use of geomechanical models in the earliest stages of a project is the case of a long, thin ridge that formed a natural dam at a reservoir rim. Assessing the behaviour of the ridge under a large water load was a key decision to accepting the project. The maximum pool level, and hence the height of the dam, were also dictated by ridge stability considerations.

Evaluating the deformations and safety of the ridge was open to discussion: it was decided to build a geomechanical model. The ridge's model is shown in Fig. 2, it was tested for different water elevations on the upstream face and for different uplift conditions along the joints i.e. for different arrangements of the grouting and drainage works. The model results, shown in Fig. 3, combined with the results of mathematical analyses, were used to select the maximum safe height of the water in the reservoir and to design the grouting and drainage systems.

It may be interesting to show how the ridge actually behaved under the water load. The deformations at the top were obtained with an inversed pendulum 200 m long. They proved remarkably close to the model results.

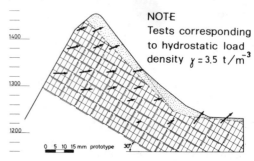

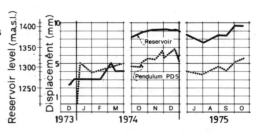

Fig. 3 (above) - Comparison of horizontal displacements measured in a geomechanical model and in the prototype rock ridge

Fig. 2 (left) - Geomechanical model of a natural ridge to assess deformations and stability under water loads and seepage

## 2.3 Large scale testing of construction materials

Whenever we deal, as we usually do in gravel fills or rockfills, with materials the grading, maximum size and actual packing of which govern the behaviour of the mass (including permeability, volume change during shear and contraction upon wetting), conventional laboratory equipment cannot reproduce prototype conditions. Very large testing equipment capable of overcoming these difficulties exist at a few testing centers around the world. Seldom can they be set up at the project site.

Seeing that compressibility and permeability are the engineering characteristics of greater relevance to today's practice and design procedures, testing rockfill materials under stresses possibly exceeding or at least comparable with those of the dam, becomes essential. Under very high vertical stresses and substantial confinement, the strain behaviour of different zones in a high dam may actually undergo striking reversals. A good example comes from Chicoasen dam. Fig. 4 shows how, under vertical stresses larger than       , the limestone rockfill of the shell becomes more compressible than the clay of the core.

Permeability is equally strictly dependent on grading and should be obtained on samples large enough to contain the prototype grading and with a cross sectional area sufficient to assume that the statistical distribution of the individual particles in the specimen is nearly that of the actual fill.

Among the very large testing facilities presently in use, we can mention the solid wall 1 130 mm diameter, 1 000 mm high, 600 Mg maximum load oedometer in use at the Comisión Federal de Electricidad of Mexico and the 600 mm diameter, 1 200 mm high, 250 Mg maximum load stacked ring oedometer developed by ISMES, Bergamo, Italy. The stacked ring oedometer shown in Fig. 5 allows the tangential strains to be measured and the lateral pressures to be derived. In its latest versions, it also allows running permeability tests under back pressure. Interestingly enough, a limited number of experimental results from the

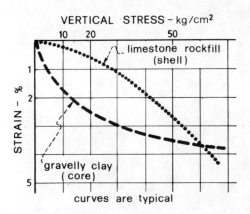

Fig. 4 - Compressibility under high loads of limestone rockfill shell and gravelly clay core (from S. D. Wilson)

Fig. 5 - One of the stacked rings oedometer presently in use at the field laboratory of a large dam

stacked rings oedometer and from conventional, small size, triaxial tests seem to indicate that the measured radial pressures can be related to the friction angle of the material using the same relationship

$$\emptyset = \arcsen (1 - \frac{r}{a})$$ derived for

normally consolidated clays.

## 2.4 Preconstruction instrumentation

The complexity of any damsite is such that it is not possible to know all its details. It is even less likely that a comprehensive behavioural model of a natural foundation is derived from a few experimental data. An alternative or rather complementary approach is to base the relevant design decisions on the overall behaviour of the site. Observing and monitoring phenomena likely to be the result of natural conditions in their real complexity all at play with their relative weight, may provide the best support to the customary experimental determinations.

Preconstruction instrumentation is intended to perform exactly this and should consist of a set of well selected and properly installed instruments to be placed as soon as possible and to be monitored for sufficient time during design; monitoring should be continued during construction, experimental impounding and normal operation.

As an example of useful application of preconstruction instrumentation, we can mention a reservoir slope facing a spillway intake. The slope was steep, formed by intensely tectonized and highly weathered graphytic schists. It was therefore inherently weak and showed signs of creep movements. The slope was instrumented at the outset of design activities, well before construction. Several long base, invar wire precision extensometers were installed in adits. The recorded

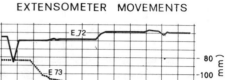
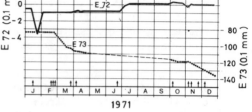

Fig. 6 - Creep movements of a reservoir slope detected during early design stages thanks to "preconstruction instrumentation"

movements, partly shown in Fig. 6, soon appeared as being more than creep but luckily confined to the outer portion of the slope. Extensometer data were confirmed by slope indicator observations.

The relevance of the movements, irrefutably proven by preconstruction instrumentation convinced the designers to introduce an important

Fig. 7 - A 150 m high rockfill introduced during design to buttress a creeping slope facing the spillway

change in the project layout: it consisted of a rockfill 150 m high placed to buttress the whole slope in front of the spillway. The completed buttress is shown in Fig.7. Its size and the associated construction difficulties made it a major work. Luckily, the timely warning of preconstruction instrumentation helped to introduce it early enough so that it could be used as a controlled rock waste. Appropriate provisions could be built into the contract documents which minimized extra costs and avoided claims.

2.5 Preconstruction works

For a site of special complexity, the Construct to Design approach may have to be expanded up to building first stage works or parts of the final structure to test the adequacy of the solution proposed.

This is quite typical, and often the only sound way of proceeding, when dealing with karstified foundations. Karstic limestone exists over large areas of South East Asia like Thailand, Malaysia, Indonesia. Over large extensions of Europe, from Lebanon to the South of France through Turkey, Yugoslavia and Italy, limestone is widespread and karstified to some extent.

To convey this point better, let us make one example only. A large dam was under design in a karstic area of Lebanon. As often in karstic land, the river was too small to fill the reservoir and the project had to include a diversion weir and a tunnel tapping a river in a nearby valley. A wrong assessment of the watertightness of the reservoir or a mistaken assumption about the possibility of efficiently sealing the damsite would have left three structures (a dam, a tunnel and a diversion weir) as useless monuments.

It was then decided to build first a low dam, with a shorter and lighter grout curtain, and to carry out a trial impounding with a water head nearing a third of the final one as shown in Fig. 8. The low dam could be built with the product of the excavations required for the final structure. The low dam could be waterproofed quite simply and economically, by a thin polymer membrane connected to the peripheral concrete beam to become later the

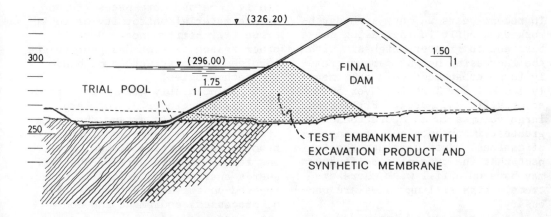

Fig. 8 - A low dam intended as a large scale testing work to pond a potentially leaky reservoir before engaging in the construction of the main dam. The testing dam is built with the product of the excavations for the final dam and waterproofed with a recoverable synthetic membrane

tie-in of the bituminous concrete facing of the high dam. After testing to satisfaction the reservoir and foundation with the low dam, this will be demolished (the membrane can be reused elsewhere) and the high dam built together with the rest of the project with a reasonable degree of confidence.

Introducing the Construct to Design concept had an important influence both on the layout and on the type of the final dam: the choice of an upstream bituminous facing was, in a way, a forced one.

Even if it is unlikely that conditions similar to these will repeat themselves, there certainly exist projects where anticipating some of the construction activities could be beneficial if not necessary. Karstic foundations are a typical case. It is often advisable to treat a karstic foundation to satisfaction before the rest of the project is set in motion. The otherwise associated risks may be that of unpredictable delays, cost overruns and even abandoning the whole dam unused.

3 DESIGN TO BUILD

In recent years we have seen earth-rock dams built in increasing numbers and to greater heights. With the increasing height came extremely large embankment volumes: nearly 100 $Mm^3$ for Tarbela, over 50 $Mm^3$ at Oroville and Nurek. Filling such large volumes is an operation of the greatest importance and of many implications. Just consider the impact that the type of specification may have on a fill work where extremely high filling rates are necessary.

If a "performance" (or end result) specification is adopted, the order to remove a given zone of the fill, because it does not meet the specified standards may come too late when the rejected fill has already been covered, or be such as to disrupt the entire filling process if this has been organized upon a continuous operation (e.g. conveyor belts).

On the other hand, if a "prescriptive" (or equipment and method) specification is chosen, modifying the characteristics of fill which, when placed according to specs is not satisfactory, may become a tremendous task if the Contractor has to provide new equipment or to modify his installations (i.e. processing plants).

It may not always be necessary to direct the Contractor to remove and replace the defective work: this course of action may result in delays and loss of revenue which would be more costly to the Owner than the diminished value of the work. However, and this is particularly true for large dams, even a minor portion of the structure built below standard may present an unacceptable threat.

The designer must therefore involve himself with the problems and alternatives of construction and provide the Contractor with as much information and insight on his idea of how best to build the structure.

Processing the materials of the dam is more and more necessary to ensure material uniformity or to improve fill performance. Here is another reason for the designer to get involved in construction methods and technologies.

It is undeniably within the trends of today's engineering for large earth-rock dams to extend the scope of design so as to orient the selection of construction technologies. The result of this part of the designer's analysis and effort is often imposed on the Contractor in terms of processes, equipment and construction sequences.

All this should be recognized as a necessary development of the Art and not dismissed as an undue intrusion in others' jobs and responsibilities. It should be accepted by all concerned parties: the Owner, Contractors and not last, by soil engineers themselves. It is not by chance that some of the finest and outstanding earth-rock dams of the last decade have been built by Authorities that design, build and monitor their own dams thus implementing to the highest extent what we may call here a Design to Build concept.

DESIGN TO BUILD means extending the customary process of computing, drafting and specifying each part of a dam, to figuring and defining ways and means of building it. Design to build means making sure that what is asked for can be built in the first place and working out one reasonably sound and economic construction method. Strictly speaking, at least for very large or unusually difficult dams, construction costs and programs should be defined only through such an exercise and no cost or time estimate should be considered a serious professional one if obtained otherwise. The writer can recall at least one dam where by refusing to consider at design stage the problems inherent to construction, led to the walk-out of two successive Contractors.

Again an example may be the best way to present further this concept. The proposed San Roque dam in the Philippines is a gravel fill 215 m high requiring an overall fill volume in excess of 45 $Mm^3$: nearly 80% of the fill is processed river alluvium. Processing as Fig. 9 shows, ranges from just scalping and crushing all boulders larger than 300 mm to a full classification and recombination of different sand and gravel classes to produce buffer and transition zones. Processing is maximum for the core which is obtained as a mix of 35% clay and 65% of sand free gravel. After blending, the clay-gravel mix is placed in a stock yard, wetted with a sprinkler system and allowed to cure for six months.

To make sure that all operations

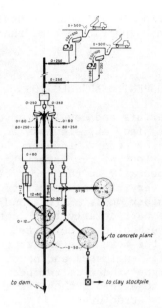

Fig. 9 - Schematic diagram of equipment, operations and products of processing plant for San Roque dam

were consistent with the materials available and could be performed with minimum waste and within expected times, the design also had to cover the processing plant and the material conveyance system. This was also necessary to define accurately enough the different unit costs. A comparative study showed that a belt conveyor system could efficiently move the 35 $Mm^3$ of alluvium required for the dam from the borrow to the processing plant and onto the fill. The same study showed that to complete the fill in 4 years, dumper hauling would cost 20% more.

Whenever dealing with very large embankment volumes, efficient construction methods and appropriate technologies become the key to success as they allow both shorter construction times and a saving in the project cost which is implicitly very high. If the dam is not conceived with this in mind, applying the best technology may prove impractical, or even impossible.

When confronted with embankment volumes larger than 15 $Mm^3$, designers should always consider design to build an essential part of their duties.

## 4 NON CONVENTIONAL SOLUTIONS

New materials and novel solutions occupy a place of relevance in dam engineering today. Industry is making a variety of new materials available especially in the areas of very low and very high permeability, they add to older ones. Stress and strain analysis particularly the Finite Elements Method, provide a real insight even in the most complex structures. Yet, soil engineers and particularly dam designers have so far been reluctant to expoloit these advantages for their structures.

### 4.1 New materials

In recent projects of earth-rock dams there has actually been an increase in the use of non-soil materials and geomembranes. It will be enough to mention concrete faced rockfills and the success of bituminous concrete for thin, flexible, impervious elements. It may be interesting to recall steel faced rockfills. Fig. 10 shows one such dam, designed some 15 years ago which has been operating successfully ever since. Worth noting is that the dam is at 3 600 m a.s.l. and has endured countless, large and rapid temperature variations and a few earthquakes with no noticeable detriment.

Fig. 11 shows a 1.5 mm thick hypalon membrane holding a pressure equivalent to 100 m of water. The support is a definitely rough one: angular rock chunks from 35 to 50 mm in size of crushed basalt. Fig. 12 shows a spun mat of synthetic fibres used as a filter between two materials the gradings of which were incompatible. This solution was successfully adopted at Phitsanulok dam on the Nan river in Thailand between a fine sandfill and its rocky armour.

Besides their characteristics of extremely low (or extremely high) permeability and of large deformability (up to 300%), synthetic materials offer to the designer a vital advantage: a quality standard and a uniformity that can very rarely be

Fig. 10 - Aquada Blanca steel faced rockfill is an example of the use of non soil materials in an earth-rock dam

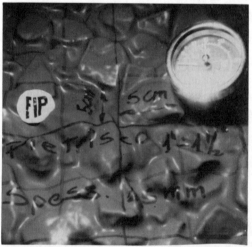

Fig. 11 - A synthetic membrane 1.5 mm thick on 2" rock fragments under an equivalent load of 100 m of water is seen here during a long duration test

Fig. 12 - A spun mat made with synthetic fibers used to filter materials of otherwise uncompatible grading

provided by a Civil Contractor. By deciding to use non-soil materials (including steel) the designer can set a lower standard for civil works because he is actually expecting that the vital contribution to ensure the dam safety, be provided by materials produced in a chemical plant or in a steel mill.

4.2 Unconventional design

Advances in design practice based on behavioural analysis enable dam engineers to adopt solutions that could not be considered only a few years ago. Adopting unconventional design may be the only key to projects which would otherwise be unfeasible.

A typical case of this sort is met where a relatively soft formation is followed by a much stronger one: classical are the sequences schists-limestone of flysch-limestone. Fig. 13 shows the solution adopted after two earlier and more conventional designs failed. The site is a deep and nearly vertical gorge less than 5 m wide, cut in sandstone just upstream, the ground is gentler, formed by flysch. Placing an earth-rock dam in the gorge was posing excessive problems: excavations, arching, seepage control. Accepting unconventional design led to a half dam with a markedly sloping core tying into the impervious flysch blanketing the sandstone. Shear displacements on horizontal planes in the core across the gorge gap, could be quantified and dealt with.

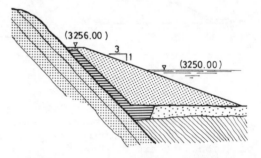

Fig. 13 - Unconventional design allowed a dam to be built where two previous attempts along standard concepts had proved nearly impossible to build

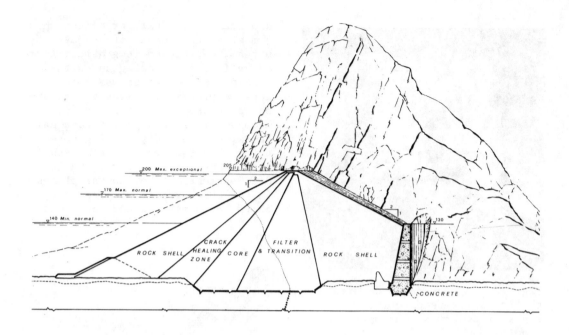

Fig. 14 - Unconventional design proposed damming a narrow gorge of competent but karstified rock. Placing the core upstream allows tying in an impervious formation. The gorge helps a check structure in the form of a concrete arc

This same concept has been proposed at another site where geomorphological conditions were quite similar. There, the gorge was much wider and cut through highly karstic limestone. In addition, the possibility existed of reservoir surges so great that the dam might be overtopped. The solution proposed, shown in Fig. 14, is similar to the previous one, but takes advantage of the extremely competent rock of the gorge where a concrete, arch-shaped toe wall, will check the crest washout and prevent the dam from being destroyed to any substantial depth below maximum normal pool level.

As a conclusive remark, we can say that earth-rock dam engineering today is living through an extremely interesting period: it is called upon to build the largest water retaining structures ever made by man and to find solutions to extremely difficult situations. Adequate tools and new materials are becoming available: it takes responsible and innovative engineers to make full use of them and to further develop the Art of building with natural materials.

# Earthquake-resistant design of earth dams

H.BOLTON SEED
*University of California, Berkeley, USA*

SYNOPSIS. Lessons gained from observations of the field performance of earth dams during earthquakes are reviewed and used to illustrate the primary problems of concern. Defensive design measures which may be taken to mitigate the various hazards are reviewed and illustrated. Analytical approaches for evaluating seismic stability and the deformations of earth dams during earthquakes are discussed together with recent developments which facilitate their implementation in special cases; situations which require careful consideration of special effects such as the three-dimensionality of the dam-valley system and pore pressure re-distribution following an earthquake are discussed and illustrated.

INTRODUCTION

Since the near-failure of the Lower San Fernando Dam during an earthquake just north of Los Angeles in 1971--an event which necessitated the immediate evacuations of over 80,000 people whose lives were endangered--the design of earth dams to resist earthquake effects has assumed a position of much greater significance among design engineers. Prior to this event it was generally believed that earth dams were inherently resistant to earthquake shaking, but the major slide in the Lower San Fernando Dam (see Fig. 1), which involved the upstream shell, the crest of the dam, and 30 ft of the downstream slope, with a resulting loss of 30 ft of freeboard, provided dramatic evidence that this is not necessarily so. As a result, regulatory agencies became more stringent in their requirements for demonstration of adequate seismic stability, and design engineers responded by developing new and more convincing design approaches than had previously been used. Thus the past 10 years have seen a major change in interest and attitude towards this aspect of design.

The problem is not of limited interest. Many parts of the world are subjected to the potentially hazardous effects of earthquakes and this includes many countries in Southeast Asia--Japan, the Philippines, Java, Sumatra, Burma, India, Pakistan and China, for example, are all areas of potentially high seismicity--and stringent earthquake-resistant design criteria have already been adopted for the design of Tarbela Dam in Pakistan, Koyna Dam in India and Tseng-wen Dam in Taiwan, among others.

As in all aspects of geotechnical engineering, the initial starting point in developing an understanding of the problem lies in observations of the field performance of structures during actual earthquakes. The significant lessons to be gained from such studies are summarized briefly in the following section.

LESSONS FROM FIELD PERFORMANCE OF DAMS DURING EARTHQUAKES

The sequence of events associated with the performance of Hebgen Dam in the Hebgen Lake earthquake of 1959 first brought to the attention of design engineers the wide variety of damage which may result from earthquake shaking (Sherard, 1967). The various possibilities are listed in Table 1 and virtually all of these, with the exception of sliding along the base, were evidenced to some degree in the behavior of Hebgen Dam during this particular earthquake. A careful study of the behavior of this dam is an object lesson in earthquake-

Fig. 1. View of Lower San Fernando Dam after Upstream Slope Slide in Earthquake of Feb. 9, 1971.

resistant design for engineers interested in this field.

Other studies of failures and non-failures of dams shaken by strong earthquake motions also provide valuable insights into types of behavior, however (Seed et al., 1978). Of major importance for example is a review of the performance of earth dams which existed in close proximity to the San Andreas Fault in the San Francisco earthquake of 1906. At that time there were 33 dams within 35 miles of the fault (and 15 within five miles of the fault) on which a magnitude 8-1/4 earthquake occurred, so there can be little doubt that all of these dams were subjected to strong shaking for a prolonged period of time (over 1 min). Based on recent correlations of ground motions with distance in California earthquakes, it seems reasonably sure that all of these dams were subjected to ground motions having peak accelerations greater than 0.25g and those within five miles of the

Table 1. Possible Ways in which an Earthquake May Cause Failure of an Earth Dam.

---

1. Disruption of dam by major fault movement in foundation
2. Loss of freeboard due to differential tectonic ground movements
3. Slope failures induced by ground motions
4. Loss of freeboard due to slope failures or soil compaction
5. Sliding of dam on weak foundation materials
6. Piping failure through cracks induced by ground motions
7. Overtopping of dam due to seiches in reservoir
8. Overtopping of dam due to slides or rockfalls into reservoir
9. Failure of spillway or outlet works

---

fault to motions with peak accelerations greater than about 0.6g. Yet significantly none of these old dams suffered any significant damage and there was certainly no evidence of slope instability. It is not possible to attribute this to the use of flat slopes for the dams (slopes varied typically from 1 on 2 to 1 on 3) or to the high quality of construction (most of the dams were not compacted with rollers but by moving livestock or by teams and wagons). However, a significant characteristic of all of the dams is that they were constructed of clayey soils on rock or clayey soil foundations. Only two of the dams were built largely of sand and for these structures the sand was apparently not saturated. It is reasonable to conclude therefore that dams built of clayey materials seem to exhibit high resistance against slope failures during earthquakes-- but the field data provide no information on the possible behavior of saturated sands.

This important conclusion is reinforced by the study by Akiba and Semba (1941) of the performance of dams in the 1939 Ojika earthquake in Japan. As a result of this earthquake 12 cases of complete dam failures occurred together with about 40 cases of reported slope failures. The main conclusions of this study were

(a) There were very few cases of dam failures during the earthquake shaking, most of the failures occurring either a few hours or up to 24 hours after the earthquake.

(b) The majority of the damaged and failed embankments consisted of sandy soils and no complete failures occurred in embankments constructed of clay soils.

(c) Even at short distances from the epicenter, there were no complete failures of embankments constructed of clay soils; however, at greater epicentral distances there was a heavy concentration of completely failed embankments composed of sandy soils.

It is clear that these observations tend to confirm those concerning dams built of clay soils in the 1906 San Francisco earthquake. In addition the Ojika earthquake experience provides clear evidence of the vastly superior stability of embankments constructed of clay soils under strong seismic loading conditions over those constructed of saturated sands--a fact fully in accord with Terzaghi's insightful considerations on this question. Moreover, the performance record clearly suggests that the critical period for an embankment dam subjected to earthquake shaking is not only the period of shaking itself, but also a period of hours following an earthquake, possibly because piping may occur through cracks induced by the earthquake motions or because slope failures may result from pore pressure redistribution.

More recently, important earthquakes in which dam performance was observed include: the 1968 Tokachi-Oki earthquake in which a large number of slope failures occurred at shaking levels of the order of 0.2g in dams constructed of loose volcanic sand (Moriya, 1974) and the 1971 San Fernando earthquake in which slope failures occurred in the Upper and Lower San Fernando Dams, both having sand shells (Seed et al., 1975b), while generally excellent performance was observed in 25 rolled earth fill dams at shaking levels between 0.2 and 0.4g (Seed et al., 1978). The latter event, together with the performance of hydraulic fill dams in Russia (Ambraseys, 1960), also showed that even dams constructed with hydraulically deposited sand shells can withstand levels of shaking up to about 0.2g from magnitude 6-1/2 earthquakes without detrimental effects.

The slide movements in the San Fernando Dams were of special importance since in both cases, field observations showed that the shaking induced by the earthquake caused a dramatic increase in pore-water pressures in the shells of the dams and, in the case of the Lower San Fernando Dam, a condition of liquefaction which led to a major slide resembling a flow slide, in the upstream shell. In both cases the slide movements were apparently associated with a loss of strength associated with these pore water pressure increases.

The general conclusions (Seed et al., 1978) which seem to follow from a close study of embankment dam performance during earthquakes are as follows

(a) Hydraulic fill dams have been found to be vulnerable to failures under unfavorable conditions and one of the particularly unfavorable conditions would be expected to be the shaking produced by strong earthquakes. However, many hydraulic fill dams have performed well for many years and when they are built with reasonable slopes on good foundations they can apparently survive moderately strong shaking--with accelerations up to about 0.2g from magnitude 6-1/2 earthquakes with no harmful effects.

(b) Virtually any well-built dam on a firm foundation can withstand moderate earthquake shaking, say with peak accelerations of about 0.2g with no detrimental effects.

(c) Dams constructed of clay soils on clay or rock foundations have withstood extremely strong shaking ranging from 0.35 to 0.8g from a magnitude 8-1/4 earthquake with no apparent damage.

(d) Two rockfill dams have withstood moderately strong shaking with no significant damage and if the rockfill is kept dry by means of a concrete facing, such dams should be able to withstand extremely strong shaking with only small deformations.

(e) Dams which have suffered complete failure or slope failures as a result of earthquake shaking seem to have been constructed primarily with saturated sand shells or on saturated sand foundations and these types of dams require careful attention to ensure their seismic safety.

(f) Since there is ample field evidence that well-built dams can withstand moderate shaking with peak accelerations up to at least 0.2g with no harmful effects, we should not waste our time and money analyzing this type of problem--rather we should concentrate our efforts on those dams likely to present problems either because of strong shaking involving accelerations well in excess of 0.2g or because they incorporate large bodies of cohesionless materials (usually sands) which, if saturated, may lose most of their strength during earthquake shaking and thereby lead to undesirable movements.

(g) For dams constructed of saturated cohesionless soils and subjected to strong shaking, a primary cause of damage or failure is the build-up of pore water pressures in the embankment and the possible loss of strength which may accrue as a result of these pore pressures. Methods of stability analysis which do not take these pore pressure increases and associated loss of strength into account are not likely to provide a reliable basis for evaluating field performance.

DEFENSIVE DESIGN MEASURES

It may be noted that most of the potential problems which may develop as a result of earthquake action do not require analytical treatment but simply the application of commonsense defensive measures to prevent deleterious effects. Thus to prevent a dam being disrupted by a fault movement in the foundation may simply require the identification of potentially active faults and the selection of a site where such faults do not exist. Similarly the potential for settlement, slumping or tectonic movements, all of which could lead to loss of freeboard, can be ameliorated by the provision of additional freeboard so that the loss of some portion would not have serious consequences. In short, many of the potentially harmful effects of earthquakes on earth and rockfill dams can be eliminated by adopting defensive measures which render the effects non-harmful. A list of such defensive measures would include the following (Seed, 1979):

(a) Allow ample freeboard to allow for settlement, slumping or fault movements.

(b) Use wide transition zones of material not vulnerable to cracking.

(c) Use chimney drains near the central portion of embankment.

(d) Provide ample drainage zones to allow for possible flow of water through cracks.

(e) Use wide core zones of plastic materials not vulnerable to cracking.

(f) Use a well-graded filter zone upstream of the core to serve as a crack-stopper.

(g) Provide crest details which will prevent erosion in the event of overtopping.

(h) Flare the embankment core at abutment contacts.

(i) Locate the core to minimize the degree of saturation of materials.

(j) Stabilize slopes around the reservoir rim to prevent slides into the reservoir.

(k) Provide special details if danger of fault movement in foundation.

This list should not by any means be considered all-inclusive. Occasionally special situations will require or provide the opportunity for unique defensive measures such as the double dam system which protects against release of water from the Los Angeles Dam. The dam itself is designed to withstand probably the strongest earthquake criteria ever established but in the very remote possibility of a release of water, the people living downstream are protected by a second dam half a mile downstream from the first which stores no water and is only required to function in the remote chance the main dam releases water. The space between the two dams is maintained as a park area. Many engineers may consider this resorting to extreme defensive measures but a combination of public pressures, political considerations and ready availability of both the downstream dam and park space dictated a highly acceptable solution to many concerned residents. I personally consider this system to provide one of the safest downstream environments of all dams with which I am acquainted.

On the other hand, situations may develop where, in spite of the utmost care in planning, actual events may change professional evaluation of the potential activity of faults. This would appear to be the case for the proposed Auburn Dam site in California. When plans for this dam were first developed in the 1960's, the area was considered seismically quiet and the dam-site immune from active faults. The occurrence of the Oroville earthquake in 1975 (State of California Department of Water Resources, 1977), some 60 miles to the north, led to a re-evaluation of this situation and the determination that potentially active faults exist very close to or possibly even across the proposed damsite. Clearly this has led to a re-evaluation of the desirability of constructing a thin concrete arch dam at this location. Such occurrences point up the need for prudence in evaluating the potential seismicity of any dam-site.

Defensive measures, especially the use of wide filters and transition zones, provide a major contribution to earthquake-resistant design and should be the first consideration by the prudent engineer in arriving at a solution to problems posed by the possibility of earthquake effects.

At the same time it is necessary to recognize that all reasonable steps should be taken to ensure that sliding such as that which occurred at the Lower San Fernando Dam, the Sheffield Dam or a number of dams in Japan does not invalidate the beneficial effects of defensive measures; and many members of the public, regulatory agencies and even leading dam engineers will readily acknowledge that relatively sophisticated analyses are warranted in many cases to provide guidance on the possibility of slide movements developing during earthquakes and their possible extent. Accordingly it is of interest to review the approaches taken to check on this possibility and evaluate their significance in the design procedure--if for no other reason than to provide some insight into their effects in producing safe designs. Early attempts in this direction are discussed in the following pages.

PSEUDO-STATIC ANALYSIS PROCEDURES

For the past 40 years or more, the standard method of evaluating the safety of earth dams against sliding during earthquakes has been the so-called pseudo-static method of analysis in which the effects of an earthquake on a potential slide mass are represented by an equivalent static horizontal force determined as the product of a seismic coefficient, $k$, or $n_g$, and the weight of the potential slide mass, as illustrated in Figure 2. Attempts by the author to determine the originator of this method have proved singularly unsuccessful but the earliest written version that I have found appears in a classical paper by Terzaghi (1950). Terzaghi described the method in the following words.

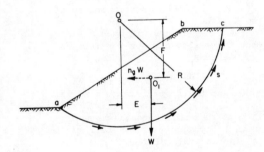

Fig. 2. Pseudo-static Method for Seismic Stability Analysis of Embankments.
(after Terzaghi, 1950)

"An earthquake with an acceleration equivalent $n_g$ produces a mass force acting in a horizontal direction of intensity $n_g$ per unit of weight of the earth. The resultant of this mass force, $n_g W$, passes like the weight W, through the center of gravity $O_1$ of the slice abc. It acts at a lever arm with length F and increases the moment which tends to produce a rotation of the slice abc about the axis O by $n_g FW$. Hence the earthquake reduces the factor of safety of the slope with respect to sliding from $G_s$, equation (1) to

$$G_s' = \frac{slR}{EW + n_g FW} \quad (2)$$

"The numerical value of $n_g$ depends on the intensity of the earthquake. Independent estimates (Freeman, 1932) have led to the following approximate values

Severe earthquakes, Rossi-Forel scale IX    $n_g = 0.1$

Violent, destructive, Rossi-Forel scale X    $n_g = 0.25$

Catastrophic    $n_g = 0.5$

The earthquake of San Francisco in 1906 was violent and destructive (Rossi-Forel scale X), corresponding to $n_g = 0.25$.

"Equation (2) is based on the simplifying assumptions that the horizontal acceleration $n_g g$ acts permanently on the slope material and in one direction only. Therefore the concept it conveys of earthquake effects on slopes is very inaccurate, to say the least. Theoretically a value of $G_s' = 1$ would mean a slide, but in reality a slope may remain stable in spite of $G_s'$ being smaller than unity and it may fail at a value of $G_s' > 1$, depending on the character of the slope-forming material.

"The most stable materials are clays with a low degree of sensitivity, in a plastic state (Terzaghi and Peck, 1948, p. 31), dense sand either above or below the water table, and loose sand above the water table. The most sensitive materials are slightly cemented grain aggregates such as loess and submerged or partly submerged loose sand...."

In spite of Terzaghi's profound influence on virtually all aspects of soil mechanics, the seismic design of earth dams appears to have been one area where his advice went unheeded and literally hundreds of dams of all types have had their seismic stability evaluated using this approach coupled with seismic coefficients substantially smaller in magnitude than those which he advocated. In the United States, for example, seismic coefficients have typically ranged from 0.05 to 0.15 even in areas such as California where the strongest imaginable earthquakes may well occur; in Japan values have characteristically been less than about 0.2. Similar values have been used in highly seismic regions throughout the world, as shown by the design criteria listed in Table 2, and engineers were apparently convinced that such low values were all that were required to ensure an adequate level of seismic stability. No special consideration seems to have been given to the nature of the slope-forming or foundation materials and if the computed factor of safety was larger than unity, it has generally been concluded that the seismic stability question has been satisfactorily resolved.

This is certainly not in accordance with Terzaghi's conception of embankment behavior, specifically expressed in the words: "Theoretically a value of $G_s' = 1$ would mean a slide but in reality a slope may remain stable in spite of $G_s'$ being smaller than unity and it may fail at a value of $G_s' > 1$, depending on the character of the slope-forming materials." This statement clearly indicates that in Terzaghi's opinion a slope may be stable or unstable even if the computed factor of safety is greater than 1, but the optimistic position has generally held sway--that is, dams have been considered to have had adequate seismic stability so long as $G_s'$ remained equal to or greater than unity regardless of the nature of the slope-forming material.

A more detailed discussion of the reasons for the widespread use of this method has been presented elsewhere (Seed, 1979). A careful study of its merits, based on studies of earthquake-induced slides, shows that the method does not always predict failure, where failures have been found to occur, in embankments consisting of sandy soils or constructed on sandy foundations which show a marked loss of strength due to earthquake shaking (see for example, the failure of the Sheffield Dam in Fig. 3). Thus it cannot be used reliably for evaluating the possible performance of these types of dams (Seed, et al., 1967, 1979).

On the other hand, based on a method of analyzing embankment deformations suggested by Newmark (1965) and applicable to soils which show no significant loss of strength due to earthquake shaking, (usually clayey soils, dry sands and some very dense cohesionless materials), it may be shown that (in cases where the crest acceleration does not exceed about 0.75g), deformations of such embankments will usually be acceptably small if the embankment

Table 2. Design Criteria for Selected Earth Dams (ICOLD Report)

| Dam | Country | Horizontal Seismic Coefficient | Minimum Factor of Safety |
|---|---|---|---|
| Aviemore | New Zealand | 0.1 | 1.5 |
| Bersemisnoi | Canada | 0.1 | 1.25 |
| Digma | Chile | 0.1 | 1.15 |
| Globocica | Yugoslavia | 0.1 | 1.0 |
| Karamauri | Turkey | 0.1 | 1.2 |
| Kisenyama | Japan | 0.12 | 1.15 |
| Mica | Canada | 0.1 | 1.25 |
| Misakubo | Japan | 0.12 | -- |
| Netzahualcoyote | Mexico | 0.15 | 1.36 |
| Oroville | USA | 0.1 | 1.2 |
| Paloma | Chile | 0.12 to 0.2 | 1.25 to 1.1 |
| Ramganga | India | 0.12 | 1.2 |
| Tercan | Turkey | 0.15 | 1.2 |
| Yeso | Chile | 0.12 | 1.5 |

Fig. 3. Failure of Sheffield Dam in Santa Barbara Earthquake of 1925.

can be shown to have a factor of safety of about 1.15 in a pseudo-static analysis performed using a seismic coefficient of 0.15 (Seed, 1979).

This means that in deciding whether or not to use the pseudo-static method of analysis in any given case, the critical decision to be made by the design engineer is whether the soil is likely to be vulnerable to excessive strength loss or pore pressure development or not. This can be determined by tests. However, both field and laboratory experience indicates that clayey soils, dry sands and in some cases dense saturated sands will not lose substantial resistance to deformation as a result of earthquake or simulated earthquake loading and thus pseudo-static analyses will generally provide an acceptable method of ensuring adequate performance for embankments constructed of these types of soil. In cases of doubt, however, a careful laboratory study will invariably provide the information from which an appropriate engineering decision concerning the applicability of the method can be made. It should also be noted that even some soils which might be vulnerable to the development of large pore pressures and some strength loss under conditions of strong shaking may show little evidence of these effects under less intense shaking, in which case the principles discussed above would still be applicable.

The Newmark (1965) type of analysis leading to evaluations of slope displacements by double integration of that portion of the induced acceleration response of a potential slide mass exceeding the computed yield acceleration for the mass (see Figure 4), represented a major step forward in both the philosophy of evaluating embankment performance during earthquakes and in suggesting the means for its implementation. It is an interesting by-product that the use of this improved method based on the prediction of embankment displacements should ultimately lead to the conclusion that for some types of soils and conditions (i.e., those which do not build up large pore pressures or cause significant strength loss due to earthquake shaking and associated displacements), the old pseudo-static analysis often did an adequate job of limiting displacements in the first place and thereby provided, in many cases, an entirely adequate design procedure.

BEHAVIOR OF SATURATED COHESIONLESS SOILS

In contrast to the behavior of clayey soils and very dense cohesionless soils, loose to medium dense cohesionless soils do not exhibit any clearly defined yield strength and their behavior is complicated by the possibility of large pore pressure build-ups and redistribution during and following an earthquake. Similarly the stress-strain relationships for a sample of loose sand, before and after cyclic loading may be quite different.

In materials of this type, the pore pressure build-up during cyclic loading clearly affects the resistance to deformation, both under further cyclic stresses

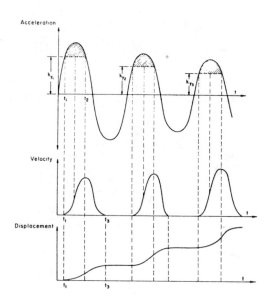

Fig. 4. Newmark Method of Evaluating Embankment Displacements.

and under static stresses. It is difficult to establish a well-defined value of yield stress for such soils and both the generation and the redistribution of the large pore pressures developed during cyclic loading will affect the seismic stability, the induced deformations and possibly the post-earthquake behavior. This was made readily apparent by the failure of the Sheffield Dam in 1925 and the subsequent analysis of its stability (Seed et al., 1969). Predicted deformations, neglecting any pore pressure build-up would have been only a few inches, yet the dam failed completely. Similar results were subsequently observed in analyses of the Upper and Lower San Fernando Dams in the 1971 San Fernando earthquake. Accordingly, an analysis technique is required which takes into account the pore pressures generated by the earthquake shaking and their potential effects. A procedure for effecting this is described below.

SEED-LEE-IDRISS ANALYSIS PROCEDURE

In recognition of the limitations of the pseudo-static analysis approach and the difficulties of evaluating a yield stress criterion for many saturated cohesionless soils, the writer developed an alternative approach to the evaluation of deformations in earth dams (Seed, 1966). The details of the general procedure have undergone many improvements since that time (Seed et al., 1975a) primarily through the development and application of finite element procedures with the aid of Doctors I. M. Idriss, J. M. Duncan, F. I. Makdisi, N. Serff, J. R. Booker, M. S. Rahman, and Professor W.D.L. Finn, but also through the development of improved testing procedures developed mainly with the help of Professor K. L. Lee, and Doctors P. DeAlba, R. M. Pyke and N. Banerjee

In spite of the improvements, however, the basic principles of the procedure have remained unchanged and involve a series of steps which might be summarized simply as follows:

(a) Determine the cross-section of the dam to be used for analysis.

(b) Determine, with the cooperation of geologists and seismologists, the maximum time history of base excitation to which the dam and its foundation might be subjected.

(c) Determine, as accurately as possible, the stresses existing in the embankment before the earthquake; this is probably done most effectively at the present time using finite element analysis procedures.

(d) Determine the dynamic properties of the soils comprising the dam, such as shear modulus, damping characteristics, bulk modulus or Poisson's ratio, which determine its response to dynamic excitation. Since the material characteristics are nonlinear, it is also necessary to determine how the properties vary with strain.

(e) Compute, using an appropriate dynamic finite element analysis procedure, the stresses induced in the embankment by the selected base excitation.

(f) Subject representative samples of the embankment materials to the combined effects of the initial static stresses and the superimposed dynamic stresses and determine their effects in terms of the generation of pore water pressures and the development of strains. Perform a sufficient number of these tests to permit similar evaluations to be made, by interpolation, for all elements comprising the embankment.

(g) From the knowledge of the pore pressures generated by the earthquake, the soil deformation characteristics and the strength characteristics, evaluate the factor of safety against failure of the embankment either during or following the earthquake.

(h) If the embankment is found to be safe against failure, use the strains induced by the combined effects of static and dynamic loads to assess the overall deformations of the embankment.

(i) Be sure to incorporate the requisite amount of judgment in each of steps (a) to (h) as well as in the final assessment of probable performance, being guided by a thorough knowledge of typical soil characteristics, the essential details of finite element analysis procedures, and a detailed knowledge of the past performance of embankments in other earthquakes.

This procedure may seem rather long and cumbersome but it also seems to incorporate the essential steps in evaluating such a complex problem as the response of earth dams to earthquake effects.

It lends itself naturally, however, to somewhat simplified versions of the method, which have often been used for reasons of time and economy (e.g., Finn, 1967; Klohn et al., 1978; Lee and Walters, 1972; Lee, 1978; Leps et al., 1978a and b; Vrymoed and Galzacia, 1978; etc.). The ultimate simplification is, of course, the total elimination of all analysis procedures and a simple evaluation, based on a knowledge of the materials comprising the dam and the judgment resulting from conducting many previous analyses and observing the performance of existing dams. However, it should be noted that each of the steps is an essential element of the procedure and if one of them is performed incorrectly, the results of the analysis may be grossly misleading. In such cases, where the job cannot be done properly, it may be better not to do it at all rather than to be misled by the erroneous results which may ensue. It is for this reason that judgment is necessary at each step in the development.

In the most modern versions of the method, the assessment of pore water pressures during and following the earthquake shaking may involve studies of simultaneous pore pressure generation and dissipation using appropriate computer programs (Booker et al., 1976; Finn et al., 1978) and the evaluation of the final configuration of the structure using a strain-harmonizing technique, again involving finite element procedures (Lee et al., 1974; Serff et al., 1976).

The particular procedure used in any given case should depend on the complexity of the case being considered, the margin of safety provided for the level of earthquake shaking likely to develop, and the judgment and experience of the engineer responsible for the study.

An interesting example of the use of this method to analyze a slope failure is provided by the analysis of the Lower San Fernando Dam (Seed, 1979). The computed response of this dam to the earthquake ground motions is shown in Fig. 5 with the dark area indicating the zones where the residual pore water pressure at the conclusion of the earthquake was equal to 100%. A stability analysis of this section, shown in Figure 6, clearly indicates that for these conditions a slope failure would develop.

An example of a seismic stability analysis where failure did not occur is illustrated by the computed response of the Upper San Fernando Dam in the same earthquake. Again extensive zones of high pore-water pressure were developed within the embankment as a result of the earthquake shaking but they were not sufficiently extensive to cause a failure, as illustrated in Figures 7 and 8. In this case the embankment suffered significant deformations, the crest moving downstream about 5 ft. Deformation analyses based on the method described above predicted a movement of about 3.8 ft.

As with all analytical procedures used in geotechnical engineering, the method should only be used if it is found to work--that is, if it provides reasonable evaluations of behavior for cases where the behavior has been or can be observed. In all, the general procedure described above has been used to study the performance of eight dams whose performance during earthquakes is known. Two of these had major slides, one underwent large deformations, one underwent small deformations and four had no discernible damage. The behavior predicted by the analysis was similar to that observed in the field in each case, and while it is true that each of these cases was in fact studied after the event involved, it seems that the procedure has the capability of giving considerable insight into the possible behavior of embankments subjected to earthquake effects. For this reason, presumably, it has been adopted in studies of many dams throughout the world as a

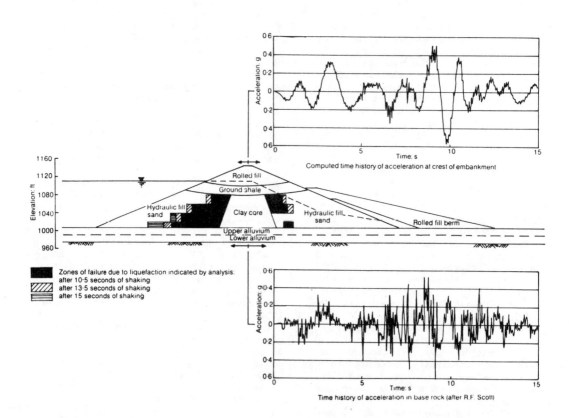

Fig. 5. Analysis of Response of Lower Dam during San Fernando Earthquake to Base Motions Determined from Seismoscope Record.

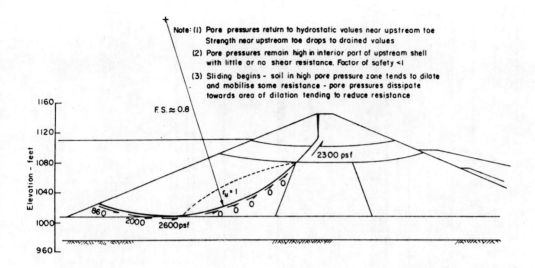

Fig. 6. Stability of Lower San Fernando Dam a Short Time after Earthquake Motions Stop.

guide to final assessment of their probable performance during earthquakes (e.g., Seed et al., 1969, 1973, 1975b; Gordon et al., 1974; Kramer et al., 1975; Marcuson et al., 1977; Makdisi et al., 1978; Sadiah et al., 1978; State of California Department of Water Resources, 1979; etc.).

If the permeability of the shell material for a dam becomes sufficiently high, say of the order of the 1 cm/s, then it may be impossible for an earthquake to cause any build-up of pore pressures in the embankment since the pore pressures can dissipate by drainage as rapidly as the earthquake can generate them by shaking. A good example of such a situation is the upstream shell of Dartmouth Dam (Seed, 1974). This 650 ft high rock fill structure (see cross-section in Figure 9) has highly pervious shells having a 10% size of about 2.5 cm and a permeability coefficient possibly of the order of 100 cm/s. In such a case analysis shows that the pore pressure build-up in 10 s of earthquake shaking would be a negligible proportion of the initial effective overburden pressure (Seed, 1979) even for an earthquake which might hypothetically be considered to produce shaking of sufficient intensity to generate a pore pressure ratio of 100% throughout the shell if it were undrained. Clearly in a case such as this, the upstream shell can for practical purposes be considered to be fully drained during any reasonable period of earthquake shaking and stability can be evaluated on this basis.

A similar conclusion has been drawn in the evaluation of the seismic stability of the Watauga Dam (Sadigh et al., 1978) and the high permeability of the shells can be considered a major factor in maintaining a high degree of seismic stability in all highly pervious rock fills.

NEW DEVELOPMENTS IN EVALUATIONS OF SEISMIC STABILITY OF EMBANKMENTS

During the past few years there have been several new developments in methods of evaluating the seismic stability of earth dams. These include the following:

1. Use of Penetration Test Data in Lieu of Cyclic Load Test Data

One of the most widely used methods of evaluating the liquefaction potential of a sand deposit under level ground conditions is to compare the cyclic shear stresses induced by an earthquake with those which, on the basis of field experience for soils of known penetration resistance, have or have not liquefied during earthquakes. The use of this procedure requires only the determination of the standard penetration resistance of a sand deposit, and avoids judgments on the part of the engineer concerning the representativeness of soil

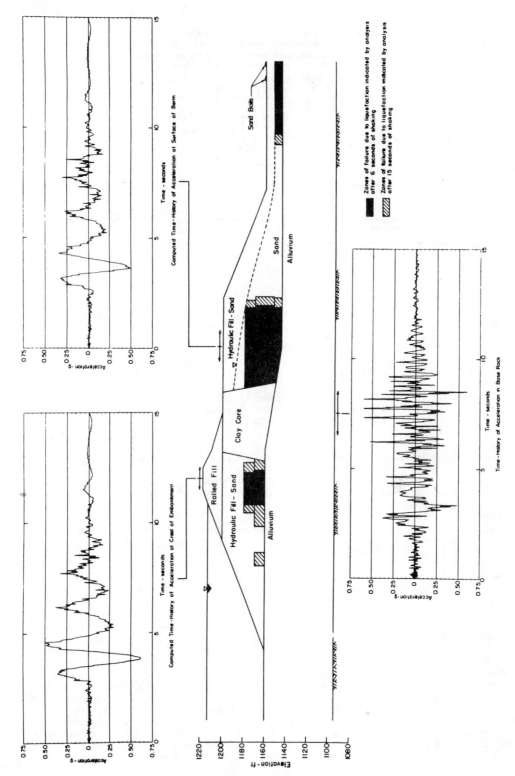

Fig. 7. Analysis of Response of Upper San Fernando Dam During San Fernando Earthquake

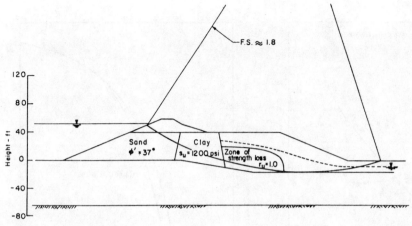

Fig. 8. Stability Analysis of Upper San Fernando Dam for Conditions Soon After Earthquake.

samples which might have been tested or the influence of sample disturbance on the results of such tests. The correlation charts most widely used in North America are shown in Figures 10 and 11 (Seed, 1979) and involves the use of a standardized penetration resistance $N_1 = C_N \cdot N$ where $C_N$ is a function of the overburden pressure at which the N-value is measured (see Figure 12).

It has long been recognized that when sand is subjected to initial static shear stresses before earthquake loading, as naturally occurs under the sloping surfaces of an earth dam, the resistance of the sand to pore pressure generation is increased. Typical results showing the effect of an initial static stress on horizontal planes on the cyclic stress required to cause a pore pressure ratio of 100% in tests on simple shear samples are shown in Figure 13. Similar results are obtained in cyclic loading triaxial compression tests conducted on samples consolidated initially under different principal stress ratios.

The form of the results shown in Figure 13 is most useful for analysis purposes since they show the magnitude of the cyclic stress ratio $\tau_c/\sigma_o'$ which must be superimposed on the initial static stresses acting on a soil element in order to cause a pore pressure ratio of 100%, for different values of the initial static stress ratio $\alpha = \tau_{fc}/\sigma_{fc}$. Tests on a number of sands have shown that the value of $\tau_c/\sigma_o'$ under these conditions increases with the value of $\alpha$ approximately as shown in Figure 14. Thus if the cyclic stress ratio required to cause a pore pressure ratio of 100% for the case where $\alpha = 0$ is known, the cyclic stress ratio causing the same pore pressure condition can be obtained from the relationship

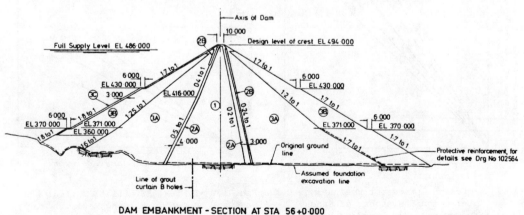

Fig. 9. Typical Cross-Section Through Dartmouth Dam.

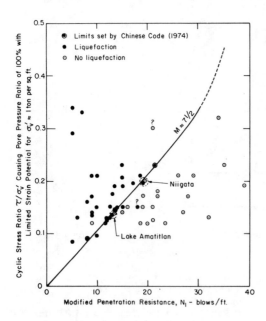

Fig. 10. Correlation Between Field Liquefaction Behavior of Sands for Level Ground Conditions and Penetration Resistance (Earthquakes with Magnitude $\simeq$ 7-1/2).

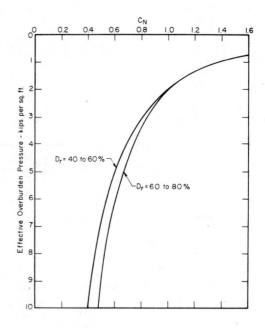

Fig. 12. Recommended Curves for Determination of $C_N$ Based on Averages for WES Tests.

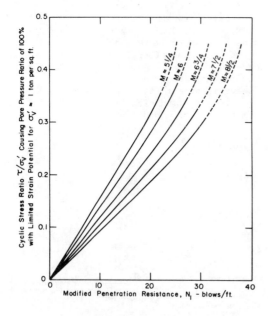

Fig. 11. Design Curves for Evaluating Field Liquefaction Resistance of Sands Under Level Ground from Standard Penetration Test Data.

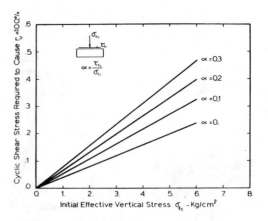

Fig. 13. Typical Effects of Initial Static Stresses on Cyclic Loading Resistance of Sands.

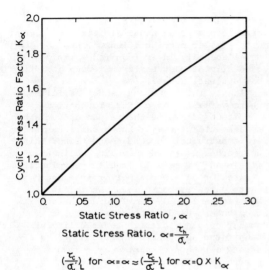

Fig. 14. Typical Chart for Evaluating Effect of Initial Cyclic Stresses on Cyclic Loading Resistance of Sands.

$$\left(\frac{\tau_c}{\sigma_o'}\right)_{\ell-\alpha} = \left(\frac{\tau_c}{\sigma_o'}\right)_{\ell-\alpha=0} \times K_\alpha$$

where $K_\alpha$ can be read off directly from Figure 14. Since the values of $\left(\frac{\tau_c}{\sigma_o'}\right)_{\ell-\alpha=0}$ can be determined from charts and as Figure 11, relating this parameter to the results of the standard penetration test, based on experiences at sites where liquefaction has or has not occurred during earthquakes, this method provides a simple but direct means of relating past field performance to probable future performance under given earthquake loading conditions.

It should be noted, however, that charts such as those shown in Figures 10 and 11 are based on sites where liquefaction has occurred under fairly small overburden pressures, typically less than 1.5 tons per sq ft and the charts are therefore applicable only to these conditions. However, as the overburden pressure on a soil increases, the cyclic stress ratio required to cause a pore pressure ratio of 100% decreases and therefore a correction must normally be made for this effect. Typically the reduction in cyclic stress ratio with increase in confining pressure approximates the values shown in Figure 5.

Thus a complete procedure for obtaining the cyclic shear stress required to cause a pore pressure increase of 100% could involve only the following sequence of steps:

1. Determine the standard penetration resistance, N, of the sand at a given point

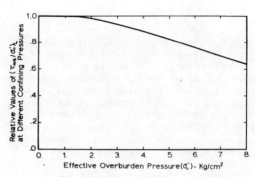

Fig. 15. Typical Reduction in Cyclic Stress Ratio Causing Liquefaction with Increase in Initial Confining Pressure.

of interest in a dam or its foundation.

2. Convert the measured N value to the value it would have at a confining pressure of 1 ton/sq ft using the relationship

$$N_1 = C_N \cdot N$$

where $C_N$ is a function of the effective overburden pressure as shown in Figure 12 (Seed, 1979).

3. From the value of $N_1$ thus determined, read off the corresponding value of the cyclic stress ratio required to cause a pore pressure ratio of 100% for cases where $\alpha = 0$ from charts such as that shown in Figure 11.

4. Determine the pre-earthquake static stresses $\tau_{hv}$ and $\sigma_v'$ at the point under consideration. Hence determine the value of $\alpha = \frac{\tau_{hv}}{\sigma_v'}$ for the point.

5. From Figure 14, determine the value of $K_\alpha$ corresponding to the known value of $\alpha$ and hence determine the value of

$$\left(\frac{\tau_c}{\sigma_v'}\right)_{\ell-\alpha} = \left(\frac{\tau_c}{\sigma_v'}\right)_{\ell-\alpha=0} \times K_\alpha$$

6. If the effective vertical stress is greater than 1.5 tons/sq ft, reduce the value of $\left(\frac{\tau_c}{\sigma_v'}\right)_{\ell-\alpha}$ obtained in step 5 by the correction factors shown in Figure 15.

7. For the final value of $\left(\frac{\tau_c}{\sigma_v'}\right)_{\ell-\alpha}$ obtained in step 6, determine the cyclic shear stress required to cause a pore pressure increase of 100% from the equation

$$\tau_c = \left(\frac{\tau_c}{\sigma_v'}\right)_{\ell-\alpha} \times \sigma_v'$$

The main advantages of this approach to cyclic resistance evaluation are that it is based on field performance data, it shows the variations in resistance from point to point, it avoids basing an analysis on a limited number of tests on hopefully representative samples and it avoids any issues associated with sample disturbance during the boring, transportation and handling of samples in the laboratory.

The main disadvantages of the approach are well-known difficulties of evaluating the values of standard penetration test data in the light of the factors known to influence the test results: e.g., method of lifting the hammer, method of supporting the walls of the hole, diameter of the drill hole, length of the drill stem, type of anvil used in the striking mechanism, etc.

However, these limitations of the standard penetration test do not preclude a meaningful evaluation, and the directness of the method has a major appeal to many engineers. It has already been used on several dams for evaluating the characteristics of sand deposits in the foundation (Khilnani and Byrne, 1981, Wahler and Associates, 1981) and will no doubt be used more frequently in the future. A typical example is shown in Figure 16, which shows the computed cyclic shear stresses induced by the design earthquake (Magnitude 6-1/2 at a distance of 24 miles) in the foundation deposits for the Camanche Dam in California and the resistance to cyclic loading determined from standard penetration test data following the procedure discussed above. It was concluded that the foundation sands had adequate capacity to support the dam, except near the toes of the embankment where soil liquefaction would cause minor sloughing of the outer parts of the shells but without any major damaging effects to the embankment.

2. Use of Comparative Procedures to Evaluate Seismic Stability

With the increasing number of dams for which (1) properties and performance under earthquake shaking of known intensities have been established and (2) properties and evaluations of seismic stability have been made, it is becoming possible to evaluate the probable performance of other dams by simple comparison with available performance or evaluation data for embankments constructed under similar conditions with similar configurations. Thus no detailed analysis is required--simply a good file of past performance and analytical evaluation data.

Clearly this approach cannot be followed where variable and unusual foundation conditions dominate the response, but where embankments with well known characteristics are constructed on good foundations, this approach can provide a valid and inexpensive means of seismic stability evaluation.

Data for dams whose performance during earthquakes is known is shown in Table 3 and Figure 17. Cyclic loading characteristics under standard conditions for a number of other dams whose stability has been found acceptable under conditions of strong earthquake shaking are shown in Figure 18. These and other data can greatly simplify stability evaluations and will also find increasing use in the future.

3. Use of Three Dimensional Response Analyses

It has been conventional practice in the past to evaluate the dynamic response of an embankment by making plane-strain analyses of several representative sections through the dam. For cases where the valley width is several times the height this approach provides a good engineering approximation but when a high dam is built in a relatively narrow canyon, plane strain analyses can give grossly misleading results concerning embankment response. Thus while much effort is always directed towards simplification of the evaluation procedure, it is sometimes necessary to use more sophisticated analyses in the interests of improved understanding and economy.

An interesting case in point is the interpretation of the records of motion at the Oroville Dam in California which crosses a V-shaped valley and has a crest to height ratio of about 7. During the 1975 Oroville earthquake, records of earthquake-induced motions were made at the crest of the dam and at its base. These records can be used to determine the dynamic response characteristics of the cobble and gravel shells for this 750 ft high embankment. For such materials the shear moduli can be represented with a high degree of accuracy by an equation of the form

$$G = 1000 \, K_2 \cdot (\sigma_m')^{1/2}$$

where $K_2$ is a material characteristic having

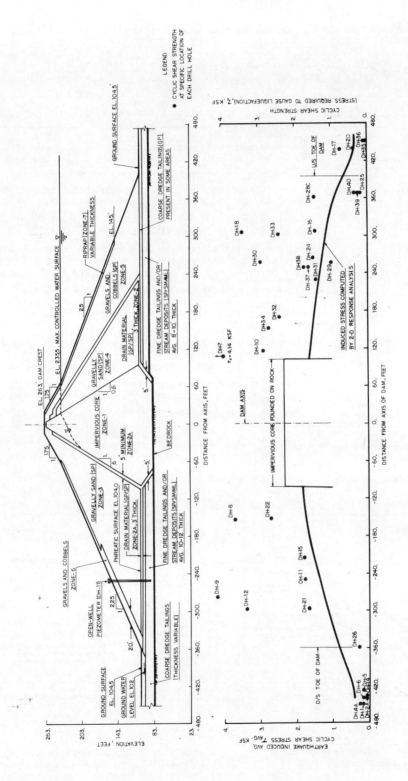

Fig. 16. Analysis of Stability of Foundation Sands at Camanche Dam Using Standard Penetration Test Data

Table 3. Performance During Earthquakes of Five Dams with Sandy Shells and Central Clay Cores.

(see material characteristics in Fig. 17)

| Dam | $(K_2)^*_{max}$ | Field Performance |
|---|---|---|
| Chabot | 65 | Survived magnitude 8-1/4 earthquake at distance of 20 miles with no apparent damage; would apparently survive magnitude 7 earthquake at distance of 2 miles. |
| Lower Franklin | 50 | Survived magnitude 6-1/2 earthquake at distance of about 20 miles ($a_{max} \simeq 0.2g$) with no apparent damage. |
| Fairmont | 50 | Survived magnitude 6-1/2 earthquake at distance of about 20 miles ($a_{max} \simeq 0.2g$) with no apparent damage. |
| Lower San Fernando | 45 | Major upstream slide including upper 30 ft of dam due to magnitude 6-1/2 earthquake at distance of about 5 miles ($a_{max} \simeq 0.55g$); performance and analysis indicate dam would have survived same earthquake at distance of 20 miles without significant damage but would have failed catastrophically for magnitude 8-1/4 earthquake at distance of 20 miles. |
| Upper San Fernando | 30 | Downstream slide movement of crest of dam due to magnitude 6-1/2 earthquake at distance of about 5 miles ($a_{max} \simeq 0.55g$); this embankment had an extremely wide downstream berm which together with its lower value of $(K_2)_{max}$ no doubt contributed to its better performance than the Lower San Fernando Dam. |

*Shear modulus $G = 1000\ K_2\ (\sigma_m')^{1/2}$

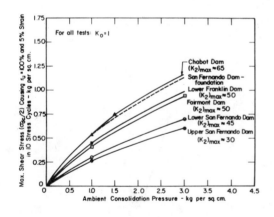

Fig. 17. Cyclic Loading Resistance of Shell Materials for Selected Dams (see Table 3).

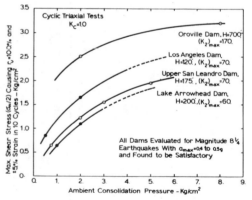

Fig. 18. Cyclic Loading Characteristics of Shell Materials for Several Dams Shown by Analyses to be Capable of Withstanding Strong Earthquake Shaking.

a maximum value $(K_2)_{max}$ at very low strain levels of the order of $10^{-4}$ percent and $\sigma_m'$ is the effective mean principal stress. A plane strain analysis of the embankment response in this case would lead to about 100% error in evaluation of material characteristics compared with a three-dimensional evaluation of embankment response.

For high dams in narrow canyons three dimensional analyses of response are therefore a necessary tool in the development of improved understanding of embankment behavior and therefore of seismic stability evaluations.

4. Re-distribution of Pore Water Pressures and Post Earthquake Stability Evaluations

The delayed failure of one of the Mochi-Koshi tailings dams in the Near Izo-Oshima Japanese earthquake in 1978 (Marcuson et al., 1979) serves as a forceful reminder that the post-earthquake stability of an embankment dam may in some cases be as serious a problem as the during-earthquake stability. This condition results from a re-distribution of pore water pressures in the embankment following an earthquake, and while in many cases, this redistribution will have beneficial effects on stability, it may in some cases have deleterious effects (Seed, 1979). Consideration of post-earthquake stability is thus now being considered an important aspect of seismic design and it imposes new and difficult demands on the skill of the design engineer.

CONCLUSIONS

In the preceding pages I have tried to review the field experiences, design concepts and analytical techniques currently available to assist the design engineer in making a reliable evaluation of the ability of a dam to resist the potentially damaging effects of earthquake shaking. Much progress in this field has been made during the past 10 years--both with regard to our understanding of seismicity and techniques for assessing the magnitude and locations of earthquakes--so that we can eliminate the surprises which so often have accompanied their occurrence in the past--and with regard to the development of design measures and embankment evaluation procedures which enable us to anticipate and mitigate the damaging effects of earthquakes.

It is important to keep the problem of earthquake-resistant design in perspective. Many parts of the world are not influenced significantly by these events and not all types of dams suffer major damage even if they are subjected to strong earthquake shaking. Thus the number of earthquake-induced failures of engineered dams is relatively small. However where the potential for strong earthquake shaking exists it is important that the possibility of failure be considered adequately and handled effectively even if it is only by a careful non-analytical assessment of hazards and the incorporation of design details to eliminate their effects. Analyses are a valuable supplementary tool to aid the engineer in achieving this goal. No matter how it is accomplished, however, the hazard potential of many dams makes it essential that adequate safety be achieved by one means or another.

REFERENCES

Akiba, M. & H. Semba 1941, The earthquake and its influence on reservoirs in Akita prefecture, J.Agric.Engng.Soc. Japan, 13, No. 1.

Ambraseys, N.N. 1960, On the seismic behaviour of earth dams, Proc. 2nd World Conf., Earthquake Engng., Tokyo.

Booker, J.R., M.S. Rahman & H.B. Seed 1976, GADFLEA, A computer program for the analysis of pore pressure generation and dissipation during cyclic or earthquake loading, Report No. EERC 76-24, Earthquake Engineering Research Center, University of California, Berkeley.

Finn, W.D.L. 1967, Behaviour of dams during earthquakes, Proc. 9th Int.Cong. on Large Dams, Istanbul, 4, 355-367.

Finn, W.D.L., K.W. Lee, C.H. Maartman & R. Lo 1978, Cyclic pore pressures under anisotropic conditions, ASCE, Geotech. Engrg. Div., Specialty Conf. on Earthquake Engineering and Soil Dynamics, Pasadena.

Gordon, B.B., D.J. Dayton & K. Sadigh 1974, Seismic stability of Upper San Leandro Dam, J.Geotech.Engng.Div., ASCE, 100, No. GT5, 523-545.

Khilnani, K.S. & P.M. Byrne 1981, Evaluation of seismic stability of foundation soils under Revelstoke earthfill dam, Proc.Int.Conf. on Recent Advances in Geotechnical Eearthquake Engineering & Soil Dynamics, April 26-May 3, St. Louis, MO, Vol. II, 837-842.

Klohn, E.J., C.H. Maartman, R.C.Y. Lo & W.D.L. Finn 1978, Simplified seismic analysis for tailings dams, Proc. ASCE Geotech.Engng.Div. Spec.Conf. on Eq.Engrg. & Soil Dynamics, Pasadena, 540-556.

Kramer, R.W., R.B. MacDonald, D.A. Tiedmann & A. Viksne 1975, Dynamic analysis of Tsengwen Dam, Taiwan, Republic of China, U.S. Dept. of the Interior, Bureau of Reclamation.

Lee, K.L. 1974, Seismic permanent deformation in earth dams, Report No. UCLA-ENG-7497, School of Engr. & Appl. Science, University of California at Los Angeles.

Lee, K.L. 1978, Seismic stability considerations for tailings dams adjacent to San Andreas Fault, Proc. 1st Central American Conf. on Earthquake Engineering, San Salvador.

Lee, K.L. & H.G. Walters 1972, Earthquake induced cracking of Dry Canyon Dam, ASCE Annual and National Environmental Engrg. Meeting, Houston, TX, Preprint No. 1794.

Leps, T.M., A.G. Strassburger & R.L. Meehan 1978a, Seismic stability of hydraulic fill dams--Part 1, Water Power & Dam Construction, October 27-36.

Leps, T.M., A.G. Strassburger & R.L. Meehan 1978b, Seismic stability of hydraulic fill dams--Part 2, Water Power & Dam Construction, October, 43-58.

Makdisi, F.K., H.B. Seed & I.M. Idriss 1978, Analysis of Chabot Dam during the 1906 earthquake, Proc., ASCE Geotech. Engrg. Div. Specialty Conf. on Earthquake Engineering & Soil Dynamics, Pasadena, 569-587.

Marcuson, W.F.III, R.F. Ballard Jr. & R.H. Ledbetter 1979, Liquefaction failure of tailings dams resulting from the Near Izu Oshima earthquake, 14 & 15 January, 1978, Proc. 6th Panamerican Conf. Soil Mechanics & Foundation Engineering, Lima.

Marcuson, W.F.III, E.L. Krinitzky & E.R. 1977, Earthquake analysis of Fort Peck Dam, Montana, Proc. 9th Int. Conf. Soil Mech. & Found. Engrg., Tokyo.

Moriya, M. 1974, Damages caused to small earthfill dams for irrigation in Armori Prefecture by Tokachi offshore earthquake, Report prepared for Committee on Lessons from Incidents of Dam Failures, Intl. Commission on Large Dams, 1974.

Newmark, N.M. 1965, Effects of earthquakes on dams and embankments, Geotechnique, 15, No. 2, 139-160.

Sadigh, K., I.M. Idriss & R.R. Youngs 1978, Drainage effects on seismic stability of rockfill dams, Proc. ASCE Geotech. Engng. Div. Specialty Conf. on Earthquake Engrg. & Soil Dynamics, Pasadena, 802-818.

Seed, H.B. 1966, A method for earthquake resistant design of earth dams, J.Soil Mech.Fdn.Div., ASCE, 92, No. SM1, 13-41.

Seed, H.B. 1979, Considerations in Earthquake-Resistant Design of Earth and Rockfill Dams, Geotechnique, Vol.XXIX, No. 3, September.

Seed, H.B. 1979, Soil liquefaction and cyclic mobility evaluation for level ground during earthquakes, J. Geotech. Engng. Div., ASCE, Vol. 105, No. GT2, February, 201-255.

Seed, H.B., J.M. Duncan & I.M. Idriss 1975a, Criteria and methods for static and dynamic analysis of earth dams. In Criteria and assumptions for numerical analysis of dams (Naylor, D.J., K. Stagg & O.C. Zienkiewicz, Eds.), Swansea: University College, 564-588.

Seed, H.B., I.M. Idriss, K.L. Lee & F.I. Makdisi 1975b, Dynamic analysis of the slide in the Lower San Fernando Dam during the earthquake of February 9, 1971, J. Geotech. Engng. Div., ASCE, 101, No. GT9, pp. 889-911.

Seed, H.B., K.L. Lee & I.M. Idriss 1969, Analysis of Sheffield Dam failure, J. Soil Mech. Fdn. Div., ASCE, 95, No. SM6, 1453-1490.

Seed, H.B., K.L. Lee, I.M. Idriss & F.I. Makdisi 1973, Analysis of slides in the San Fernando Dams during the earthquake of February 9, 1971, Report No. EERC 73-2, Earthquake Engineering Research Center, University of California, Berkeley.

Seed, H.B., K.L. Lee, I.M. Idriss & F.I. Makdisi 1975c, The slides in the San Fernando Dams during the earthquake of February 9, 1971, J. Geotech. Engng. Div., ASCE, 101, No. GT7, 651-668.

Seed, H.B., F.I. Makdisi & P. DeAlba 1978, Performance of earth dams during earthquakes, J.Geotech. Engng. Div., ASCE, 104, No. GT7.

Serff, N., H.B. Seed, F.I. Makdisi & C.K. Chang 1976, Earthquake induced deformations of earth dams, Report No. EERC 76-4, Earthquake Engr. Res. Center, University of California, Berkeley.

Sherard, J.L. 1967, Earthquake consideration in earth dam design, J. Soil Mech. Div., ASCE, 93, No. SM4, 377-401.

State of Calif. Dept. of Water Resources 1979, The Aug. 1, 1975 Oroville Earthquake investigations, DWR Bull. No. 203-278.

Terzaghi, K. 1950, Mechanisms of landslides, The geological survey of America, Engr. Geology (Berkey) Volume.

Vrymoed, J.L. & E.R. Galzascia 1978, Simplified determination of dynamic stresses in earth dams, Proc. ASCE Geotech.Engng.Div., Specialty Conf. on Earthquake Engrg. & Soil Dynamics, Pasadena, 991-1006.

Wahler & Associates 1981, Seismic reevaluation of Camanche Reservoir main dam," Report prepared for East Bay Municipal Utility District, Oakland, CA, January.

*Symposium on Problems and Practice of Dam Engineering / Bangkok / 1-15 December 1980*

# Comparative behaviors of similar compacted earthrock dams in basalt geology in Brazil

VICTOR F.B.DE MELLO
*Consulting Engineer, Sao Paulo, Brazil*

SYNOPSIS. One important problem in higher compacted earth-rock dams concerns compatibility of settlements between core and rockfill. For such compressibility behavior, adjustment coefficients between laboratory and field are sought. It is shown that both the very clayey cores and the sound angular compacted rockfill exhibit nominal compaction precompression followed by virgin compressibility. The magnitudes of settlements in both materials have proved very similar. Statistical correlations are offered for preconsolidation pressure and compression indices of undisturbed block sample specimens of compacted clays. Construction pore pressures are very low.

1. INTRODUCTION

The question of compatibility of settlement deformations in zoned embankment dams is well recognized as one of the problems that significantly condition satisfactory embankment dam design and behavior. Moreover, there is every theoretical reason to anticipate that such a problem should be amenable to direct and rapid adjustments of the necessary testing and consequent interpreted stress-strain parameters: settlements derive from integrations of compressibilities of ΔH Soil elements, and are therefore definitely associated with statistics of averages. The purpose of the present paper is, however, to further advance my own earlier indications on how significant an adjustment has to be made in deriving the appropriate mental model and geotechnical design-prediction parameters because of historic misconceptions on the significance of compaction, and because of premature crystallization of conventional soil mechanics on would-be standardized routine test procedures, interpretations, and design computations.

In fig. 1 I submit once again (cf. Tokyo ICSMFE 1977, Vol. III) a schematic visualization of what I have called the Experience Cycle in Civil Engineering. No matter what degree of crudences of the "data" on which we have to base our design decisions, such decisions really begin by establishing a "physical model" intended to fulfill "equal to or better than" the desired functional behavior. That is why I emphasize that Civil Engineering is based on "prescriptions", themselves based on predictions of what will not happen, rather than restrictively on difficult predictions of what exactly will happen (Rankine Lecture 1977). For instance, on the basis of the earliest identification and Index Tests we already associate certain likely behaviors with the word "clay" and "rockfill": and thereupon from the most preliminary phase of investigation - identification, would not hesitate to "design" a presumed earthcore-rockfill dam section at sites blessed with a red porous clay of presumably high plasticity for a core, and a sound dense basalt from required excavations for compacted rockfill shells.

Such was the condition in 1969 when the Salto Osorio dam was submitted for construction bidding. The Consulting Board had considerable concern about the anticipated high compressibilities of the clay core in comparison with the rockfill shells, and directly decided to change the central core to an inclined-core section. A posteriori finite element analyses and geotechnical tests were engaged (cf. ICOLD, Madrid 1973). Suffice it to mention that the reasonable parameters proposed by experienced specialists were E values of

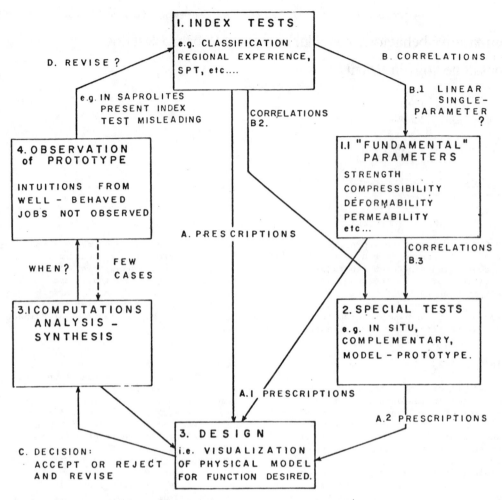

FIG. 1 EXPERIENCE CYCLE IN CIVIL ENGINEERING

about 80 kg/cm² for the clay and 1000 kg/cm² for the basalt. Careful geotechnical testing was undertaken (cf. steps 1.1 and 2 in Fig. 1) and obviously the case called for careful monitoring of the prototype behavior (step 4). In rapid succession two similar higher dams, Salto Santiago 80m, and Itauba, about 100m at deepest points, followed the same design trend. The data on these three dams are further enhanced, as regards the behavior of the sound dense well compacted basalt, by the monitored data from the Foz do Areia 160m concrete--face dam, in many respects a record--breaking case. Figs. 15 and 16 summarize the design crossections of these four dams used herein to confirm my views on needed revisions on interpretation of behavior of compacted embankment dams.

2. REVISING THE PREVAILING MENTAL MODEL ON THE NATURE OF COMPACTION

It is quite comprehensible that the historical mental model on the nature of compaction should have been, in analogy with the understanding of dense vs. loose sands, a hypothesis of developing through densification a material homogeneously improved. All the publications (predominantly from laboratory research) through the 1950's and 1960's, attributed to each compaction condition (defined by percent compaction PC%, and water content deviation Δw% from optimum) a presumed intrinsic new quality, such as schematically represented in Fig. 2 as regards shear strength. Even meticulous and sophisticated research on "structure of compacted clays" as dependent on compaction parameters (truly second-order effects in much civil engineering construction) always associated all specimens of a given set of compaction indices as pertaining to a common statistical universe. I have repeatedly emphasized that in unsaturated clays and rockfills (materials with significant hysteresis of absorption of compressive energy) compaction really represents the introduction of an apparent preconsolidation effect, with little influence on the $C_c$ behavior of the materials at pressures beyond the preconsolidation pressure, i.e. in the nominal virgin compression range. In Fig. 3a are reproduced a few of the many laboratory tests conducted with regard to different residual clay-silt-sand soils using specimens molded at different (PC%, $\Delta$w%) compaction indices: bearing in mind inevitable statistical dispersions the conclusion repeatedly extracted has been that within the narrow range of compaction indices within which a dam is constructed, we may accept that $C_c$ is for practical purposes independent of (PC%, $\Delta$w%) for each soil.

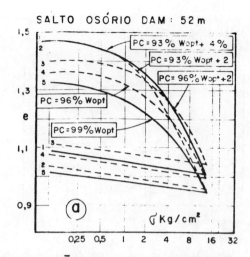

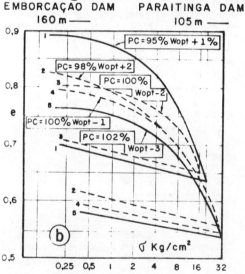

FIG.3 COMPACTION IMPLIES PRECOMPRESSION WITH PRACTICALLY NO EFFECT ON VIRGIN COMPRESSION INDEX.

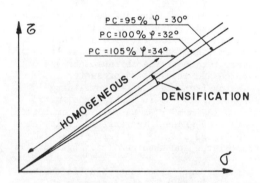

FIG.2 HISTORICAL MENTAL MODEL OF COMPACTION EFFECT

Returning for a moment to Fig. 1 I wish to emphasize that on looking back at the historical development of soil mechanics (as of any other technology) it is quite comprehensible that both Index Tests (for identification and classification) (de Mello 1979) and also the principal tests for Fundamental Parameters, and even the further Special Tests, should have arisen one by one, without much regard to compatibilities between them. Also, it is inexorable that each test is derived from a preconceived mental model of how the specimen behavior will be incorporated into the integrated prototype behavior. It is inexorable, therefore, that the historical tests should require considerable revision and readjustment, both for compatibilities between them, and for adequate estimation of the behavior anticipated for new projects. It is firstly sad that in many quarters the need for such recycling revision has neither been promoted nor even perceived. Much more regretable, however, is it when additional indices are lightly suggested without recourse to the context of theoretical and practical knowledge already accumulated: Fig. 4 exemplifies some immediate conceptual questions that surround a recently proposed index, the Dispersion Index, regarding a problem that is in some parts emphasized as afflicting compacted clay dams.

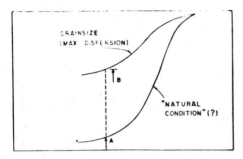

FIG 4 EXAMPLE OF CONFUSING PROPOSAL OF INDICES FOR SOIL BEHAVIOR

2.1 Estimates of compaction preconsolidation pressure $\sigma_c$ from oedometer tests on undisturbed block samples from various dams

The results of oedometer tests on 167 block samples from various Brazilian dams have been analyzed statistically for correlations between PC% and the corresponding preconsolidation pressure. The simple conclusions are summarized in Fig. 5. The correlation depends somewhat on the classification of the soil, as indicated by the maximum Proctor dry density. As can be readily understood the correlation is not good in the sandier materials. The same regressions are furnished in the lower graph in a manner that should make it easier to estimate for any given maximum Proctor dry density and corresponding PC%, what to expect as a compaction preconsolidation pressure.

In the two graphs of Fig. 5 we have plotted the zones corresponding to 14 tests obtained more recently in the compaction of the clay cores of the Salto Osorio and Salto Santiago earth-rock dams employing the presumably very clayey basalt red porous clay.

2.2 Evidence of pressure distributed into the lifts by the tire of compacting equipment

No data can be found on pressure transmitted by sheepfoot and tamping rollers to the lifts during the passes of compaction equipment. In fact, all or most pressure cell measurements in earth dams have preferred to protect cells from damage, and, having been installed in pits opened after fill has risen a couple of meters, have served the questionable purpose of checking themselves as recorders of overburden pressure increments. One can reason however, in average, that like effects associate with like causes. Therefore it suffices to examine experimental data collected (in the field of pavements) from tire pressures, analogous to pneumatic rollers.

Some published data are reproduced in Figs. 6 and 7, and require little comment since the experimental evidence fully confirms theoretical predictions. It is of interest to note the obvious increase of applied and transmitted pressures in changing from yielding to unyielding support (Fig. 7): in other words, with increasing number of passes and compaction, there is a gradual increase of rigity of support and also of the very pressures transmitted and absorbed (cf. Figs. 8a, 9a, 9b).

Another very important point is the much more significant PC gradient across the lift in sandy material than in clayey soil, and in dry compaction compared with wet compaction (Fig. 6b).

2.3 Schematic visualization of compaction hysteresis and bearing capacity limit of compactability in clayey soils

Fig. 8 configurates schematically the obvious mental model that emerges regarding absorption of compactive energy. In clay the compression-expansion hysteresis is well recognized to be very significant because of plate-like grains and change of structural arrangement. Note that an interesting index of how much capillary negative pore pressure should develop in a given clay should be associated with the difference in the expansion behaviors on stress release (as the roller moves away) with and without access to free water (surround-

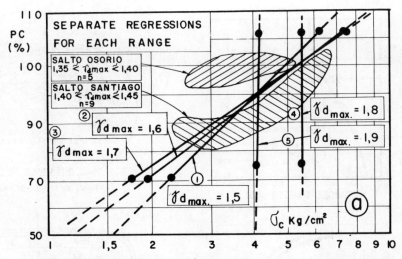

① $\log \sigma_c = -5.35 + 0.017\ PC + 2.92\ \gamma_{d.max.}$  CORREL. COEF.  $r = 0.75$, $n = 45$
② $\log \sigma_c = 0.25 + 0.022\ PC - 1.07\ \gamma_{d.max.}$   $r = 0.53$, $n = 52$
③ $\log \sigma_c = 4.67 + 0.024\ PC + 1.77\ \gamma_{d.max.}$   $r = 0.63$, $n = 61$
④ $\log \sigma_c = 4.53 + 0.00029\ PC - 2.12\ \gamma_{d.max.}$   $r = 0.53$, $n = 29$
⑤ $\log \sigma_c = 1.50 + 0.00031\ PC - 0.48\ \gamma_{d.max.}$   $r = 0.29$, $n = 50$

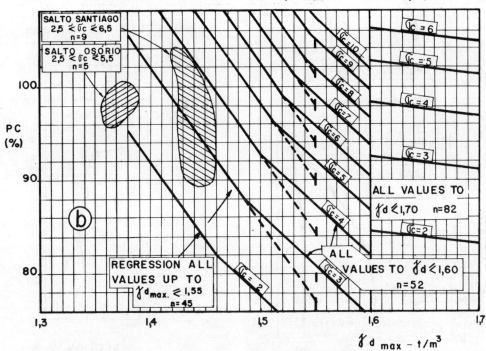

FIG.5  BASES FOR ESTIMATING PRECONSOLIDATION PRESSURES OF COMPACTED SOILS

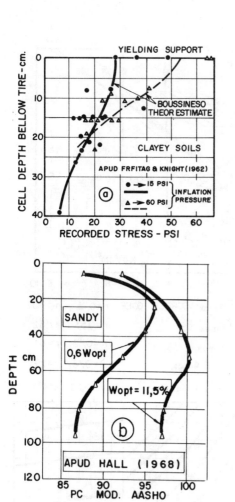

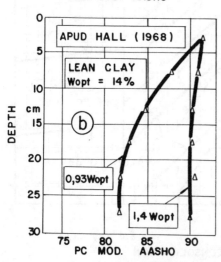

FIG.6 PRESSURE DISTRIBUTIONS WITHIN LIFT DUE TO PNEUMATIC COMPACTION

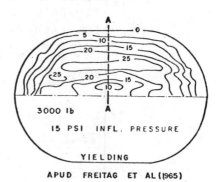

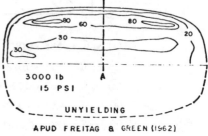

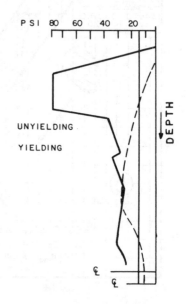

FIG.7 APPLIED AND TRANSMITTED PRESSURES

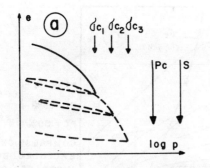

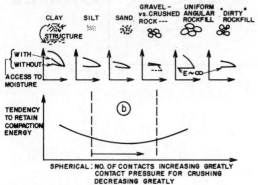

FIG.8 SCHEMATIC COMPARISON OF COMPACTION GRAIN STRUCTURE AND COMPRESSIBILITY HYSTERESIS

ing the specimen in oedometer tests).

Silts, sands and rounded gravels are materials in which there is least absorption of compactive prestress. In the case of coarser granular material the compressions are due to the very significant difference between (point to point) intergranular stresses (as differentiated from the nominal average effective stress obtained by definition and dogma): therefore a crushed aggregate is obviously more compressible than a gravel of same grainsize, and also obviously a clean angular sound rockfill should be more compressible than crushed aggregate of the same rock. Also under the same reasonings it is quite understandable that a "dirty rockfill" should be less compressible than the corresponding clean angular rockfill.

Finally, it should be of considerable interest to note (for finite element analyses etc...) that in the coarse angular materials of compressibility generated by crushing of point contacts, the E value for stress release may be essentially infinite.

Some of the very significant observations in modern compaction of clayey materials are: the fundamental importance of densities and degrees of saturation S% of the clay nuclei in situ (in the borrow pit); this is schematically illustrated in Fig. 10.

Moreover, one faces the problems of bearing capacity both of earthmoving equipment and of compactability of lifts; and the difference between laboratory compaction within metal molds, and the compressive-shearing compaction of a lift in the field (Fig.11c).

Such observations have become more and more salient as equipment greatly increased in capacity and weight, and as one advanced in the use of much more clayey (and/or less unsaturated) borrow materials. Excavations in borrow pits presently often advance through materials much denser than the Proctor maxima, and at S% values absolutely incompatible with any compaction (one only compacts the air voids, principally the macro-airpores): and one always deals with clay nuclei and not with the "totally disintegrated" particle sizes. Ironically the $\Delta w\%$ index would indicate a need for adding water: in a dense, close--to-saturated clay nucleus, the mechanical action would be remoulding and not compaction.

The data from the Salto Osorio, Salto Santiago, Itauba claycore material have been used in Figs. 9, 11a, 11b to demonstrate that the existing theoretical and experimental background can be readily used for such evaluations of compaction and bearing capacity compactability limits. The effective stress envelope from Fig. 9a has been transformed into total stress equations (Fig. 11a) by use of the pore pressure data for two different S% from Fig. 9c. In Fig. 11b we reproduce a set of curves of surface plate bearing capacity $\sigma_{rut}$ (de Mello, 1969). Finally against such a background we can plot the probable trajectories of compaction of the two different lifts, recognizing that in each lift the S% increases (and $\phi_a$ decreases) with increasing passes.

2.4 Statistical estimation of virgin compression $C_c$ of compacted clay-silt--sand materials

Having discussed and estimated the compaction pre-consolidation $\sigma_c$, we must now establish the $C_c$ values for the purpose of extrapolations of behaviors beyond the precompression benefits of the compaction PC%.

The same oedometer tests on undisturbed block samples were used for statistical regressions that are summarized in Figs. 12 and 13. All points are plotted, and appear to configurate too wide a dispersion: it must be repeated, however, that we are principally interested in the average, and the respective coefficients of correlation are very good. The first obvious attempt

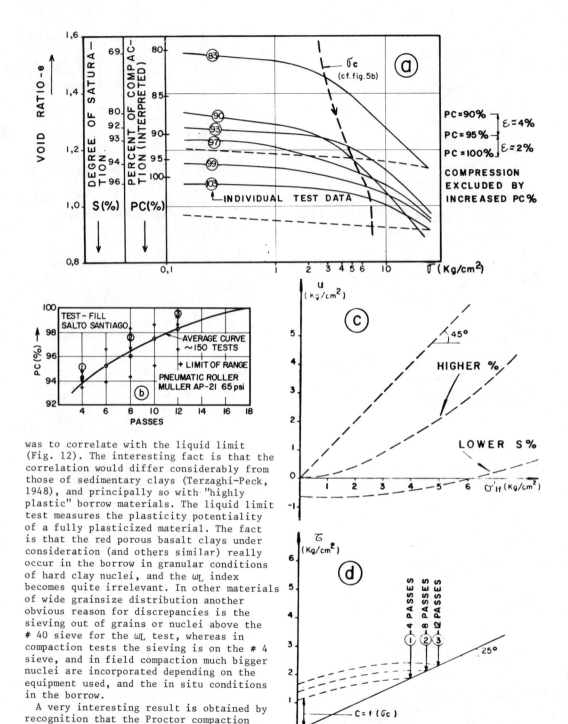

was to correlate with the liquid limit (Fig. 12). The interesting fact is that the correlation would differ considerably from those of sedimentary clays (Terzaghi-Peck, 1948), and principally so with "highly plastic" borrow materials. The liquid limit test measures the plasticity potentiality of a fully plasticized material. The fact is that the red porous basalt clays under consideration (and others similar) really occur in the borrow in granular conditions of hard clay nuclei, and the $\omega_L$ index becomes quite irrelevant. In other materials of wide grainsize distribution another obvious reason for discrepancies is the sieving out of grains or nuclei above the # 40 sieve for the $\omega_L$ test, whereas in compaction tests the sieving is on the # 4 sieve, and in field compaction much bigger nuclei are incorporated depending on the equipment used, and the in situ conditions in the borrow.

A very interesting result is obtained by recognition that the Proctor compaction test itself is an appropriate index for classification of soil types. Excellent coefficients of correlation were obtained (Fig. 13) for estimation of $C_c$ values on undisturbed block sample oedometer tests on the basis of Proctor compaction maximum

FIG.9 COMPACTION SEQUENCE WITH INCREASED PASSES, ESTIMATED CHANGE OF STRENGTH ENVELOPE

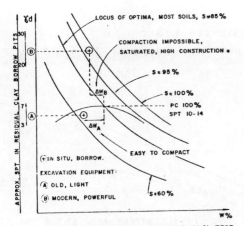

FIG.10 SCHEMATIC EXPLANATION OF WHY THE ΔW% TEST OF BORROW PITS IS INSUFFICIENT FOR INDICATING COMPACTABILITY OF CLAYS

dry density (de Mello, 1977).

The problems of comparative E values of compressibility-settlement in earthrock dams are summarily indicated in Fig. 14. Below compaction preconsolidation pressures materials tend to have been practically homogenized by the absorbed compactive energy: differences accentuate in the virgin compression range. In the recompression range the "compressible materials" (exclude dense sands, etc..., cf. Fig. 8b) E values are not very different. It is important to recognize this problem since simple finite element analyses seduce one into the impressions of linear extrapolations. Precedents of satisfactory behavior of low dams cannot be lightly extrapolated to similar dams of much greater heights.

## 3. COMPARATIVE BEHAVIORS OF DAMS

Fig. 15 gives the basic data on cross--sections of three earth rock dams employing the very clayey red porous clay compacted core, and the very sound angular quarry basalt compacted rockfill. Fig. 16 gives the corresponding crosssection of the concrete-face rock-fill dam of much greater height employing exactly similar compacted rockfill.

### 3.1 Need to compute overburden pressures including some Influence factor $I \lesssim 1.00$, and interest in plotting on log $\sigma$ scale

Practically all publications on behavior of embankment dams have automatically assumed that increments of overburden stress $\gamma z$ are applied and transmitted as if pertaining to infinite loaded area conditions, I=1.00, and yet are limited strictly to the material vertically above the point under consideration. If slopes are flat and the dam construction does not incorporate varied construction phasing, such an assumption does not become patently unreasonable. In fact, however, as the fill rises, a given mid-point between vertically positioned settlement gages near the bottom of the dam only receives $I\gamma z$ pressure increments, and the least that can be done is to use the elastic model charts (Poulos and Davis, 1974) for estimating approximate I values. One of the illusions generated by such a wrong practice of interpreting and plotting construction settlement data is that at some moment $\Delta\sigma$ has stopped above a given point (Fig. 17), and settlement continues (discussed either as "consolidation" settlements" or as "secondary compressions").

Further improvements may be made iteratively in the very assumption of $\gamma z$ (presupposes no lateral shear redistributions due to differentiated tendencies to settlement) and in the mathematical model for extraction of I values. At present, however, suffice it to correct an evidently wrong practice.

In Fig. 17 it is further indicated that it is preferable to plot settlement vs. pressure data in the semilog plot that was adopted for the oedometer test, for the express purpose, entirely pragmatic, of facilitating the determination of the preconsolidation pressure.

### 3.2 Field settlement behavior compared with laboratory test data, compacted very clayey core

Figs. 18, 19, 20 and 21 summarize the comparative information on compressive vertical strains as derived from laboratory tests, both routine and sophisticated, and from settlement observation of the construction settlements.

Firstly it must be mentioned that for the purpose of approximation with current routines, the field data on $\sigma$ have not been corrected with regard to I. Fig. 22 summarizes, for comparison, typical data on some settlement gages, presented both with respect to $\gamma z$ and with respect to the nominal elastic $I\gamma z$: the principal consequence is that the log ($I\gamma z$) graph accentuates more definitely the apparent nominal preconsolidation pressure.

Secondly, it must be mentioned that overburden stresses were measured in the core (because of fears of redistributions due to incompatible deformabilities and

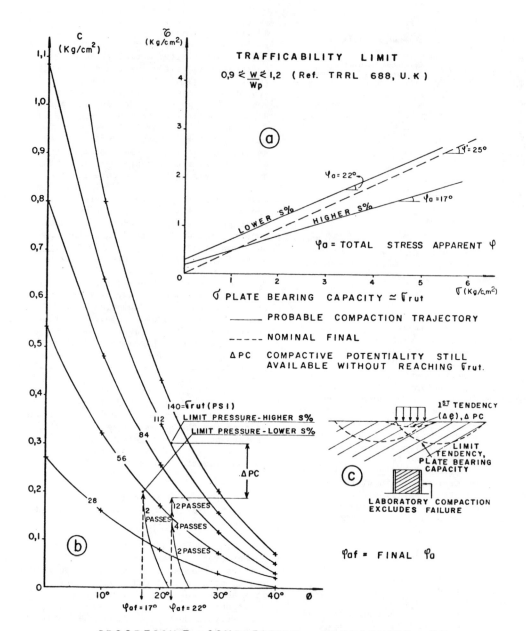

FIG. II  PROGRESSIVE COMPRESSION - COMPACTION UP TO BEARING CAPACITY. TRAFFICABILITY LIMIT

possible "hang-up" of the cores) and the pressure data attributed to the field behaviors are well confirmed. Finally, it must be repeated that only one or few points of field observation are represented in each case because all of the other points gave almost exactly the same indications: the scatter in laboratory data is apparently great, but the consisten cy of field observed behavior was so very close as to be surprising (and to render impossible packing more curves into the same drawing).

The principal conclusions from these observations stand in surprisingly emphatic confirmation of some of the points discussed regarding the need for very significant revision of the routine test

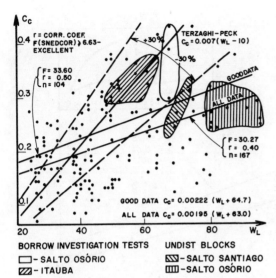

FIG.12 REGRESSIONS OF $C_c$ VS. $W_L$ UNDISTURBED BLOCK SAMPLES

procedures and corresponding parameters.
Laboratory molded specimens would seem useless, both if tested in oedometer compression and in triaxial compression. One suggestion for the oedometer test might be to mold the specimens by compacting directly into and against the oedometer ring, incorporating into the specimen an initial lateral confining stress: it would require research and adjustment. The use of specimens cut from undisturbed block samples from the compacted fill offers some improvement: that is why in our design and construction experience with compacted earth dams we prefer to use the first few thousand cubic meters of actual placement and compaction as a field compaction test for field adjustments and for extraction of undisturbed block samples. One suggestion would be to resort more to field tests (plate load tests, duly conducted and interpreted, or pressuremeter tests, etc...) rather than so-called undisturbed block samples: thereby one approaches somewhat more the in situ condition that should retain residual stresses from compaction (de Mello, 1977).

Sophisticated laboratory tests on 4" diameter undisturbed specimens from block samples do offer some improvement (cf. the $K_0$ tests and the anisotropic triaxial compression with $\sigma_1/\sigma_3$ ratio of 1.5). The predicted strains (and settlements would still be of the order of 2 to 3 times higher than the field behavior observed. It hardly seems worth the trouble: moreover there is no plausible theoretical reasoning in favour of such sophistication. Therefore, in other cases the comparison could be either less favourable or more so, purely by coincidence.

Note that the recompression strain values from oedometer tests on the undisturbed block samples come adequately close to representing the field behavior. Therefore, a technique may well be adjusted for using such simpler tests, but resorting to two, three, or more cycles of compression-decompression, before extracting the appropriate recompression E value (much as was done for building settlements on Stuttgart, cf. Schultze and collaborators).

One very remarkable observation is, of course, the fact that the clay compressibility is not, as was early feared, significantly greater than that of the clean sound dense basalt, compacted in 0.8 to 1.2m lifts.

3.3 Field settlement and E values of clean well compacted angular basalt rockfill

In the case of the Foz do Areia dam of record dimensions every care was exercised to improve the E values through compaction, watering, varied lift thicknesses etc... The limitation was of the weight of the vibratory roller (roughly 10 tons., static). The field compression data were so consistent as to appear having been faked: dispersions were frequently less than 2% around the mean.

Fig. 22 summarizes some of the typical settlement data which, once again very clearly demonstrate the precompression effect. If such a reasonable conclusion is really proven, for the cases of higher dams what would be required is to use heavier vibratory rollers with proportionally higher impact so as to increase the crushing of angular intergranular contacts during compaction.

Finally, in closing this presentation it is of interest to compare the behaviors of variations of E for the core and for the rockfill, with increasing $\sigma$. Firstly the difference is not at all one in the range of $E_{clay} \approx 80$ kg/cm$^2$ to $E_{rock} \approx 1000$ kg/cm$^2$, but much closer to similar. Secondly, due to more perceptible "jumps" in the early stages of rockfill compressions there is considerable initial scatter, even using secant E values (average from start): scatter would be much higher at start if employing $d_s/d\sigma$ values for E. Finally, beyond a certain loading the rockfill compressions become quite smooth, and, surprisingly, reach, in the cases presented, greater compressibilities than the very clay beyond a certain pressure (Fig. 23).

There is at present no means for

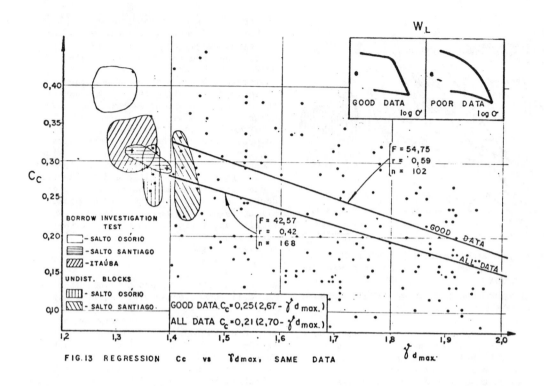

FIG.13 REGRESSION $C_c$ vs $\gamma_{dmax}$, SAME DATA

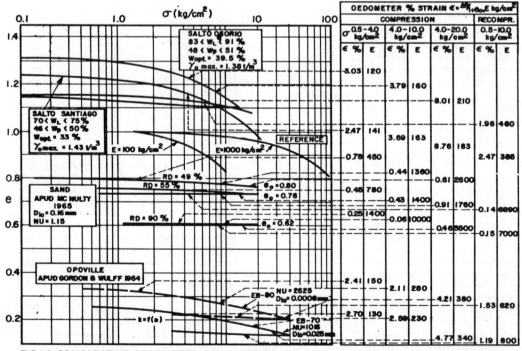

FIG.14 COMPARATIVE OEDOMETER COMPRESSIBILITY STRAINS IN WIDELY DIFFERENT MATERIALS

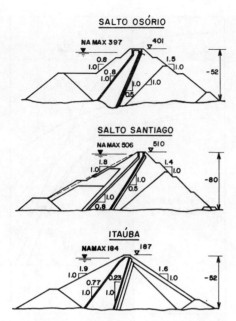

FIG.16 SOUND CLEAN BASALT COMPACTED ROCKFILL DAM WITH CONCRETE FACE

evaluating if the smaller Erock vs. Eclay beyond pressures of the order of 6 to 7 kg/cm$^2$ might be associated with compaction preconsolidation or precompression values of about the same magnitudes (possibly by coincidence).

3.4 Construction pore pressure behavior

Consistent with the compressibility - expansivity behavior above discussed as inherent in the process of compaction of the clay-core material, one also finds the real field vs. laboratory behavior on construction pore pressure. Figs. 24 and 25 summarize some of the data.

FIG.15 THREE SIMILAR EARTH-ROCK SECTIONS

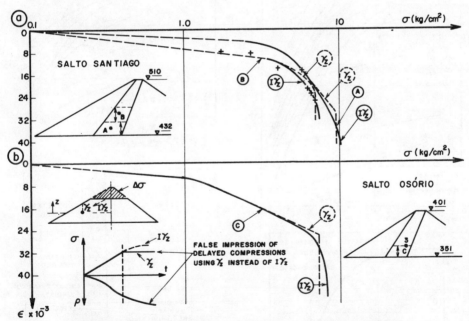

FIG.17 PRESSURE SETTLEMENT CURVES PREFERABLE USE OF LOG PRESSURE AND CORRECT $I\gamma_z$

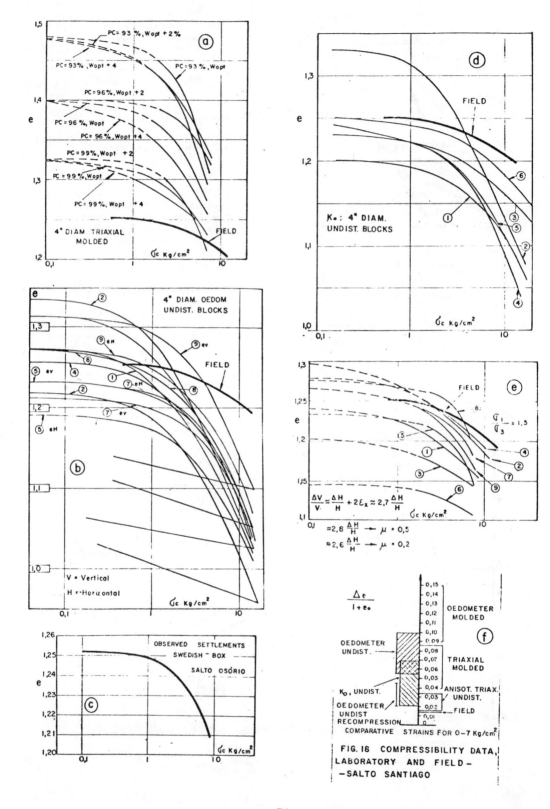

FIG. 18 COMPRESSIBILITY DATA, LABORATORY AND FIELD — SALTO SANTIAGO

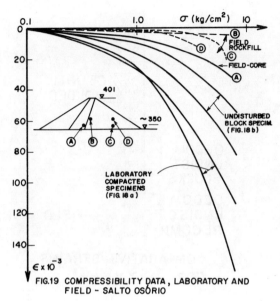

FIG.19 COMPRESSIBILITY DATA, LABORATORY AND FIELD - SALTO OSÓRIO

Both the early USBR (1930-1955) field observed data, hampered by a big consistent error of high initial u values due to soaking around the cell, and the indications from near-saturated borrow pits subjected more to remolding than to compaction (cf. some European and British experience), have spread an impression of great problems due to construction pore pressures. In fact, the limits of trafficability quite exclude such a possibility. Truly it has been and continues to be very difficult to measure negative pore pressures, that are inexorable right after compaction of a clayey material (duly unsaturated): it is difficult in the laboratory, and much more so in the field. In field observations even with the best of constant volume cells, and respective installations, the early measurements tend to be erratic "around zero": only beyond a certain overburden pressure do the readings begin to develop the consistent trend of concave-upward curve of increasing u and increasing rate of change of u with σ. The best way to estimate initial conditions in situ would be by extrapolating backwards, with due cognizance of the varying expansivity with stress release as the roller moves away.

The typical detailed data of trends of u vs. σ from the Salto Osorio field observations have been presented (de Mello, 1975): positive pressures only begin to build up, at a small rate, beyond an overburden pressure of the order of 4 kg/cm$^2$.

Moreover, as summarized in Figs. 24 and 25, all the test data are most unrealistically pessimistic. The impression that a very clayey material should give high construction u values is obviously unrealistic. One should reason on the basis of the "equivalent suction" values necessary to give the same trafficability, in all materials: with a low $\phi'$ associated with a very clayey material, there has to be a compensation of a high "equivalent suction".

In a separate paper we shall discuss how inappropriate it is to employ such ratios as $\bar{B}$ or $r_u$, wherein early values "explode to infinity" merely because σ begins from zero, and also how unnecessary, and detrimental to economy of shallow circle or low embankment analyses, it has been to employ a straight line u vs. σ variation starting from the origin.

4. SUMMARY

4.1 The experience cycle in the cases configurated demonstrates that conventional classification index tests do not lead to any valid predictions of behavior.

4.2 Compaction behavior of clayey materials depends very much on the condition of the clay chunks and/or nuclei in situ, in the borrow, and principally on S% and porosimetry (frequency of macropores). Conventional compaction indices PC% and Δω% may often lead to grave errors of planning of plant and of prediction of behavior.

4.3 Both clays and sound angular clean rockfills present definite evidence of nominal preconsolidation pressures corresponding to the compactive energy absorbed.

4.4 For the estimation of settlement behavior of compacted clay fills one should resort to adjusted field and laboratory tests, pertaining to field compacted volumes. Statistical regressions are offered for $\sigma_c$ and $C_c$ based on undisturbed block samples. The best classification index appears to be the Proctor maximum dry density.

4.5 Field compressibilities are much smaller than those of the regressions mentioned in 4.4, probably because of residual stresses from compaction, that benefit in situ conditions. Recompression curve settlements may be an avenue for

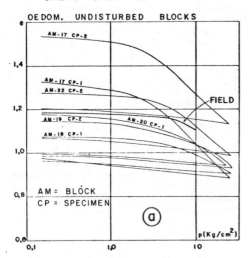

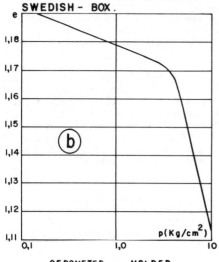

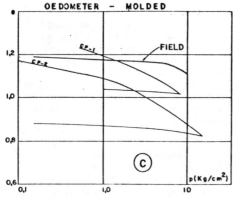

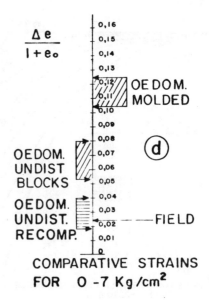

FIG.20 COMPRESSIBILITY DATA, LABORATORY AND FIELD — SALTO SANTIAGO

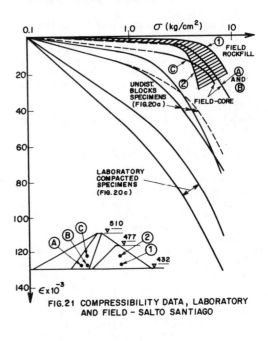

FIG.21 COMPRESSIBILITY DATA, LABORATORY AND FIELD – SALTO SANTIAGO

suitable adjustment between prediction and observation.

4.6 The guarantee of trafficability for modern heavy earthmoving and compaction

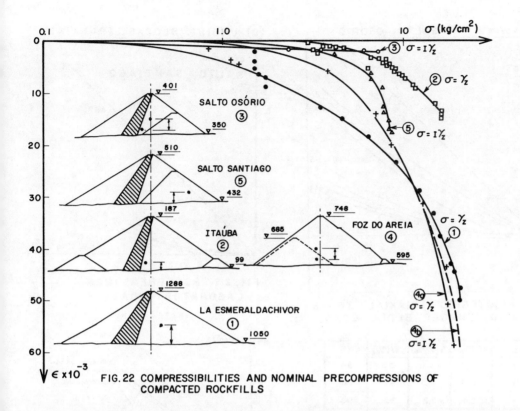

FIG. 22 COMPRESSIBILITIES AND NOMINAL PRECOMPRESSIONS OF COMPACTED ROCKFILLS

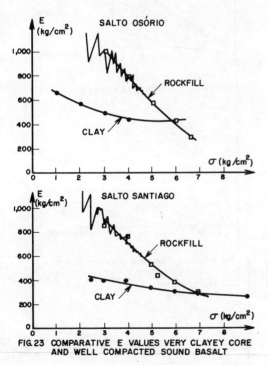

FIG. 23 COMPARATIVE E VALUES VERY CLAYEY CORE AND WELL COMPACTED SOUND BASALT

equipment is generally a sufficient guarantee that there will be no problem from construction pore pressures.

4.7 Many of the automatically accepted dictates of testing, parameters, and computations are herein shown to require rather significant revision. Precedents will not be sufficient to guarantee success if we are called to extrapolate, and herewith recognize that the mental models for such analysis and possible extrapolation are not really valid.

5. REFERENCES

de Mello, V.F.B. (1969), "Foundations of Buildings in Clay", Proceedings of the Seventh International Conference on Soil Mechanics and Foundation Engineering, Mexico, vol. 3, pp. 127-139.

de Mello, V.F.B. (1973), Eleventh International Congress on Large Dams, Madrid, Spain, vol. 5, pp. 394-406.

de Mello, V.F.B. (1975), "Some Lessons from Unsuspected, Real and Fictitious Problems

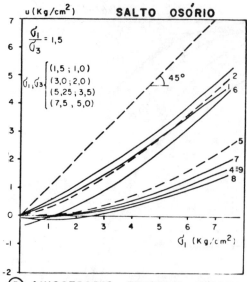

(a) ANISOTROPIC TRIAXIAL TESTS UNDISTURBED BLOCK SPEC.

| AM | PC (%) | ΔW (%) | W (%) | S (%) |
|---|---|---|---|---|
| 1 | 94,4 | +4,3 | 39 | 94 |
| 2 | 95,4 | +3,3 | 41 | 96 |
| 4 | 100,4 | +2,2 | 39 | 93 |
| 5 | 100 | -0,5 | 38 | 93 |
| 6 | 101 | +0,9 | 37 | 97 |
| 7 | 99,8 | +0,1 | 36 | 83 |
| 8 | 99,2 | +1,0 | 38 | 91 |
| 9 | 97,8 | +1,1 | 39 | 92 |

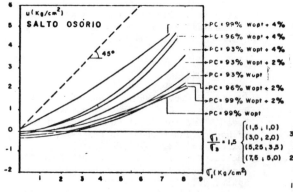

(b) ANISOTROPIC TRIAXIAL TESTS LABORATORY COMPACTED SPEC.

(c) UNDIST. BLOCKS TRIAXIAL $\overline{CU}$

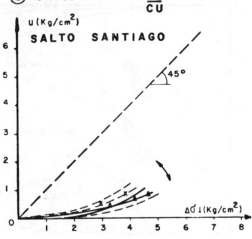

FIG. 24 POREPRESSURES. LABORATORY DATA

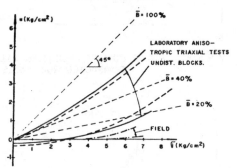

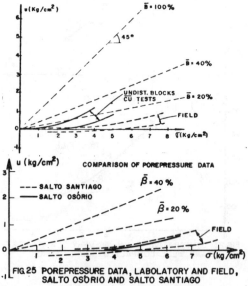

FIG. 25 POREPRESSURE DATA, LABOLATORY AND FIELD, SALTO OSÓRIO AND SALTO SANTIAGO

in Earth Dam Engineering in Brazil, "Proceedings, Sixth Regional Conference for Africa on Soil Mechanics & Foundation Engineering, Durban, South Africa, vol.2, pp. 285-304.

de Mello, V.F.B. (1977), "Seventeenth Rankine Lecture: Reflections on Design Decisions of Practical Significance to Embankment Dams", Geotechnique, London, I.C.E., 27 (3), pp. 281-354.

de Mello, V.F.B. (1977 b), Proceedings of the Nineth International Conference on Soil Mechanics and Foundation Engineering, Tokyo, vol. 3, pp. 364-368; 369; 381; 389.

de Mello, V.F.B. (1979), "Soil Classification and Site Investigation", Proceedings of the $3^{rd}$ International Conference, Applications of Statistics & Probability in Soil & Structural Engineering, Sydney, Australia, vol. 3, pp. 123-144.

Poulos, M.G. & Davis, E.H. (1974), Elastic Solutions for Soil and Rock Mechanics, New York, John Wiley & Sons.

Terzaghi, K. & Peck, R.B. (1948), Soil Mechanics in Engineering Practice, New York, John Wiley & Sons.

# Some problems and revisions regarding slope stability assessment in embankment dams

VICTOR F.B.DE MELLO
*Consulting Engineer, Sao Paulo, Brazil*

SYNOPSIS. In assessing slope stability of embankment dams, current practice of limit equilibrium calculations may be accepted as established, temporarily. Such factors of safety FS are nominal. The need is to establish histograms of FS values vs. varying behaviour, setting aside the right-wrong dichotomy at $1.00 \lessgtr FS \lessgtr 1.00$, and to search for meaningful acceptability criteria. A distinction is recognized between Factors of Safety FS and Factors of Guarantee FG. The conventional stability calculations of construction period, full reservoir, rapid drawdown, and of dumped and compacted rockfills are discussed, exposing the needs and procedures for significant revisions in present practices.

INTRODUCTION

In some of my latter papers (including one of the accompanying ones herein presented), I have attempted to emphasize that one begins by sorting out carefully which are the types of failures that one should visualize physically associatable with a major civil engineering project such as a dam, and should thereupon carefully distinguish between viable design philosophies for the different cases. Much depends on the statistics of truly repetitive conditions from which we derive our "laws" and the implicit or explicit histograms of probable behavior.

In the present summary paper I shall concentrate on the problems of slope instability and sliding failure analyses.

1. DISTINCTIONS BETWEEN NOMINAL FACTORS OF SAFETY

It is more than comprehensible that engineers should concentrate attention initially on failures. Human cognizance of the continuum comes from perception of when it ceases, the recognition of the discontinuity: one does not feel health, does not notice one's members except when they begin to hurt or to fail in the continuous functions of the silent majority. Slopes slide, some of them inexorably, in their allotted geological time. But how do we distinguish between a slow sliding movement that merely causes cracks and acceptable damages, and the more rapid movement that "endangers" lives and "totally disrupts" property? The cognizance of slope sliding started being associated with the latter rapid and major movement. Understandably the analysis of slope slides, both rapid and of major volumes, and the analysis of Factors of Safety FS against such sliding absorbed a major proportion of the efforts of the geotechnical engineer and of dam engineering.

Let us herein accept that the analytical methods of limit equilibrium, and of corresponding computations, are reasonably established as working tools of the geotechnical engineer. Even if we wish to discard such working hypotheses and vindicate more modern working methods of stress--strain distribution analyses, let us assume that they also have been distilled into a comfortably established routine engineering tool. It is the purpose herein to emphasize, however, that no matter how good our methods of investigation-testing--computation, all our FS are but "nominal", and the problems of DECISION (yes-no, acceptable-unacceptable) continue to be an arbitrary discontinuity within the continuum of reality. At some point in the histogram of Percent Probabilities PP% "Failures" (?), or better, PP% of Satisfaction Indices SI (de Mello, 1977) vs. varying nominal FS values, a designer must sever with the yes-no guillotine.

The problem would become one of discussing acceptable FS values. The behavior of a sliding mass should be reasonably conditioned by statistics of averages within the big volumes and surfaces at play. Why then is it that we are yet totally lacking in histograms of "behavior indices" of slope movements vs. nominal FS? The fault is surely not in Nature, but in our own mental model of rigid-plastic limit equilibrium, that would transform the problem into one of adjusting our analyses of failed slopes to the hypothetical condition of $FS \lessgtr 1.00$, having always stumbled on the presumption of dealing with "true" FS values, as if engineering were science, and both were deterministic. Moreover failures have always been analyzed a posteriori, under all the psychological and technical conditioning that this implies, with no real observations from the plane of failure at the time of failure movements.

In this paper I shall restrict my comments to the discussion of: (a) the nature of Facture of Safety FS vs. Factor of Guarantee FG (de Mello, 1979); (b) the conventional stability analyses of soil mechanics as applied to earth and earth-rock dam slopes, and the need to adjust them to reality in order to accumulate observational experience on the histograms desired of SI vs. FS or FG; ($b_1$) the need to adjust tests; ($b_2$) the need to adjust some of the analyses. The discussion of concepts regarding dam failure and embankment dam slope failure must be faced independently of any common slope sliding failure because of the disproportionate risks at stake. Practising dam engineers may be right in their disagreement with theoreticians when they insist that "a dam cannot fail" (i.e. cannot be permitted to visualize a risk of catastrophic sliding failure). Definitely the downstream slope of a reservoir should never be permitted to fail rapidly. Probabilistic calculations are an illusion. We should definitely resort to a physical change of statistical universe, so that the probability of the event should be guaranteed to be zero (de Mello, 1977).

## 2. FUNDAMENTAL PRINCIPLES OF SLOPE DESIGN AND OF STATISTICAL DEFINITION OF ACCEPTABLE INDICES OF BEHAVIOR

2.1 In the companion paper I expatiate on what I propose as the most fundamental principles of good theorizable design (de Mello, 1977), particularly relevant in embankment dam engineering calculations. Principle of the pretest, wherein one ensures that construction conditions should be more critical than the operative ones (FG vs. FS); further, the principle of humility in changes of conditions, wherein one avoids rapid changes of conditions disproportionate with experience and/or the status quo; thirdly, the all-important aim that foreseeable changes with time be in the favourable trend, however minute. The fundamental consequence is that in good design often a stability computation under critical operational conditions may well have been turned quite unnecessary: that is when wisdom supercedes knowledge. However, recognizedly we still lack statistical knowledge of acceptable indices of slope behavior, and of indices of acceptability of such behavior.

2.2 By convention we define FS as the ratio of:

$$\frac{\text{Predicted Resistances R } (\pm e_r)}{\text{Predicted Stresses S } (\pm e_s)}$$

$\pm e$ = dispersion

If during construction we have established satisfactory stable (elastic?) behavior up to a given stress level $S_c$ (c = construction), we know deterministically that $R \gtrsim (FS)R_c$. Now, if we have enforced that $S_c \gtrsim S_o$ (o = operational-time) and reasonably anticipate that $R_{ot} \gtrsim R_c$, then we cannot continue to use the ratio

$$\frac{R_{ot} (\pm e_r)}{S_{ot} (\pm e_s)}$$

as a value FS. I have proposed to call such a different nominal factor of safety as FG = Factor of Guarantee. Obviously one may accept FG << FS without risk of dissatisfaction: e.g. FG = 1.1 might well prove satisfactory in a material and condition that would require FS = 1.5.

2.3 In Fig. 1 I attempt to demonstrate that whereas Soil Engineering has generally considered only one definition of Factor of Safety, FS, it can be important to recognize three distinct Factors (considering only the differentiation of statistical dispersions around Resistances, R, without any further delving into the histograms of acting stresses S). Factor of Safety FS is a routine definition. I have chosen to call Factor of Guarantee FG the situation wherein by some lower rejection criterion I have assured myself that the histogram of Resistances can only be higher than some value already pretested or guaranteed. Obviously a value FG = 1.5 constitutes a much greater assurance of success than

FS = 1.5. In order to clarify the concepts it may be convenient to exemplify with regard to piles, with which familiarity is greatest, and softground tunnels, in which execution effects are of greatest moment.

A pile jacked down under 60 tons to absolute stoppage of penetration/settlement has FG = 2 if used for a working load of 30 tons: if the estimated resistance is 60 tons it has the conventional FS = 2. Setting aside the discussions on dynamic vs. static resistances of piles and cases of sensitive clays, driven piles checked by "refusal" observations can well be said to imply factors FG. In contrast, a bored pile would suffer from two disadvantages in its load-settlement behavior. Firstly, it would never have been pretested, and therefore one might conclude that it is affected by the FS (poorer than FG). Secondly, upon closer examination we should reason that it is even worse than that. All efforts of advancement of Soil Mechanics are towards minimizing sampling and testing disturbances and better representing in situ soil parameters (intact soil elements). In reality the assessed intact parameters would establish an upper rejection criterion, since the soil affecting bored-pile behavior represents a histogram of resistances always lower, to varying degrees, truncated at the upper value. A situation diametrically opposite to that of FG, with the lower rejection criterion. One could denomite the new ratio of averages (Resistances/ /Stresses) a Factor of Insurance FI: insurance is against something essentially inevitable, that should be attenuated.

The basic fact is that FI < FS < FG and depending on the dispersions of the histograms the differences may be very significant. If projects continue to be designed generally for (nominal) FS = 1.5 without recognition of this significant difference, all structures in which FI is at stake will record a much greater degree of troubles, while structures in which FG is at stake will incorporate an unnecessarily higher degree of safety. Tunnels and bored piles involve execution effects that only deteriorate in situ parameters (resistance, deformation) and therefore involve FI conditions. In the case of dams, if we allow flownets upon first filling to alter significantly the stability conditions of the downstream zone, we may be inviting FS conditions rather than the desirable FG situation of pretested behavior. More over, if the long-term flownets generate uplifts on tensile stresses, which can only deteriorate the strengths achieved as ascertained, we may be inviting most unfavourably the conditions associated with FI.

There is yet another important point to emphasize: the distinction between conditions which permit applying averages (as above), and those that involve localized situations corresponding to somewhere along the ends of histograms. That is, confidence limits and factors of safety might be related to "individual events on the histogram" (or fractiles) rather than on the median. Such is, for instance, the situation of instability of localized pockets along a bentonite-stabilized bored pile before concreting: after the concreting, the rigidity of the concrete guarantees applicability of the average over the profile. Similarly in a tunneling open face, during excavation localized instability may well be at play, corresponding to much more unfavourable conditions: behavior behind a steel face-plate of shield tunneling, or around a lining, can well be accepted to be averaged, which implies an inevitable benefit in comparison with localized worse conditions. A steel face--plate of shield tunneling, or around a lining, can well be accepted to be averaged, which implies an inevitable benefit in comparison with localized worse conditions. In the case of dams it has been argued (de Mello, 1977) that sliding mass instability can be treated on the basis of averages but locally generated piping failures could be characteristically associated with extreme-value statistics. In this paper I shall concentrate on problems associated with averages.

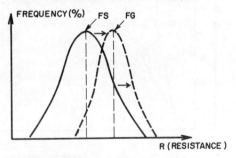

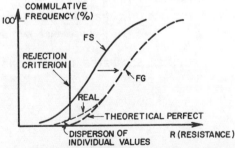

FIG.1 PROPOSED DISTINCTION BETWEE FACTOR OF SAFETY (FS) AND FACTOR OF GUARANTEE (FG)

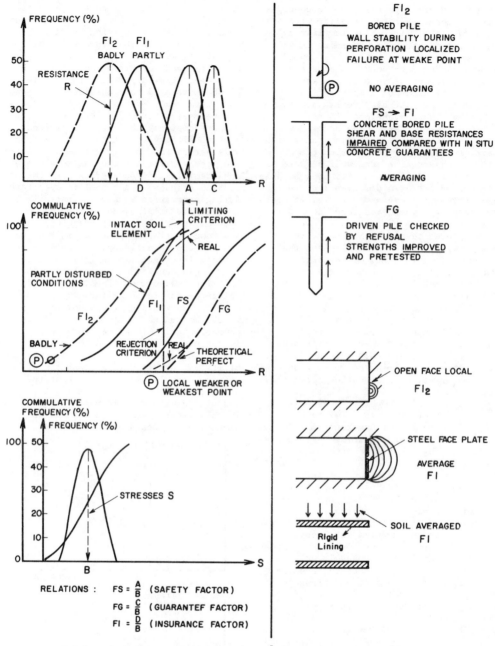

FIG.2 SUGGESTED DISTINCTIONS IN "FACTOR OF SAFETY"

2.4 In "complete" reasonable stress-strain-
-time path trajectory (Taylor) reasoning
and testing, we must not neglect to consider

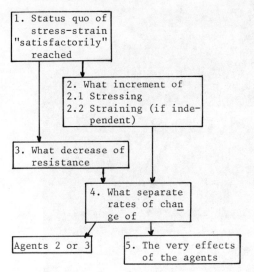

Note that internal stresses (1) are most
frequently different from the simply adopted geostatic assumptions of early soil
mechanics. Moreover, straining (2.2) is
sometimes quite independent of load-stress-
ing (e.g. collapse of structure). Further,
the onset of failures can be due to any of
the agents (2) and (3). Finally, what
matters much is not merely the rate of
onset (4) of the agents (2.3), but also,
the rate of onset of the effect (5) of the
agent, since it is well-known (e.g.
"viscous" and other complex rheologies)
that the rates of causes and rates of
effects are not similar.

Thus in estimating pore pressure development along a potential sliding plane due
to a change of flownet pore pressures, it
is not sufficient to consider, for the
effective stress analysis, the u value as
that corresponding to the new flownet. One
should consider the u value as composed of
two parts: the first is the hydrodynamic
flownet pore pressure; the second is an
incremental excess (positive or negative)
pore pressure due to tendencies to variation of volume ($\Delta V$) that would accompany
the incremental straining (normal and
shearing). This $\Delta u = f(\Delta V)$s depends of
course on an estimate of the incremental
stresses anticipated (always assumed on the
pessimistic side), but it depends also on
an estimate of the anticipated rate of
change of stresses, and principally rate
of change of strains. The latter important
consideration is why liquefaction slides
and mud flows are foreboded when the
stress-strain curves show a sharp post-
-peak drop, and the incremental shearing
is highly contractive.

In recognizing the above we merely
emphasize the recognition of how nominal
are our procedures of sampling-testing-
-computing (stability). They will always
continue to be so in our engineering
endeavours.

2.5 That is why I proposed (de Mello, 1977)
that we should establish an operational
Satisfaction Index SI for assessing the
behaviors of slopes associated with
different FS (and/or FG values). The
importance is to use in each same statistical universe (same embankment) several
different slopes, to accumulate theorizable
observational data (e.g. on "plastic
incremental movements" compared with
pseudo-elastic "stable" reference values).
The importance is to collect hundreds,
thousands of such pairs of data (SI vs. FG)
so as to establish the necessary histograms.
Then we will be in a position to apply our
acceptance - DECISION truncations, rati-
onally and economically.

Statisticians conversant with its
mathematics will kindly develop the
relationships between the conventional FS
and the newly proposed additional factors
FG and FI in function of the histogram
truncations. In good dam design, if
consequences of risk are high we want to
be dealing with FG conditions: thus, in
much of the following text I shall limit
myself to mention of FG. However one must
emphasize that FS conditions may well be
at play in most cases of conventional
designs, and if so the corresponding
computed value for satisfaction must be
decidedly higher than if FG conditions
were guaranteed (under pretested situations).

3. SLOPE STABILITY IN DUMPED AND COMPACTED
   ROCK-FILL

In my Rankine Lecture the subject was
somewhat discussed, to emphasize that: (a)
infinite slope analysis is an extremely
conservative lower bound and could well
accept a FG ≈ 1.00 +; (b) "stability is
automatically self-tested as the fill
rises at its constant slope" (i.e. we are
dealing with FG and not FS); (c) there are
advantages of deterministic u = 0 to permit
vely low FG; (d) there should be advantages
of locked-in prestress (in crushed angular
contacts) whereby we should count on greater
stability than implied by conventional
computations.

In furthering the subject herein the
following facts are emphasized, summarizing

PHOTO - 1

a vast number of observations on rock-fill and corresponding aggregate stockpiles (heights 30-45m) of very big projects. For interesting comparisons, specific data are presented on the sound dense angular basalt quarried in the Salto Santiago and Foz do Areia Projects.

The following photos (1 to 7) represent (a) comparative end-dumped vs. bottom-excavated repose-slopes of basalt rock stockpiles on which preliminary statistical data on face angles were carefully surveyed; (b) the 48m high 1:1 compacted rockfill slope of the upstream cofferdam

PHOTO - 2

PHOTO - 3

incorporated into the 78m high earth-rock dam of Salto Santiago; (c) the final 1:1.4 downstream slope of the Salto Santiago dam; (d) the finished 1:1.25 downstream slope of the 160m compacted concrete-face rockfill dam of Foz do Areia.

(1) The stable slopes (angles of repose) were surveyed in detail in minimum stretches involving more than about 15 big-size rocks. The histogram for the end--pushed loose rock may be considered conditioned by the most unstable surface rock having to stop from a moving position. (Fig. 3)

(2) In comparison, the excavated slopes show two distinct trends, a steep stretch

PHOTO - 4

PHOTO - 5

(even partly subvertical), dominated by the more stable rocks having to be moved out of their interlocked rest ("static" vs. dynamic friction?), and the lower stretch comprising mixed excavation-slope and rolled--slope material.

(3) In comparison with smaller granular material, we deal with a histogram that is not so tight as in "uniform sand laboratory tests".

(4) However, even in loose end-dumped angular rock stockpiles there is a definite strength gain from prestress.

(5) The rhetorical question posed is, which $\phi'$ aver should prevail in nominal stability analyses, that of slopes a-fill-

PHOTO - 6

PHOTO - 7

ing, or that of slopes excavated from the bottom, after benefit of prestress?

(6) Considering the very significant prestress contributed by compaction of rockfill in lifts, how much steeper can we go without any risk of unsatisfactory behavior?

(7) In consonance with the Rankine Lecture suggestion, the vertical and horizontal movements of points on the downstream slope were carefully recorded as the compacted rockfill dam rose by its final increment of height. (Fig. 4). Can we say that the tendencies of movements are such as upon progressing would model a slope sliding failure?

(8) Note that this rockfill was relatively uniform and compressible, not suggestive of the most stable interlocking rigid blocks. Face instability is satisfactorily tended by arranging bigger blocks as markers for offset and lift thickness.

## 4. CONSTRUCTION PERIOD STABILITY, CLAYEY MATERIALS

In the companion paper it is emphasized that under modern heavy earthwork equipment the conditioning factor is trafficability, and that in any well-designed dam having a chimney filter (cf. Rankine Lecture) it should be difficult to conceive of compactable conditions facing end-of-construction instability. There is inexorably an effective stress preconsolidation cohesion even if only short-term) for the short--term condition considered. Moreover, there is the initial negative pore pressure, and not the presumed $u_c$ vs. $\gamma z$ (c = construction) diagrams insinuated by the early (and most recent but faulty) laboratory tests and field observations (USBR). Finally, the constant $r_u$ or $\bar{B}$ coefficient assumed for simplifying computations is quite unnecessary, and misleading regarding greater instability for shallower circles (really benefited by suction and cohesion). (Fig. 5)

In the zone downstream DS of the chimney, affecting the all-important DS stability, we should prefer generating some $u_c$ in order to assure ourselves of the pre-test principle and satisfactory FG, and especially its inexorable improvement with $u_c$ dissipation with time (cf. Rankine Lecture). Should any $u_c$ develop higher than desired, we may use at will the intermittent $u_c$ - ADJUSTERS comprising dry layers (functioning as "blotting-paper" non-exiting filters), without fear of layering permeabilities. The same expedient may be used to advantage also in the upstream US zone of the dam, excluding what would be equivalent to a "core": in the "core zone" it is necessary to avoid unfavourable permeability $k_h \gg k_v$ effects on flownet, and is desirable to have a high $u_c$, perferably close to the $u_{fn}$ (fn = flownet, full reservoir) so as to minimize the change of conditions in the

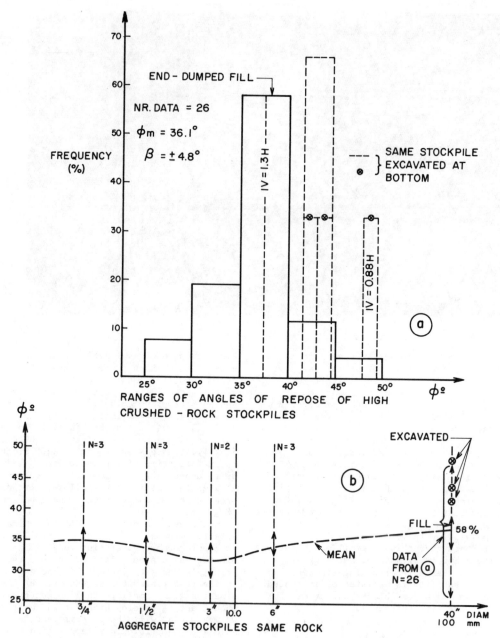

FIG.3 BEHAVIOUR OF ANGULAR ROCK SLOPES AT F.S. = 1.00

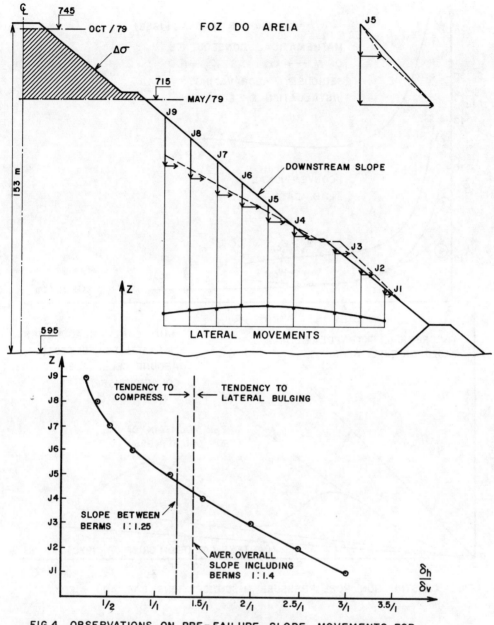

FIG. 4 OBSERVATIONS ON PRE-FAILURE SLOPE MOVEMENTS FOR "SATISFACTION INDICS" (cf. RANKINE LECTURE) vs. NORMAL F.S (160 m ROCKFILL)

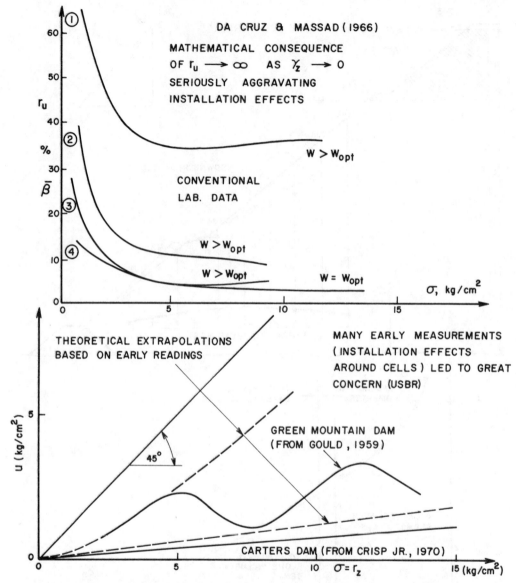

FIG.5 CONSTRUCTION PORE PRESSURE MISCONCEPTIONS, LAB. AND FIELD

core on first filling.

As is easily proven and well known, a core may have quite high construction pressures without any impairment of end-of-construction US slope stability. If necessary, stability analyses can well be run, using appropriate estimates of internal stresses, cohesion, negative and positive $u_c$, and the effective stress envelope but the acceptable (and even desirable design aim) FS should be close to 1.0 for adequate "pre-testing". The analyses are merely to facilitate Bayesian insertion of succesive $u_c$ observations for continually improved assessment of FS, while deformation measurements furnish indications on SI.

Since one cannot anticipate the coincidence of achieving $u_c \approx U_{fn}$ it is of interest to discuss in which direction the tolerance should be more favorable, $u_c > u_{fn}$ or $u_c < u_{fn}$. The ideal situation would be to have had $u_c$ developed to values higher than $u_{fn}$ but dissipated to a value slightly lower than the $u_{fn}$. Thereupon the soil will have been pretested to values of u higher than necessary, and will be behaving within the precompressed or preconsolidated range wherein changes of void ratio and of behaviors with change of stress are small.

## 5. DOWNSTREAM STABILITY, FULL RESERVOIR

### 5.1 First filling

The interesting problems of rapid vs. slow filling will not be discussed herein. We shall firstly assume a traditional critical "permanent flownet".

One would imagine that finally this will be the all-important case in which a conventional stability analysis is indispensable. Truly, however, quite to the contrary. Indeed, the hypothesis of such a failure is so unthinkable that one can only accept the wisdom of a design wherein the establishment of critical full reservoir conditions will not cause u values, $u_{res}$ (res = reservoir), higher than the $u_c$ already satisfactorily borne (cf. Rankine Lecture). This is fundamental. And with appropriate design of position of chimney, controlling $u_{fn}$, and appropriate control of compaction parameters, controlling $u_c$, it is quite simple to achieve this wise design situation that dispenses such additional stability calculation. With $\Delta u = u_{res} - u_c$ = negative (modestly) the change of stability from end-of-construction to first filling can only be an increase ($\Delta FG$ = positive). ||Note. There are other conditions that may similarly be reasoned to be satisfactory ||.

Many a dam has behaved satisfactorily without any inkling of such principles. However, absence of evidence is not evidence of absence. The only way to guarantee zero probability of downstream sliding failure is to have pretested the FG $> 1^{++}$, and to have $\Delta FG$ $^+$ve. Any number of design sections and conditions may be rapidly sketched showing how to compare $u_c$ vs. maximum possible $u_{res}$: it is a simple exercise, and requires no qualms of decision.

What is the most unfavourable $u_{res}$ possible? The flownet hypothesis $u_{fn}$? Under which hypotheses (cf. Rankine Lecture), all higly dependent themselves on other hypotheses? Would it then be right to assert that the maximum maximorum attainable would be the full $u_{hyd}$ (hyd = hydrostatic)? Most specialists might still claim so. I dare emphasize that even this claim is, in principle, wrong, because it would not incorporate the influence of rates of change and of possible rheological consequences, and rates of consequence. If a rapid reservoir filling, and unfortunate upper limit, leads to a positive $\Delta u = u_{fn} - u_c$, or $u_{res} - u_c$, and thus there is a rapid drop in stability, there can be a rapid strain and consequent $\Delta V$ due to shear around the sliding surface. If we guarantee a dilatant rheology, we would be secure, but if there were a tendency to compression, we would have an increase $\Delta u_s$ (incremental excess pore pressure due to contractive tendencies as already discussed) such that the real maximum u at play in the sliding stability could be $u_{res} + \Delta u_s$. It is imperative to seek the right geomechanical and rheological statistical universe, that may dispense the full reservoir stability analysis as such, by assurance of much better known truncated histogram analyses of changes of conditions.

### 5.2 Long term stability

For a long-term DS slope stability once again the only appropriate analysis is by mentally postulating what changes of loadings and resistances can tend to occur to affect the already established first-filling FG. It is really absurd to think of a slope analysis "from scratch", because of the much much greater probabilistic imprecision than by Bayesian analysis of posterior probabilities as superposed on the prior (cf. Rankine Lecture). There is no difference in principle in using the probability changes of u from $u_{res}$ ff (ff = first filling) to $u_{res}$ lt (lt = long--term), in manner similar to the roughly suggested progressive adjustment of $u_c$ as a quantification of the Observational Method.

Would $u_{res}$ tend to increase with time? Depends on how consolidation, secondary compressions, tensile cracking, etc... would change relative permeabilities. We must arrange for strength to increase with time, and u to decrease with time: both trends are associated with tendency towards compression (desirably modest). Strength may further gain from favourable cementing and thixotropic effects. It isn't at all difficult to design to guarantee such compressions, such that after $FG_{res}$ ff $> 1^{++}$ we guarantee a $\Delta FG$ + ve.

It would seem that the fear of long-term

instability in natural slopes has unduly influenced dam design: whereas long-term instability is inexorably a problem in a natural slope at FS ≈ 1.00, in a good dam design of FG > $1.0^{++}$ and + ve ΔFG, there should be scant probability of such a problem. Note, however, that much depends on the shear strains and brittle stress-strain, and strain rates: repeating concisely what has been emphasized, it is not valid to be content with using effective stress envelope and merely $u_{res}$, when there may well be a Δu = f(Δ strain).

One loading condition that would seem to thwart the postulated simplicity would be the seismic one. I cannot herein advance into this additional case. Suffice is to emphasize, however, that even the probabilities of a catastrophic seism should not be considered independently of the probabilities of occurrences of smaller seisms. Except for the extreme event of the very first seism being that of maximum maximorum probable intensity, the occurrences of smaller events could and should be considered with regard to cumulative improvement of conditions by successive compressions (cyclic). It would be ideal to aim at a slightly dilative instantaneous behavior for intensities higher than some moderately rare recurrence level.

## 6. UPSTREAM SLOPE, INSTANTANEOUS DRAWDOWN

This is, indeed, a topic in which both the theorizing, and the conventional practices, are blatantly wrong. And, as a result, it would appear that upstream US earth slopes are significantly overdesigned. There have been cases such as the emptying of the Tarbela reservoir (≈ 4m per day) on the 10 hr emptying of the 58m deep Euclides da Cunha reservoir (Jan 77) after overtopping failure, in which nothing budged. The following photos (8,9) of the rapid and subvertical erosion scar clearly indicate the sliding that can be said to have been generated on this steep face by the instantaneous drawdown condition to which it was exposed: one must note, however, that along this face the condition is much more exacting than merely that of a rapid drawdown, because besides instantaneous removal of the water pressure diagram there is the removal of the earth-pressure diagram also.

Thus, as regards slope stability analysis and design of upstream slopes subject to rapid drawdown it is not merely a case of permitting lowering FS to 1.1 (as presently often applied), and has little to do with the fact that drawdown is never quite "instantaneous" and that some drainage lowering of the phreatic might be included. It is a case in which the very mental model appears fraught with inconsistencies from the start (cf. Rankine Lecture): if one does rightly lower FS but does it on a wrong mental model, there can well occur some undesirable surprises.

In consonance with routines of examination of limiting critical hypotheses, I shall herein limit myself to considering the hypothetical absolutely saturated embankment. We well know how high are the backpressures necessary to saturate triaxial specimens (6 to 12 kg/cm$^2$), especially if the air micropores are first reduced in diameter by the soil consolidation under confining stresses. Thus it is obvious that in modest dams and/or shallow sliding circles, the compacted material would not be saturated. The principles below summarized restrict themselves to considering tendencies to change of σ' (assuming saturated incompressible pore fluid): in a generalized extension, we will have to consider tendencies to change both of σ' and of u (besides, of course, the incremental shear stress and strain rate $\Delta u_s$, and any sophistications such as rotations of principal stresses, etc...).

The fundamental errors of concept may be summarized as the following:

(a) The dichotomy could not possibly be between:

| "FREELY-DRAINING FILLS" | vs. | "COMPRESSIBLE FILLS" |
|---|---|---|
| (Terzaghi-Peck, 1948, accepting flownets, rapid drawdown RDD) | | (Pore pressures generated due to $\Delta^\sigma t$ on dam face and $r_u$ concept, $r_u$ ≈ 1.0) |

Obviously draining vs. non-draining has no obligation to any assumption on compressibilities. Likewise, incompressible vs. compressible has no obligation of direct deterministic association with drainability or rates of drainage.

(b) Except in extremely differentiated materials, there is no black-white distinction of intrinsic qualities of draining vs. undraining, compressible vs. incompressible, as regards materials and conditions thereof. The Portuguese Spanish languages (and others?) have a conceptually important distinction in the verbs ESTAR (to be temporarily, to behave as being) and SER (to be in essence, to be permanently). Firstly no material possesses a homogeneous tendency (to compress, to drain, or whatever) over the entire upstream body of a dam. Secondly, compressibility (etc...) is a temporary behavior problem (ESTAR): a material will not be have as compressible (even if it is generally or frequently

PHOTO - 8

compressible) if it is subjected to a stress release. A soil element in one part of the critical circle may be subjected to stress release, and want to behave as dilatant, while another exactly similar soil element in another part of the same circle may be subjected to a stress increment, and want to compress.

As a curious extreme example to emphasize the conceptual point we could say that in a material that is (SER) homogeneous and pervious functioning under a flownet, each flowline (surface) behaves as (ESTAR) impervious, since no iota of water from one side or other of the surface crosses it (it is immaterial that it does not "wish"

PHOTO - 9

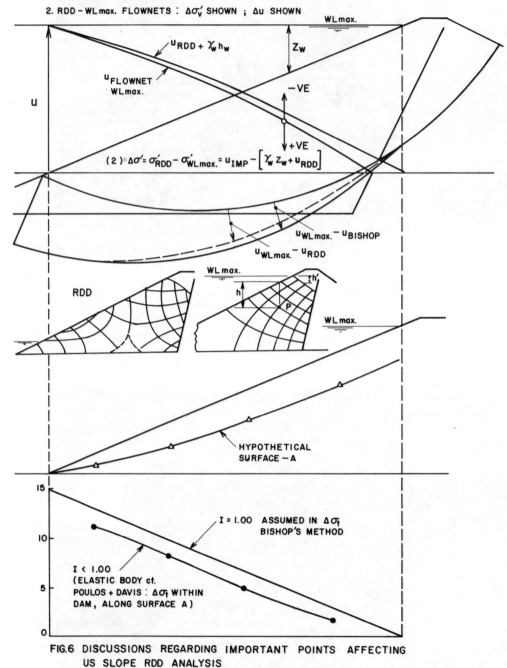

FIG.6 DISCUSSIONS REGARDING IMPORTANT POINTS AFFECTING US SLOPE RDD ANALYSIS

to cross the line). All materials are compressible, pervious, etc..; it is a question firstly of algebra (tendency to expand vs. tendency to compress) and secondly of degree (compress more, or less).

(c) Thirdly, there are some over-simplifications in the $\sigma_t$ and $r_u$ or $B$ procedure that in no way permit incorporating the advantages (or disadvantages) of the most fundamental design weapon of a good dam design, which is the chimney filter and the drainage (u - controlling) features (be they exiting or non-exiting drains, cf. Rankine Lecture). The degree of significance of such consequences varies considerably in different design crosssections.

The principle proposed is simple, and quite consistent with the hypothesis of perfectly saturated (incompressible) pore fluid. Tendency to change of flownet is "instantaneous", and what matters are instantaneous pore pressures changes (compare with rheological model of Terzaghi consolidation theory). Therefore we can draw the two flownets, $WL_{max}$ and RDD; the positions of the drainage features are duly incorporated. Once these internal flownet pore pressure conditions are established, based on the rapidly changed boundary conditions, it is quite simple to check what would be the tendencies to change $\Delta\sigma'v$ that determine whether or not the soil elements would wish to behave as compressible. If there is a tendency to compression, obviously there should be a corresponding increased transient pore pressure (due to $\Delta V$). The stability analysis should be based on the RDD flownet (valid for incompressible pores) complemented (at worst) by the $\Delta u$ due to the compressive $\Delta\sigma'$ (and any consequent shearing $\Delta V$). In other words, the principle is that routine flownets presume incompressible pores and incompressible fluid: therefore, one checks first what would be the tendency, to compress or to expand, in an instantaneously incompressible assumption. Thereupon one immediately concludes whether the material will behave as compressible or behave as dilatant, but in any case superposed on the background of the saturated instantaneous RDD flownet. Fig. 6 exemplifies this principle of stability analysis that is quite generally applicable.

Once again instability conditions are best analysed with regard to changes of conditions superposed on the earlier proven slope obliquities of stress under $WL_{max}$. Moreover, the occurrence of frequent partial pool drawdowns can be introduced with the appropriate analysis of whether or not the trend is toward gradual small tendencies to compress (and increase strength). The ideal design of crosssection and compacted material would aim at slight compressions under the frequent smaller drawdown episodes, and, hopefully, a tendency to dilate under the most critical drawdown.

However, once again, if construction $u_c$ had been satisfactorily high in comparison with the new $u_{RDD} + \Delta u$, and US stability well established, we should be dealing with a pre-tested slope stability and a perfectly satisfactory FG $\approx 1^+$.

There is considerable evidence, direct and indirect, that US slope instability due to rapid drawdown is much overestimated by presently current analyses, and that any tendency to sliding would tend to be of shallow scoop circles.

## 7. CONCLUSION

Recommendations are made that stability analyses be used as furnishing nominal FS (or preferably FG) values and that the acceptance levels of such index values be established by great number of non-failure situations, treated statistically. For more critical conditions the only satisfactory approach for guarantee of design is to consider changes of conditions, starting from pretested conditions, preferably more critical than the operational ones.

In a manner similar to the evolution of foundation design as it moved from discussing FS with regard to bearing capacity formulae, to the more fruitful approach of aiming at limiting deformations for avoiding minor damages, the field of dam slope design will only move significantly forward when such countless statistical data are collected, correlating FS or FG vs. Satisfaction Indices SI.

## 8. REFERENCES

Bishop, A.W. & Bjerrum, L. (1960), "The Relevance of the Triaxial Test to the Solution of Stability Problems", Research Conference on Shear Strength of Cohesive Soils, ASCE, pp. 437-501.

da Cruz, P.T. & Massad, F. (1966), "O Parametro B em Solos Compactados", Anais do III Congresso Brasileiro de Mecânica dos Solos, Belo Horizonte, vol. I, pp. I1-I24.

de Mello, V.F.B. (1977), "Seventeenth Rankine: Reflections on Design Decisions of Practical Significance to Embankment Dams", Géotechnique, 27 (3), Sep. 1977, pp. 281-355.

de Mello, V.F.B. (1979), "Soil Classification and Site Investigation", 3$^{rd}$

International Conference 'Applications of Statistics & Probability in Soil & Structural Engineering', Sydney, Australia, vol. III, pp. 123-144.

Poulos, H.G. & Davis, E.H. (1974), "Elastic Solutions for Soil and Rock Mechanics", New York, John Wiley & Sons, 1974.

Terzaghi, K. & Peck, R.B. (1948), "Soil Mechanics in Engineering Practice", New York, John Wiley & Sons, 1948.

ns / Bangkok / 1-15 December 1980*

# Some examples of design changes required during construction

J.R.HUNTER
*Snowy Mountains Engineering Corp., Cooma, Australia*

## 1 INTRODUCTION

The foundation and the connection of the foundation to a dam are possibly the two most important aspects in dam design. However, prior to construction the design parameters and character of the foundation and construction materials can only be assessed by exploration methods currently available. These methods can give a general indication of the conditions and the overall competency of a foundation but they cannot necessarily define all conditions which will be revealed when the foundation is excavated and which, in some cases, require modifications to the design of the dam. Similarly the properties of embankment construction materials may vary from those assumed for the design.

This paper cites six examples of dams which were constructed on reasonably firm rock foundations, and which had to be modified during construction as a result of:

- conditions exposed in the foundation excavation being different to those assumed; or

- the behaviour of embankment materials was different to that assumed in the design

The examples show that a dam designer must play a close liaison role during the construction of a project so that the design can be amended if conditions vary from those assumed. In all cases quoted in the paper the owner adopted the policy of having a Foundation Inspection Committee to approve the foundations prior to commencing dam construction. This Committee comprised representatives from design, construction and geomechanical fields to provide the necessary liaison.

## 2 FOUNDATION EXPLORATION

The methods used for exploration of the dam foundations were regional and surface geological mapping supplemented by core drilling, trenching, augering, seismic traverses and tunnels and adits. Whilst these methods can give a good indication of the subsurface conditions and are adequate for establishing the feasibility of a project and the main design parameters they cannot define all conditions pertaining to a site. The engineer and the geologist are therefore faced with the decision as to when to terminate the exploration. In all cases cited in this paper the decision was made when it was agreed that further exploration would not advance the interpretation of the site conditions significantly. There are other cases, of course, where financial and budgetary restrictions limit the amount of exploration and in these cases the possibility of changes being required during construction is more likely.

In the case of very complex and variable foundations it could be necessary to adopt an 'observation method' as adopted by Terzaghi and Leps for Vermillion Dam in the mid-1950's (Terzaghi & Leps 1958). They concluded that the foundations were so complex and erratic that the most complete investigations would still leave a wide margin of interpretation and advocated that the tenders be based on more or less arbitrary assumptions for the character of the subsoil and that the designer make provision for adequate monitoring of the construction so that any discrepancy in the assumptions could be taken into account and the design modified during construction. A similar method was used on Willard Dam in USA (Walker 1967).

It may also be possible for some organisations to do the foundation excavation under their direct control and thus reveal any potential problems before entering into costly contracts. This approach has been used successfully by several organisations in Australia. Two of the examples cited in this paper, namely Thomson and Cania Projects, illustrate the possible advantages in excavating for the dam foundations separately.

## 3 MATERIALS EXPLORATION

Materials exploration is normally done by drilling, augering, trenching and laboratory testing to determine the properties and the suitability of the materials to be used in the embankment. Trial quarries are often opened up to assess properties of rock for concrete or rockfill and wherever possible a trial embankment is tested for embankment-type dams, using equipment that the Contractor is likely to use, to assess the behaviour of the materials under construction conditions. By necessity these trial rollings are on a small scale and do not necessarily simulate the behaviour of the materials during full-scale operation in the borrow area and under the construction procedures adopted by the Contractor. In addition it is not practicable to test all materials within the borrow areas particularly those at depth. They can therefore only be regarded as an indication of the likely behaviour of the materials within the embankment.

## 4 TOOMA DAM - SNOWY MOUNTAINS SCHEME

Tooma Dam is an earth and rockfill structure 67 m high (Figure 1) and was completed in 1961 (Hunter & Hartwig 1962). Geological investigations consisted of surface mapping, bulldozer trenches and a total of seventeen diamond drill holes and water testing of all drill holes. The foundation of the dam is about 30% granite-gneiss and 70% biotite granite which was obtained after stripping topsoil vegetation and organic materials.

During the excavation for the dam foundations, open cracks were noticed on the right abutment above a large outcrop of rock which was just upstream of the cut-off trench. The cracks ranged in size from very small to up to 300 mm width and included a movement of a large mass of rock and earth. Monitoring of the suspect area confirmed that movement was occurring and further exploration at the toe of the outcrop revealed a clay seam approximately 25 mm thick dipping towards the river on which the rock mass was moving. The cracks in the abutment extended from the cut-off trench for a distance of approximately 60 m upstream. It was considered that grouting of these cracks could only be done with partial success and any cracks remaining would provide direct passage for the water to penetrate under the earthfill, at least to the cut-off trench. In addition, any further movement of the abutment during construction could cause disturbance and possible cracking of the earthfill after it has been placed in the fill. Rezoning of the embankment so as to replace the upstream portion of the earthfill with rockfill would effectively overcome this undesirable situation and the embankment was rezoned accordingly. To arrest any further movement of the right abutment and provide added stability, a stabilising fill was placed immediately against the abutment.

With the steepened upstream face of the earthfill in the amended section, the grout curtain provided in the original design was very close to the upstream toe of the new earthfill zone. The seepage path through the foundations and from the upstream face of the earthfill into the foundation downstream of the grout curtain was considered to be too short for the reservoir head and a secondary downstream grout curtain with blanket grouting between the two curtains was therefore provided.

The earthfill was a residual soil of completely weathered biotite granite. During the early placement of soil in the embankment difficulty was found in keeping the moisture content in the fill below the maximum wet limit. Although borrow pit moisture samples were dry of optimum the results of moisture tests on compacted fill was consistently wet. It was then realised that the change in gradation of soil caused by breakdown in excavation and rolling had a profound effect on the Optimum Moisture Content (O.M.C.). The whole area of the borrow

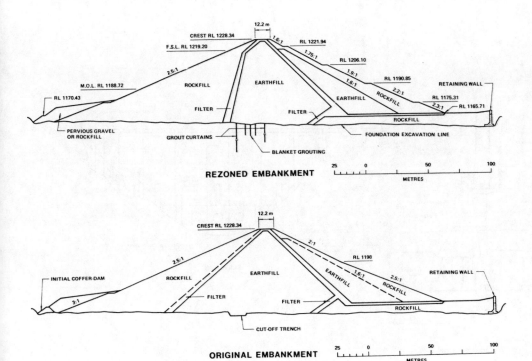

Figure 1  Tooma Dam – Cross-sections

pit was opened up so that full advantage could be taken of sun and wind to dry the soil. However, the moisture content in the embankment was still higher than the upper limit specified and after checking the stability with the increased pore pressures the upper limit for the moisture content was increased from O.M.C. +1% to O.M.C. +1.3% to facilitate construction.

As a result of this experience it became practice in the investigations for all earth borrow materials for future dams in the Snowy for the soil to be subjected to mixing in a cake mixer to simulate breakdown of the soil due to construction operations. This procedure appeared to improve the interpretation of soil behaviour but is by no means the absolute answer. Breakdown of soil during compaction applies more to residual soils than deposited soils.

5. MURRAY 2 DAM – SNOWY MOUNTAINS SCHEME

The Murray 2 Dam is a concrete arch structure approximately 43 m high. First stage completed 1965, second stage completed 1968 it consists of a 30 m high thin arch shell constructed on a relatively massive concrete foundation block which was used for diversion of the river during construction of the arch (Figure 2). To produce a symmetrical arch wall, to take up possible variations in excavation depth from that assumed in the design and to distribute the stresses transferred from the arch wall to the foundations a socle was provided between the base of the arch and the foundation.

For construction program reasons it was decided to construct the dam in two stages. The first stage being the construction of the base block with its openings for river diversion so that the discharges from an upstream power station could be diverted whilst the arch shell was being constructed. The first stage contract also included the excavation for the foundations for the socle and the abutment thrust blocks. This proved very advantageous because it revealed different conditions to those assumed and provided an excellent means of exploring the foundation before a contract was awarded for the construction of the main dam. It also allowed modifications to be made, as discussed below, without any delays or claims from the Contractor of Stage 2.

The foundations are metamorphosed sediments consisting mainly of sandstone grading to quartzite, siltstone and slate and are tightly folded into an anticline whose axis is approximately parallel to the

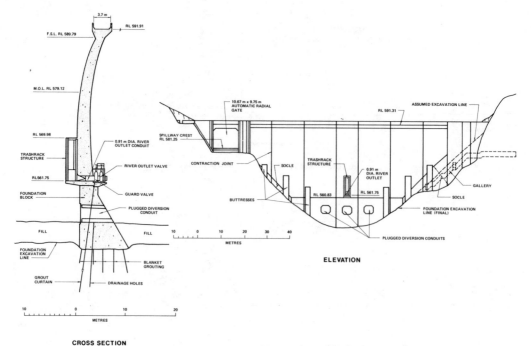

Figure 2  Murray 2 Dam - Elevation and Section

stream bed. The initial exploration was mainly aimed at establishing the feasibility of the project and the final design of the first stage construction. It consisted of surface geological mapping supplemented by diamond drilling, trenching and seismic traverses. A total of six drill holes were drilled.

On the upper right abutment a sheared zone was exposed under the thrust block during the excavation in Stage 1. Additional excavation was carried out and three additional drill holes were drilled in an attempt to ascertain the extent and importance of the seam on the abutment stability and to suggest possible remedial measures. Strength tests were done on material from the seam and stability analysis carried out on the abutment. These indicated that the foundation was stable as finally excavated provided there were no seams in the abutment beyond the ones exposed. To confirm the adequacy of the right abutment and ensure that there were no further seams or fault zones behind the one already exposed, a 15 m long adit was constructed into the abutment. This adit was later incorporated into a grouting gallery and formed part of a gallery system in the right-hand abutment thrust block.

The general consensus of opinion was that the unfavourable clay seam was small and its importance could only be recognised after foundation excavation. It was fortuitous that the construction was performed in two stages because there was no delay to the Contractor's program and the modifications could be carried out without being subjected to tight time restraints. The modifications were to increase the depth of the socle and thrust block on the right abutment and no redesign of the arch shell was required because the socle ensured that the profile of the arch shell was unchanged as shown on Figure 2.

6  THOMSON DAM - MELBOURNE AND METROPOLITAN BOARD OF WORKS - VICTORIA

Thomson Dam is a central core rockfill dam approximately 165 m high with a crest length of approximately 550 m. The Thomson River runs generally east and passes through deep valleys and gorges where side slopes are commonly $20°$ to $40°$. The topographic relief in the dam area is typically 300 m to 400 m but the full range from valley floor to the top of the plateau is about 1 200 m.

The rocks at the damsite are of

Palaeozoic Age consisting of interbedded siltstones and sandstones. These rocks are intensely folded and faulting, both thrust faulting and faulting across the bedding, is common throughout the area.

The geological investigation up to the tender design stage consisted of geological surface mapping supplemented with sub-surface exploration comprising trenching, diamond drilling and seismic traversing. A total of eighteen bulldozer trenches, forty-eight drill holes up to 160 m deep and twenty seismic traverses were undertaken prior to commencing dam foundation excavation.

These investigations identified two disturbed areas, one on each abutment. The one on the left abutment was minor. It was upstream of the toe of the dam and did not present any problem to the safety of the dam. The one on the right abutment was found to be more extensive and although not affecting the technical feasibility of the project, was recognised as presenting a serious stability problem. It extended from downstream of the axis of the dam to a point some 50 m to 100 m upstream of the proposed upstream toe of the dam. The main concerns were the amount of excavation required to get a satisfactory core foundation and the treatment required to ensure the stability of the upstream right abutment under submerged conditions.

Further exploration was considered unlikely to give any better understanding of the extent of the disturbed material or its characteristics. It was therefore decided to excavate the foundations for the core (the core trench) on the right abutment before awarding a contract for the main dam. This procedure would not seriously affect the overall program.

The excavation of the core trench required a rather extensive system of haul roads to gain access. During the excavation of these roads movements and local slumping were observed on the abutment and a monitoring system was established to determine the extent of movement and monitor the stability of the abutment.

Movements of up to 50 mm in a 2-week period were recorded and movements increased during wet weather. Expert geological advice was obtained to assist with the overall understanding of the problem. The initial assessment was that most of the disturbed area was the result of an ancient synclinal slide dipping towards the river within the ridge which forms the upstream right abutment. This slide had been reactivated by the excavation of the haul roads and had the potential of developing into a major landslide which could block the river during construction. This was unacceptable and it was decided to place a stabilising fill to support the abutment. Initially the stabilising fill (Section AA Figure 3) was only a poultice against the abutment with provision for diverting the river around the poultice because the diversion tunnel would not be operative for about 6 months.

At the same time, direct shear testing of the bedding fault material, upon which movement was occurring, was carried out. The results of this showed that the bedding fault material had a low residual shear strength in the range of $7^o$ to $12^o$. In addition ten diamond drill holes were drilled to supplement the geological surface mapping of the excavations.

From these investigations the extent and structure of the synclinal slide was determined with a reasonable degree of confidence and a geological model built to represent the likely conditions pertaining at the site and the possible modes of failure of the slide. The abutment was analysed by means of a three-dimensional mathematical model based on a conventional wedge analysis. These analyses indicated that to obtain the desired factor of safety under drawdown conditions the stabilising fill had to be extended to support the spur upstream of the toe of the dam as indicated on Figure 3.

The existence of the synclinal slide still did not account for the considerable disturbance below the basal surface of the synclinal slide in the core trench. The disturbed material had all the appearances of being part of an old landslide. Excavation was continued until a deeply scored siltstone surface was exposed which dipped towards the river, and as such it formed an obvious basal surface for a slide directly out into the river. In addition a fault cut across the core trench which divided the material above the siltstone surface into two parts. It was concluded that rock in the upper part had acted as an active wedge which forced the mass above the scored surface out into the river valley once the river had cut down far enough. The treatment was to basically remove all slide material.

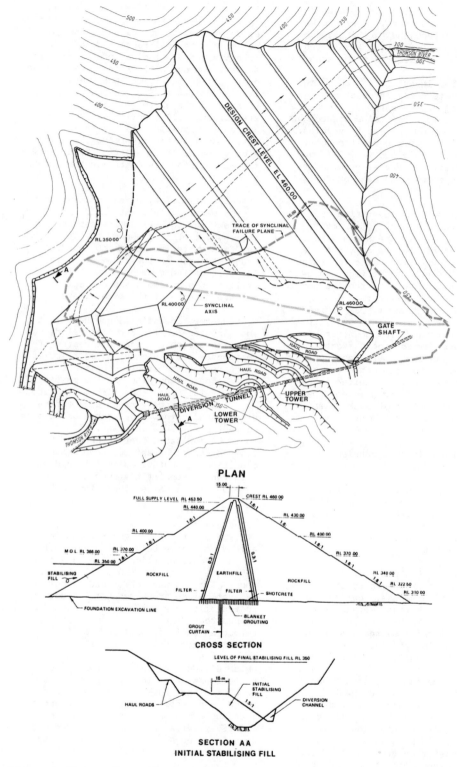

Figure 3  Thomson Dam - Plan and Sections

The disturbed rock associated with the slide on the dam axis presented no real problem as it could be maintained during construction and will finally be supported by the embankment.

Without the benefit of opening up the area for the foundation excavation the extreme complexity of the geology and the extent of landsliding would never have been revealed. If the contract had included excavation for the foundation there could have been serious delays to the contract which would have meant claims from the Contractor until the full understanding of the problems had been obtained.

## 7  JINDABYNE DAM - SNOWY MOUNTAINS SCHEME

Jindabyne Dam is a sloping core rockfill dam 68 m high, as shown in Figure 4, built in 1967 (Howard & Halliday, 1968). The foundations for the dam are granite and were explored by thirty-six diamond drill holes, four exploratory adits, four exploratory shafts and trenches together with surface geological mapping. No unforeseen conditions in the foundations were encountered and there were no significant design changes required from those set out in the original contract. However, the behaviour of the core material in the embankment was different to that anticipated from the exploration and was the cause of some concern.

The material comprising earthfill was a residual soil consisting of completely and highly weathered granite and was placed in 150 mm layers at moisture contents slightly above optimum; this was considered necessary to prevent cracking of the core. Placement commenced with moisture content between O.M.C. -0.5% to O.M.C. +1.5% with the average above O.M.C. The dam was well instrumentated and early in the construction high negative pore pressures were recorded in the hydraulic piezometers. These negative pore pressures were considered undesirable in a thin sloping core as the desire was to have as flexible a core as possible to prevent cracking due to settlement of the rockfill. In addition trenches excavated in the embankment fill gave the impression that the compacted fill was dry and brittle. A high soil suction, indicated by the high negative pore pressures, was confirmed by laboratory testing. The reduction in the apparent moisture content after placement was attributed to the migration of free moisture surrounding the soil particles to the interior of the particles themselves, as in the case of expanding clay minerals (Hosking 1974).

To achieve a more plastic core the moisture limits were progressively increased while keeping a close check on the handling and compaction characteristics of the soil and behaviour of the fill. Eventually limits of O.M.C. +1.5% to O.M.C. +3.5% were adopted with no significant loss of field dry density or change in compaction characteristics. Additional piezometers were installed to check the pore pressure build-up with the increased moisture content and this information was used to check the upstream stability of the dam. The additional wetness of the core did not affect the stability of the upstream face.

## 8  TALBINGO DAM - SNOWY MOUNTAINS SCHEME

Talbingo Dam (Hosking, 1974 and Wallace & Hilton, 1972) is a 165 m high rockfill dam with a sloping earth core completed in 1971. The foundations of the dam are predominantly rhyolite lava flows except on the upper right abutment where the rhyolites are covered with decomposed volcanic tuffs which were too extensive to remove. Dam cross-section is shown in Figure 5.

The damsite was explored by normal methods with twenty-eight diamond drill holes, thirty-six trenches, seven exploratory adits and two seismic traverses. There were no significant changes required to the design as a result of any unforeseen conditions in the foundation.

The earth borrow was an old landslide area and the material used in the embankment was mainly slope wash (CL to CH, MI to MH) and completely weathered andesite (SF clayey). The borrow area was thoroughly explored by fifty-two percussion drill holes, twenty-five power auger holes, thirteen diamond drill holes, sixty bulldozer trenches and six seismic traverses. A trial rolling was also carried out for the vibrating sheepsfoot roller.

Using the experience of a dam downstream the initial moisture placement was set in the range O.M.C. -1% to O.M.C. +1%. Based on the 'pore pressure-total stress' curves obtained from laboratory consolidation tests for this moisture content a satisfactory factor of safety was obtained for the upstream face. However, during construction pore pressure rose to values

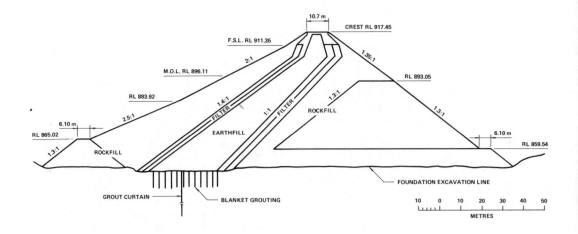

Figure 4   Jindabyne Dam — Cross-section

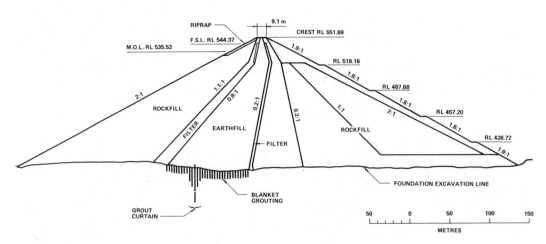

Figure 5   Talbingo Dam — Cross-section

very much in excess of those used in the design.  In fact, the material behaved as if it was fully saturated.  The placement moisture contents were made drier with the range being changed to O.M.C. −1.5% to O.M.C. +0.5% and a review was made of the stability of the upstream face for the 'end of construction' and 'initial filling' cases with various assumptions for construction pore pressure build-up in the core.  The main concern was the upstream stability particularly for the initial filling condition because had the pore pressures continued to rise at their present rate, the calculated factor of safety would have been unsatisfactory for this condition.  A close monitor was kept on the pore pressure behaviour and additional piezometers were installed in the embankment for this purpose.  The embankment stability was continually checked using recorded pore pressures and a prediction of pore pressure rise based on these values.  When the embankment neared completion the calculated factor of safety was satisfactory and no further action was required.  The dam was filled and there was no indication of any upstream instability.

## 9   CANIA DAM — QUEENSLAND WATER RESOURCES COMMISSION

The Cania Dam Project is an irrigation project on a tributary of the Burnett

River in Central Queensland. The main features are a rockfill dam 54 m high with a central impervious core, a combined diversion and outlet works on the left bank at the damsite and an unlined ungated spillway through the right abutment ridge, some 600 m from the damsite. The combined diversion and outlet works consist of a long approach channel 44 m high, and intake tower, a 140 m long reinforced-concrete conduit constructed in a trench through the damsite, an outlet structure containing the regulating valves for the irrigation releases and a long discharge channel.

In the preliminary design phase, arrangements of the outlet works on both banks were prepared and costed. Overall the cost variation was relatively small compared with the total project cost. A right bank layout was chosen basically because it gave flexibility in siting the upstream cofferdam, because it required a shorter discharge channel than any of the left bank layouts and because there appeared to be some doubt about the stability of the required excavation for the intake tower on the left bank layouts.

The Commission started actual construction work on the project in April 1978 by beginning the abutment stripping and constructing the permanent access road to the top of the right abutment.

Early in the abutment excavation on the right bank, it became apparent that there were many shears and small faults criss-crossing the foundations, several of which dipped towards the river. Later an extensive crushed seam was exposed which, if it extended down the abutment, would have been undercut by the proposed conduit excavation. Further, there was a small slide failure on the batters of the access road.

The Commission concluded that slides could occur on this seam and ones parallel to it during construction and later once the project was brought into service. While potential slides in the abutment caused by undercutting during construction could be temporarily held until the dam was built, those outside the dam limits would require extensive treatment if structures such as the tower were not to be endangered.

Fortunately, by this time, the left abutment had been almost completely stripped and a close inspection along the most likely alignment of a left bank conduit was therefore possible. In addition, three diamond cored and several percussion holes had been drilled in the areas of the tower and the outlet structure and also near the dam axis. This exploratory drilling with supplementary seismic work and the visual inspection of the stripped foundations confirmed that left bank location was better and the conduit was moved accordingly.

## 10  CONCLUSIONS

The examples discussed above and many other case histories to be found in the engineering literature all serve to illustrate certain general principles which need to be constantly in the mind of the designer and constructor of any major dam. These can be briefly summarised as:

10.1  A dam is no stronger than its foundations and therefore special care must be taken during design and construction to ensure that the foundations of the final structure are fully known.

10.2  Subsurface conditions on most sites vary considerably from point to point and subsurface exploration such as drill holes, trenches and adits can only form a relatively small proportion of the total foundation area. Therefore there is no guarantee, prior to opening up of the foundation, that adverse features will not be encountered during construction.

10.3  The designer should be aware of the effect that possible changes in foundation conditions has on the main structure and design the structure so that changes will have minimal effect, e.g., a socle to retain the profile of an arch dam.

10.4  Provision should be made for foundations to be logged as soon as possible after excavation and for monitoring the foundations in the event of adverse conditions arising during excavation.

10.5  The criteria for the foundations must be clearly specified by the designer and if any doubts exist as to whether the exposed conditions meet these criteria the situation should be assessed jointly by the designer and constructor. On many projects a Foundation Inspection Committee is formed consisting of senior members of design, construction and geomechanical disciplines to assess progressively the foundation.

10.6  Time and budgetary constraints on

site investigations for a project sometimes limit the exploration and therefore decrease the reliability of the interpretations of foundation conditions.

10.7 Where subsurface conditions are complex or the exploration has been restricted and considerable uncertainties exist consideration should be given to the desirability of sequential construction (such as separate contracts for foundation excavation and main dam construction) so that any forced changes to the design can be made before the main dam contract is awarded. Another alternative is to have conditions in the contract whereby changes can be made during contract.

10.8 Where natural (as opposed to manufactured) construction materials such as earthfill and rockfill are used the behaviour of such materials in the dam should be carefully monitored during construction to compare their behaviour with that assumed in the design and, if necessary, make changes to the construction methods to ensure that they are consistent with one another.

## 11 ACKNOWLEDGMENTS

The Author is indebted to the Snowy Mountains Hydro-electric Authority, the Melbourne and Metropolitan Board of Works and the Queensland Water Resources Commission for permission to publish the information contained in this paper and to his many colleagues who assisted in the preparation of the paper.

## 12 BIBLIOGRAPHY

Hosking, A.D. 1974, Monitoring of Earth Rockfill Dams, Australian Geomechanics Journal, G.4, No.1.

Howard, S.R. & Halliday, N.G. 1968, Design and Construction of Jindabyne Dam, Civil Engineering Transactions, Institution of Engineers, Australia, Vol. CE10, No.1, pp. 143-159.

Hunter, J.R. & Hartwig, W.P. 1962, The Design and Construction of the Tooma-Tumut Project of the Snowy Mountains Scheme, Journal of I.E. Australia, July-August, pp. 163-185.

Terzaghi, K. & Leps, T.M. 1958, Design and Performance of Vermillion Dam, California, Journal of Soil Mechanics and Foundations Division, A.S.C.E. Proceedings, Vol. 84, SM3, pp. 1728-1 - 1728-30.

Walker, F.G. 1967, Willard Dam - Behaviour of a Compressible Foundation, A.S.C.E. Proceedings, Vol. 93, SM4.

Wallace, B.J. & Hilton, J.I. 1972, Foundation Practices for Talbingo Dam, Australia, A.S.C.E. Proceedings, Vol. 98, SM10, pp. 1081-1098.

# Dam engineering activities in Japan

E.MIKUNI
*EPDC International Ltd., Tokyo, Japan*

SYNOPSIS. In the 1970's the number of fill dams exceeded concrete dams because of the fact that favorable sites with good foundation conditions became scarce and also with the development of construction equipment. However, where large flood discharge must be treated, gravity dams with large fillet are often economically superior to fill dams even if foundation conditions of the sites are poor. Roller compacted concrete method has been introduced for the construction of such dams. As for arch dams, parabolic arch has been adopted in order to enhance the stability of abutment.

Due to unfavorable climate conditions, impervious soils in Japan are generally wet and these materials are used without any improvement for small earth dams. "Kanto loam" and "shirasu", deposits of volcanic ash which had been discarded as not being suitable have been used in the construction of earth dams from the 1970's.

For large rockfill dams, coarse, weathered rocks or talus deposits are mainly used as core materials with appropriate improvement by stockpiling.

Behavior analysis of rockfill dams during embankment, water impounding and earthquakes have remarkably progressed and it has become a common practice to carry out dynamic analysis in addition to static analysis. Some of the recent study activities on deformation characteristic and results of some dynamic model tests are briefly introduced.

Embankment materials, zoning and foundation treatment of several noteworthy rockfill dams are described.

## 1. STATISTICAL REVIEW

In the register of dams in Japan (1979 issue) published by the Japanese National Committee on Large Dams (JANCOLD) there were 2,060 dams over 15 meters high at the end of 1978, and statistical compilation are given in Table 1. Similar information for dams under construction and in the planning stage are given in Table 2 and 3.

Of the approximately 1,000 dams constructed in pre-war years, about 90% of the total are earth dams for irrigation purpose built by empirical methods. Included in these dams are Kaerumata-ike dam (17 m high) which according to record was completed in the year 162 A.D. and many other dams that have been in existence for several hundreds of years. In a seismic country like Japan, it is astonishing to know that the many old dams are still intact and functioning.

In the 1950's which is the post-war reconstruction period, the expansion and reinforcement of electric facilities was taken up by the government as one of the high priority programs, and in those ten years dams for power, including multipurpose uses were built in succession. And in the 1950's the advanced technology of the western countries and large earth moving equipment were introduced into our country. Thus, in this decade large dams over 100 meters in height were constructed, and approximately 60% of the dams constructed in this period was concrete structures.

The 1960's was a decade of rapid economic growth. In this decade, the past trend of learning foreign technology changed to the pattern of developing our own technology which is continuing to this day. Together with the development of our own technology, large dams, particularly the construction of large rockfill dams which started in this decade can be characterized for the 1960's.

In the 1970's, because of the fact that favorable sites with good foundation conditions became scarce and also it became possible to construct rockfill dams economically with the development of construction equipment, the number of fill dams exceeded concrete dams. Also, by reason that oil-fired thermal plants became the major source of electric supply and that hydro plants were built as a part of multiple purpose development, the number of dams for power only drastically fell small in number.

There are 135 dams under construction, and 60% of these dams are fill structures. Noteworthy is that there are only a few dams for power only and the majority are for multipurpose uses which is the same trend as in the 1970's.

Table 1. Completed dams

a) Type

|  | TE | ER | PG | VA | CB | Total |
|---|---|---|---|---|---|---|
| -1899 | 473 | 0 | 0 | 0 | 0 | 473 |
| 1900-1940 | 405 | 0 | 119 | 1 | 5 | 530 |
| 1941-1950 | 110 | 0 | 44 | 0 | 0 | 154 |
| 1951-1960 | 129 | 4 | 166 | 12 | 2 | 313 |
| 1961-1970 | 119 | 25 | 162 | 31 | 9 | 346 |
| 1971-1978 | 64 | 67 | 106 | 6 | 1 | 244 |
| Total | 1,300 | 96 | 597 | 50 | 17 | 2,060 |

b) Purpose

|  | I | S | C | H | M | Total |
|---|---|---|---|---|---|---|
| -1899 | 473 | 0 | 0 | 0 | 0 | 473 |
| 1900-1940 | 384 | 35 | 1 | 106 | 4 | 530 |
| 1941-1950 | 110 | 7 | 0 | 29 | 8 | 154 |
| 1951-1960 | 124 | 13 | 15 | 110 | 51 | 313 |
| 1961-1970 | 110 | 24 | 33 | 87 | 92 | 346 |
| 1971-1978 | 69 | 21 | 45 | 26 | 83 | 244 |
| Total | 1,270 | 100 | 94 | 358 | 238 | 2,060 |

c) Height (m)

|  | 15-30 | 31-60 | 61-100 | 101-150 | >150 | Total |
|---|---|---|---|---|---|---|
| -1899 | 471 | 2 | 0 | 0 | 0 | 473 |
| 1900-1940 | 475 | 48 | 7 | 0 | 0 | 530 |
| 1941-1950 | 125 | 24 | 5 | 0 | 0 | 154 |
| 1951-1960 | 181 | 82 | 42 | 7 | 1 | 313 |
| 1961-1970 | 162 | 119 | 45 | 17 | 3 | 346 |
| 1971-1978 | 81 | 113 | 38 | 11 | 1 | 244 |
| Total | 1,495 | 388 | 137 | 35 | 5 | 2,060 |

d) Owner

|  | G | L.G. | P.G. | Total |
|---|---|---|---|---|
| 1969-1978 | 61 (19.3%) | 222 (70.0%) | 34 (10.7%) | 317 (100%) |

Table 2. Dams under construction

a) Type

| TE | ER | PG | VA | PG/ER | Total |
|---|---|---|---|---|---|
| 30 | 49 | 54 | 1 | 1 | 135 |

b) Purpose

| I | S | C | H | M | Total |
|---|---|---|---|---|---|
| 34 | 7 | 29 | 7 | 58 | 135 |

c) Height (m)

| 15-30 | 31-50 | 51-100 | 101-150 | >150 | Total |
|---|---|---|---|---|---|
| 20 | 60 | 46 | 6 | 1 | 135 |

d) Owner

| G. | L.G. | P.C. | Total |
|---|---|---|---|
| 35 (25.9%) | 91 (67.4%) | 9 (6.7%) | 135 (100%) |

Table 3. Dams in planning stage

a) Type

| TE | ER | PG | VA | PV/ER | Total |
|---|---|---|---|---|---|
| 59 | 92 | 144 | 2 | 5 | 302 |

b) Purpose

| I | S | C | H | M | Total |
|---|---|---|---|---|---|
| 92 | 10 | 48 | 3 | 149 | 302 |

c) Height (m)

| 15-30 | 31-50 | 51-100 | 101-150 | >150 | Total |
|---|---|---|---|---|---|
| 15 | 110 | 122 | 14 | 5 | 302 |

d) Owner

| G. | L.G. | P.G. | Total |
|---|---|---|---|
| 117 (38.7%) | 182 (60.3%) | 3 (1.0%) | 302 (100%) |

TE : earth dam  
ER : rockfill dam  
PG : concrete gravity dam  
VA : concrete arch dam  
CB : buttress dam including hollow gravity dam  
PG/ER: combination of PG and ER  
I : irrigation  
S : water supply  
C : flood control

H   : hydropower
M   : multipurpose
G   : government including Water Resources Development Corporation
L.G.: local government
P.C.: power companies including Electric Power Development Co.

Of the 302 dams in the planning stage, the majority of them are concrete gravity structure. The reason for this trend is that dams of the government and prefectures are planned first as concrete structures, but it is judged as the site investigations progress and foundation conditions are disclosed to be not favorable, a number of dams may be changed to rockfill structures.

These statistical data are as at the end of 1978. However, with the second oil crisis in 1979, importance is being directed to hydropower, particularly medium and small hydro plants. And at present, serious efforts are being given to studies on dams for power only and also multipurpose dams including power.

Large dams for important irrigation, flood control, and water supply purposes to serve areas covering two or more prefectures are planned, constructed and owned by the government. Whereas dams to serve the needs of a particular area are planned, constructed and owned by the prefecture or city concerned. The electric power companies own dams for the purpose of power only and therefore, the number of dams owned by them is relatively small in number. However, as seen in Table 4, 60% of large dams over 100 meters in height is owned by the electric companies.

## 2. CONCRETE GRAVITY DAMS

### 2.1 Dams constructed on unfavorable foundation

Dams constructed in the 1950's, that is the post-war reconstruction period, with 2 or 3 exceptions were constructed at sites with relatively good topographic and geologic conditions. In the 1960's, the economic growth period, because of the influx of the people to the large cities, there arised the rapid need of constructing dams for electric power, water supply and flood control. Because of this need, it became necessary to construct dams at sites that had been outside of consideration due to their unfavorable topographic and geologic conditions. This is one of the reasons for the transition from concrete dams to rockfill dams, and as statistics show these days fill dams are very popular.

However, in case of flood control, gravity dams are often more economical even if foundation condition is poor, because large spillway and outlet works can be built in the concrete dam body. Where there is problem in the strength of the foundation rock, fillet is provided on the upstream side to broaden the width of the dam base. Thus, the external force acting on the dam is transmitted widely and uniformly to the bedrock and adequate safely against sliding is assured.

Sameura Dam (H = 106 m, 1974) of W.R.D.C. for multipurpose is an example. The spillway has a design flood discharge of 4,700 m$^3$/sec. The foundation rock is black schist with highly developed schistosity and small folds. Shear strength and bearing capacity were inspected by insitu tests. The shape finally decided is that indicated in Fig. 1 where a large fillet is provided at the upstream toe and the downstream surface is of relatively steep slope. It was clarified by model tests and numerical analysis that local safety factors against sliding at the base for such a cross section would be uniform.

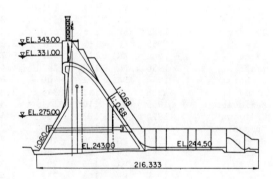

Fig.1. Sameura dam.

Another example where a fillet was adopted is Ohdo Dam (H = 96 m) under construction by Ministry of Construction. This dam is a multipurpose structure mainly for flood control, and it has a spillway to handle a maximum discharge of 7,700 m$^3$/sec. After comparative studies between a fill dam, a concrete gravity structure with fillet was designed.

Recently, there have been constructed many gravity dams with fillet in our country.

Table 4. Dams more than 100 m in height (including dams under construction)

| Name of dam | Year of Comp. | Type | Height (m) | Crest length (m) | Volume ($10^3$ m$^3$) | Storage capacity ($10^3$ m$^3$) | Purpose | Owner |
|---|---|---|---|---|---|---|---|---|
| Kurobe | 1964 | VA | 186 | 489 | 1,360 | 199,000 | H | Kansai Ele. P. Co. |
| Takase | 1978 | EE | 176 | 362 | 11,400 | 76,200 | H | Tokyo Ele. P. Co. |
| Okutadami | 1961 | PG | 157 | 480 | 1,640 | 601,000 | H | EPDC |
| Sakuma | 1956 | PG | 156 | 294 | 1,120 | 327,000 | H | EPDC |
| Nagawado | 1969 | VA | 155 | 356 | 672 | 123,000 | H | Tokyo Ele. P. Co. |
| Tedorigawa | 1979 | ER | 153 | 420 | 10,102 | 231,000 | C,S,H | EPDC |
| Ogochi | 1957 | PG | 149 | 353 | 1,680 | 189,000 | S,H | Tokyo City |
| Tagokura | 1960 | PG | 145 | 462 | 1,990 | 494,000 | H | EPDC |
| Arimine | 1961 | PG | 140 | 500 | 1,570 | 218,000 | H | Hokuriku Ele. P. Co. |
| Kusaki | 1976 | PG | 140 | 405 | 1,300 | 60,500 | C,I,S,H | WRDC |
| Kawaji | (1981) | VA | 140 | 320 | 700 | 83,000 | C,I,S | Ministry of Const. |
| Takane No. 1 | 1969 | VA | 133 | 277 | 330 | 43,600 | H | Chubu Ele. P. Co. |
| Yagisawa | 1967 | VA | 131 | 402 | 690 | 204,000 | C,I,S,H | WRDC |
| Miboro | 1960 | ER | 131 | 405 | 7,950 | 370,000 | H | EPDC |
| Hitotsuse | 1963 | VA | 130 | 416 | 557 | 261,000 | H | Kyushu Ele. P. Co. |
| Shimokubo | 1968 | PG | 129 | 626 | 1,190 | 130,000 | C,I,S,H | WRDC |
| Kuzuryu | 1968 | ER | 128 | 355 | 6,300 | 320,000 | C,H | EPDC |
| Iwaya | 1976 | ER | 128 | 366 | 5,700 | 173,500 | C,I,S,H | WRDC |
| Managawa | 1976 | VA | 128 | 362 | 490 | 115,000 | C,I,H | Ministry of Const. |
| Hatanagi No. 1 | 1962 | CB | 125 | 269 | 583 | 107,000 | H | Chubu Ele. P. Co. |
| Nanakura | 1978 | ER | 125 | 340 | 7,240 | 32,500 | H | Tokyo Ele. P. Co. |
| Takami | (1982) | ER | 120 | 427 | 5,120 | 229,000 | C,H | Hokkaido |
| Shimokotori | 1973 | ER | 119 | 289 | 3,530 | 117,440 | H | Kansai Ele. P. Co. |
| Tsuruta | 1965 | PG | 118 | 448 | 1,124 | 123,000 | C,H | Ministry of Const. |
| Kawamata | 1966 | VA | 117 | 131 | 168 | 87,600 | C,I,H | Ministry of Const. |
| Shintoyone | 1973 | VA | 117 | 311 | 348 | 53,500 | C,H | EPDC |
| Tamahara | (1982) | ER | 116 | 610 | 5,220 | 16,300 | H | Tokyo Ele. P. Co. |
| Yanase | 1965 | ER | 115 | 202 | 2,842 | 105,000 | H | EPDC |
| Sagae | (1982) | ER | 115 | 460 | 9,780 | 109,000 | C,H,I,S | Ministry of Const. |
| Ikari | 1956 | PG | 112 | 267 | 468 | 55,000 | C,I,H | Ministry of Const. |
| Ikehara | 1964 | VA | 111 | 460 | 647 | 338,000 | H | EPDC |
| Seto | 1978 | ER | 111 | 346 | 3,820 | 16,850 | H | Kansai Ele. P. Co. |
| Kamishiba | 1955 | VA | 110 | 341 | 390 | 91,600 | H | Kyushu Ele. P. Co. |
| Kajigawa | 1974 | PG | 107 | 286 | 428 | 22,500 | C | Niigata Pref. |
| Ohmachi | (1984) | PG | 107 | 338 | 775 | 33,900 | C,H,S | Ministry of Const. |
| Sameura | 1974 | PG | 106 | 400 | 1,200 | 316,000 | C,I,S,H | WRDC |
| Makio | 1961 | ER | 105 | 260 | 2,616 | 75,000 | I,S,H | WRDC |
| Koshibu | 1969 | VA | 105 | 293 | 311 | 58,000 | C,I,H | Ministry of Const. |
| Misakubo | 1969 | ER | 105 | 258 | 2,410 | 30,000 | H | EPDC |
| Ikawa | 1957 | CB | 104 | 243 | 430 | 150,000 | H | Chubu Ele. P. Co. |
| Sakamoto | 1962 | VA | 103 | 257 | 170 | 87,000 | H | EPDC |
| Shin-nariwagawa | 1968 | VA | 103 | 289 | 430 | 128,000 | S,H | Chugoku Ele. P. Co. |
| Hoheikyo | 1972 | VA | 103 | 305 | 307 | 47,100 | C,S,H | Ministry of Const. |
| Niikappu | 1974 | ER | 103 | 326 | 3,071 | 145,000 | H | Hokkaido Ele. P. Co. |
| Ohuchi | (1982) | ER | 102 | 340 | 4,000 | 19,100 | C,H | EPDC |
| Tori | 1966 | VA | 101 | 229 | 148 | 23,400 | C,I,H | Ministry of Agri. |
| Kazaya | 1960 | PG | 101 | 330 | 592 | 130,000 | H | EPDC |
| Yahagi | 1971 | VA | 100 | 323 | 256 | 80,000 | C,I,S,H | Ministry of Const. |
| Ohto | (1981) | PG | 100 | 320 | 992 | 66,000 | C,I,S,H | Ministry of Const. |

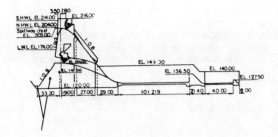

Fig.2. Ohdo dam.

## 2.2 Roller compacted concrete dams

The compressive strength of concrete for gravity dams is generally more than 250 kg/cm$^2$. For concrete gravity dams, the maximum compressive stress (kg/cm$^2$) of concrete is about 0.3 times the height (m) of a dam. Therefore, the safety factor of concrete used in gravity dams of 50 m high is more than 16 and of 100 m high more than 8 compared with the required minimum safety factor which is specified to be 4 in the design criteria established by JANCOLD. When a large fillet is constructed on the upstream of a dam, stress inside the body is smaller than a dam with a normal cross-section, and therefore the margin of compressive strength of concrete becomes larger.

On the judgment that concrete with a somewhat greater compressive strength than the maximum stress in the dam body and the foundation bedrock can be used in the dam body, roller compacted concrete method was studied from 1974 by Ministry of Construction aiming to reduce cement volume in concrete and form work, and to use common equipment such as bulldozers, dump trucks for transportation and placing of concrete.

This new method was adopted for the first time for construction of the mat concrete of Ohkawa dam (H = 78 m) under construction by Ministry of Construction. Because of the foundation bedrock condition of the dam site, a rockfill dam was more favourable, but in order to handle a large flood discharge of 5,200 m$^3$/sec a concrete gravity dam was adopted. Various shapes for the dam cross-section were studied, and a mat was finally designed as shown in Fig. 3. The concrete mat is 20 m thick and contains about 400,000 m$^3$ of concrete. First, this method was experimentally employed in the construction of the upstream cofferdam which is 19 m high and contains 10,000 m$^3$ of concrete. The height of one lift of concrete and different compaction methods were tested for 10 trial concrete mix proportions. Segregation of aggregates, uniformity of concrete compressive strength, coefficient of permeability, coefficient of elasticity and Poisson's ratio were tested and measured from cores and test pieces. From these results, the concrete mix proportion selected for the mat is given in Table 5. In the experimental concrete placing in the cofferdam, the methods of concrete quality control and treatment of joints were also studied.

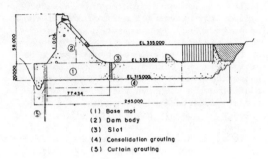

Fig.3. Ohkawa dam.

Table 5. Mix proportion of mat concrete of Ohkawa Dam

| G max (mm) | Slump (cm) | Air (%) | W/C+F (%) | F/C+F (%) | s/a (%) |
|---|---|---|---|---|---|
| 80 | -- | 1.5 ± 1 | 85 | 20 | 31.9 |

| Unit weight (kg/m$^3$) | | | | | | |
|---|---|---|---|---|---|---|
| W | C+F | S | \multicolumn{3}{c}{G (mm)} | Admixture |
| | | | 5–20 | 20–40 | 40–80 | |
| 102 | 120 | 694 | 444 | 444 | 593 | 0.3 |

At Shimajikawa Dam (H = 90 m) under construction by Ministry of Construction, the RCC method is also employed for the inner portion of dam body. Amount of cement and flyash is 84 kg/m$^3$ and 36 kg/m$^3$, respectively.

# 3. ARCH DAMS

## 3.1 Direction of arch thrust to abutment

The fact that only 50 arch dams have been constructed, which is less than 10% of the number of concrete gravity dams, is evidence of the poor geologic condition of our country. Important factors in the design of arch dams are that the stress acting in the dam body must be within permissible range and that the required stability of the foundation must be assured by properly aligning the direction of the arch thrust to the abutments. Even at relatively good dam sites, the existence of faults and predominant seams cannot be avoided in our country. A method to enhance the stability of these foundation conditions is to align the direction of thrust force to the abutments.

The relationship between the direction of thrust (T) to the abutments and central angle of arch ($\phi$) of a symmetrical circular arch dam of uniform thickness and radius was studied at the time of designing the Kawamata Dam (H = 117 m, 1966) of Ministry of Construction. The results of this study are shown in Fig. 4. It will be noted that the smaller the central angle and the larger the arch radius, the trend is that the direction of the thrust will be towards

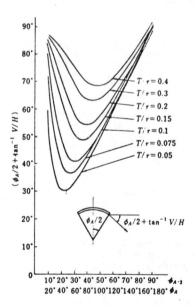

Fig.4. Relation between central angle and direction of thrust.

the abutment. In such a case, however, the arch shape would become close to a straight beam, and this will cause the development of a big tensile force near the arch crown where the bending moment is big. Arch dams, in those days, were designed with a central angle of about 110°C, but for the Kawamata Dam, a central angle of less than 90° was adopted to align the direction of thrust to the abutments, and thereby in improving the stability of the abutments.

## 3.2 Parabolic arch dams

In a parabolic arch dam, even if the central angle of arch is small, the curvature near the arch crown can be big, thereby mitigating tensile force acting near the crown. That is to say, without aggravating the stress condition in the dam body, the direction of thrust can be aligned to the abutments, and this would improve the stability of the abutments, resulting in the design of an economic structure. Yahagi Dam (H = 100 m, 1971) of Ministry of Construction is the first parabolic arch dam built in our country. In the design stage comparison was made between a parabolic arch dam (central angle 75°) and a circular one (central angle 100°). By selecting a parabolic dam, a saving of approximately 5% was achieved in concrete volume. After the Yahagi Dam, in the 1970's parabolic arch dams have been adopted in our country.

Shintoyone Dam (1973) of EPDC is a parabolic arch structure of 116.5 m high, 311 m long along the crest and the width at the center of the base is 19.6 m.(Fig.5).This dam functions as the upper storage pond for a 1,125 MW pumped-storage hydro plant which utilize the existing Sakuma reservoir as the lower storage. Natural deposit on the riverbed is shallow. The left abutment is steep and weathering of the rock is not

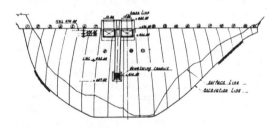

Fig.5. Shintoyone dam.

advanced, while the right abutment is a relatively gentle slope but weathering has reached fairly into the rock. In general, the bedrock is sound granite with diabase intruded in the form of dike rock. There are several small faults in the foundation rock, but they are not of a magnitude that would influence the stability of the dam. The shape of the dam is not symmetrical because of the configuration of the abutments at the dam site. In determining the shape of the dam, various in-situ rock tests were conducted to ascertain the stability of the foundation. Also, model tests, listed below, were conducted to study stress distribution, behavior during earthquake and other phenomenon of an actual structure in order to ascertain the stability of the dam.

(1) Static elasticity test
(2) Static rupture test
(3) Observation and measurement of behavior of dam with joints with some models used in the static tests
(4) Dynamic elasticity test
(5) Dynamic rupture test

Kawaji Dam (H = 140 m) under construction by Ministry of Construction is a multipurpose dam for flood control, irrigation and water supply. This dam is third highest arch structure in our country. The geology of the dam site is an alteration of diorite and tuff breccia. The left abutment is mainly diorite and the right abutment is almost wholly tuff breccia. Except for some portions, the diorite is sound and hard, and forms a good foundation for arch dam. There are several faults running parallel to the river, but by adopting a dam of parabolic arch structure, the stability and safety of the bedrock against sliding have been improved. (Fig. 6)

## 4. EARTH DAMS

### 4.1 Pore pressure during construction

Much rain and high humidity are the climatic characteristics of Japan, so that deposited soils are usually wet and natural moisture contents are generally higher than the optimum as shown in Fig. 7.

In case of small earth dams for irrigation, soil materials in the vicinity of dam sites are used without any improvement and compacted at considerably wet side from the optimum employing light rollers. In such case, neither high density nor high strength can be obtained and high pore pressure developed in embankment may threaten the safety of dam. In fact, slope sliding occurred at three dams during construction due to high pore pressure as shown in Table 6. After that, for an earth dam to be constructed with wet clayey soil, width of impervious zone or location of drain is carefully studied and designed to dissipate pore pressure as soon as possible and construction speed is controlled to avoid accumulation of pore pressure.

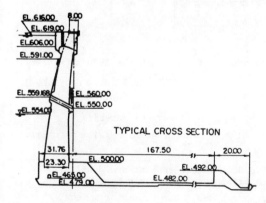

Fig.6. Kawaji dam.

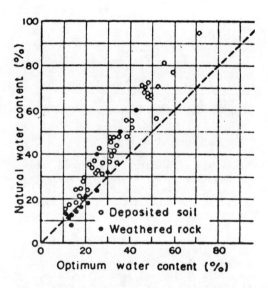

Fig.7. Moisture contents of soil at Miboro damsite.

Table 6. Sliding due to pore pressure during construction

| Dam | Height (m) | Start of embankment | Sliding |
|---|---|---|---|
| Togo | 31 | Feb. 1960 | Aug. 1960 |
| Yamakura | 41 | Aug. 1962 | Sept. 1964 |
| Shin-Kodoroku | 26 | May 1963 | July 1965 |

| Speed of embank. (m/month) | Pore* pressure at sliding (%) | Soil properties at sliding | Soil properties in design |
|---|---|---|---|
| 16/6 | 78 | $\gamma_d = 1.15$<br>$\phi = 5°30'$<br>$c = 0.12$ | $\gamma_d = 1.28$<br>$\phi = 4°50'$<br>$c = 0.5$ |
| 22.5/25 | 50 | $\gamma_d = 0.88$<br>$\phi = 5°$<br>$c = 0.37$ | $\gamma_d = 0.75$<br>$\phi = 4°50'$<br>$c = 0.39$ |
| 9/12 | 54 | $\gamma_d = 0.88$<br>$\phi = 10°$<br>$c = 0.6$ | $\gamma_d = 1.1$<br>$\phi = 22°$<br>$c = 0.4$ |

*: weight of soil

4.2 Volcanic deposited soil

As there are many volcanoes in Japan, volcanic deposited soil is widely distributed as shown in Fig. 8. Volcanic soil distributed in the central part of the Main Island is named "Kanto loam" and is known for its high moisture content and low density compared with the other clayey soil. On the other hand, "shirasu" distributed in the northern part of the Main Island is a mixture of pumice, volcanic sand and volcanic ash. A characteristic of these soils is that even high steep slope is stable under in-situ condition but mechanical properties decrease remarkably when distributed.

With the progress of soil mechanics which have been reflected in the design and construction of dams, materials that have been discarded as not being suitable are being used in the construction of dams from the 1970's. Fig. 9 gives a comparison of materials used for earth dams around 1960 and those used in the 1970's.

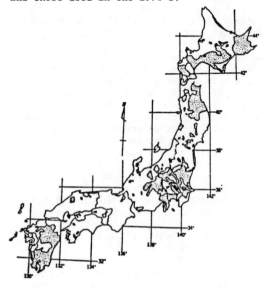

Fig.8. Distribution of volcanic deposited soil in Japan.

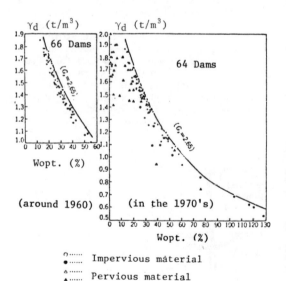

Fig.9. Moisture content and density of materials used for earth dams in Japan.

Chifuri Dam (H = 23 m, 1971) of Tochigi Prefecture is an example of an earth dam constructed of Kanto loam of high moisture content. Materials having natural moisture contents of 130 to 150% were specified to be compacted at moisture contents of 80 to

110%, and the stability of the dam was checked by employing low valves (dry density: 0.65 t/m$^3$, $\phi$ = 5°, c = 0.18 kg/cm$^2$). Compaction of the material was performed by using light tire roller of 7.5 tons weight. The dam was embanked over a period of about 5 years, and the material was dried before placing and compaction. The average moisture content and dry density were respectively 90% and 0.74 t/m$^3$. By using weathered andesite (dry density: 1.23 t/m$^3$, $\phi$ = 34°30') available near the dam site, zone "2" was constructed on the upstream side of the dam. By this method, it was possible to design the upstream gradient of zone "1" at a much steeper gradient than homogeneous dams. On the downstream side, waste materials were effectively utilized as the weighting zone. The shape of the dam is a reverse inclined core, and to avoid abnormal seepage line and concentration of flow net, drains and zone "2" were constructed on the downstream side. (Fig. 10)

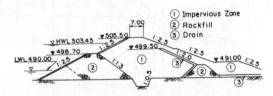

Fig.10. Chifuri dam.

"Shirasu" is a porous material which is inherent to products of volcanic eruption and therefore the density is low. The porosity of the material does not affect its shear strength, and as shown in Table 7 the angle of internal friction is relatively high. Because the density of the material is low, it is more sensitive to piping and quick sand action compared to ordinary sand. Fluidization of sandy materials caused by earthquake movement has become an important subject of study by the Japan Soil Mechanics Institute after the Niigata earthquake of 1964.

Table 7. Physical properties of "Shirasu"

| Dam | C | Proctor compaction | |
|---|---|---|---|
| | | wopt (%) | $\gamma_d$ max (t/m$^3$) |
| Ninokura | 2.3 - 2.5 | 30 | 1.0 - 1.2 |
| Matakido | 2.53 - 2.64 | 25 - 33 | 1.2 - 1.3 |
| Shiwa | 2.59 | 25 - 29 | 1.48 |

| Shear strength | | K |
|---|---|---|
| c (t/m$^2$) | $\phi$ (degree) | (cm/sec) |
| 2.0 | 33° - 0' | 10$^{-4}$ - 10$^{-5}$ |
| 6.0 | 33° - 41° | 10$^{-6}$ - 10$^{-7}$ |
| 1.8 | 50° - 30' | 7 x 10$^{-5}$ |

Fig. 11 gives an example of the results of vibration tests. It will be noted that more the material is compacted, the stability increases against fluidizing.

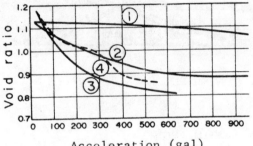

① Drained, loaded ($\sigma$ = 0.1 kg/cm$^2$)
② Undrained, loaded ($\sigma$ = 0.1 kg/cm$^2$)
③ Unloaded
④ Unloaded

——— Box size  140 x 50 x 50 mm
- - - - Box size   50 x 50 x 50 mm

Fig.11. Critical void ratio for fluidization.

Ninokura Dam (H = 37 m, 1971) of Aomori Prefecture is the first dam constructed of "shirasu". In order to prevent piping and quick sand phenomena, the material was embanked in layer of 30 cm and was thoroughly compacted with a 7-ton vibrating roller after adjusting the moisture content.

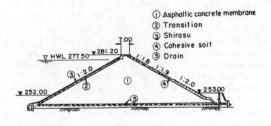

Fig.12. Ninokura dam.

Watertightness of the dam was achieved by placing an asphalt concrete membrane on the upstream face and on the downstream face cohesive materials were placed to prevent erosion by rain water. (Fig. 12)

Matakido Dam (H = 30.2 m, 1978) built on the same river basin at the Ninokura Dam, is also constructed of "shirasu". Based on the experience at Ninokura Dam, "shirasu" was spread out in layer of 20 cm and was thoroughly compacted by using vibrating roller. The dam was designed with protection zones of river deposit materials on both the upstream and downstream faces. (Fig. 13)

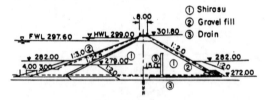

Fig.13. Matakido dam.

## 5. ROCKFILL DAMS

### 5.1 Embankment materials

a) Impervious core materials

Wet soil has sometimes been used as impervious core material of small rockfill dams for irrigation purpose. In such case, neither high density and strength nor large and fast production employing heavy equipment can be expected. This is the reason why large rockfill dams were not constructed for a long time after the 2nd World War.

Miboro Dam (H = 131 m, 1960) of EPDC is the first large rockfill dam constructed solving these problems by means of stockpiling. There were two kinds of materials available near the dam site. One is decomposed granite on the mountain slope and the other is clayey soil deposited at the foot of the same slope. Each of them could not be used along as a core material, because D.G. was too coarse and pervious, while clayey soil was too wet to compact. After making a series of soil tests using the mixed samples at various mix proportion of these materials, it was assured that, at the mix proportion of D.G. to clayey soil is about 4 to 1, excellent core material having sufficient impermeability (5 x $10^{-7}$ cm/sec) and moisture content 0.5 - 1.0% drier than the optimum (14%) could be obtained without spoiling desirable properties of D.G. such as high density, high strength and low compressibility. D.G. and clayey soil were spread alternately in thin layers at the borrow area to build stockpile of 10 to 15 m high and were automatically blended homogeneously during excavating operation to haul to the site of embankment. Thus, large scale compaction at the optimum moisture condition was realized for the total core material of about 2 million cubic meters.

After Miboro dam, for most of the large dams constructed so far, coarse materials such as weathered rock, talus deposits were used for impervious core with a certain improvement by stockpiling. Gradation curves of core materials for several dams are shown in Fig. 14.

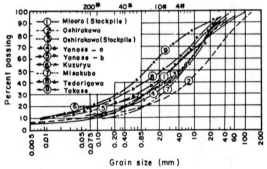

Fig.14. Gradation of core materials for large rockfill dams in Japan.

In the 1970's, three dams, Hirose (H = 75 m, 1975), Nabara (H = 86 m, 1975) and Myojin (H = 89 m, 1976) were constructed using "masa" (disintegrated granite) as the impervious core materials. "Masa" is rather pervious material because of its uniform gradation, but this problem was solved by stockpiling and compacting the material with higher energy using heavy vibrating roller.

In case of these dams, the purposes of stockpiling were (1) to crush the materials and (2) to obtain more homogeneous materials. Gradations of stockpiled materials are shown in Fig. 15.

Besides the above, stockpiles of the following purposes have been constructed at the other dams:

(1) to reduce the moisture content and to improve the gradation of wet and poor graded talus deposit by adding dry sand-gravel mixtures (Ohshirakawa dam).
(2) to homogenize the properties of materials from several borrow areas (Takase and Tedorigawa dam).

(3) to use effectively excavated materials from structure, weathered portion of rock quarry and necessary excavation for preparation works, which should be normally wasted (Misakubo dam).

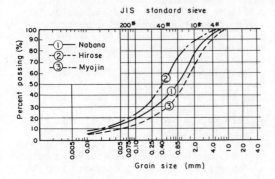

Fig.15. Gradation of "masa" used as core material.

As coarse core materials are being used in Japan, core materials are compacted at moisture condition a little (0.5 to 1.5%) wetter than the optimum in order to obtain the minimum permeability coefficient. As to the equipment for compaction, 20 tons heavy sheep's-foot rollers have been used for a layer of 20 cm thick, aiming at blending and pulverization effect during compacting operation. However, the use of vibrating rollers of 8 to 10 tons has become popular recently, because of its advantages such as (1) bigger thickness of a layer to be compacted, (2) less number of passing, (3) prevention of penetration of rain water and (4) common use of equipment for compacting the other zones. These blended and homogeneous core materials having high shear strength and low compressibility would contribute very much to the safety of dams against earthquake.

b) Filter materials

As for filter material, not only to satisfy the filter criteria, but also self-healing function for cracks in the core is considered important for earthquake resistant design. Therefore, river deposits are used as the best filter material. Where river deposits are not available, weathered quarry-run rocks are used. As the impervious core materials are coarse, wide single filter zone is generally satisfactory, if content of large size particles is not so much and segregation of large particles along the impervious core zone can be practically avoided.

40 to 50 ton pneumatic tire roller was used to compact filter material with an appropriately wetted layer of about 30 cm thick. Nowadays, heavy vibrating rollers are commonly used to compact layer thickness of about 40 cm and very dense filter with void ratio of 0.2 to 0.23 is easily obtained.

c) Rock materials

Rockfill zone in thin layers has been proven more effective to minimize settlement during construction and after completion. Recently, in addition to quarry-run rock, excavated rock materials from appurtenant structures are utilized to the maximum extent. These materials are usually well graded and can be densely compacted, so that higher shear strength will be obtained when compacted well as shown in Fig. 16, even if the materials are inferior.

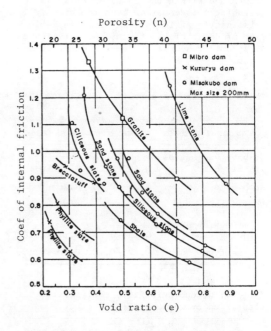

Fig.16. Relation between void ratio and tan $\phi$ of rock materials.

The thickness of layer is determined by the gradation of rock materials. For materials mainly consisting of spalls and finer rock in the transition zone between the filter and rockfill zone, the layer is generally 30 to 50 cm. The layers of 1 and

2 m thick are generally adopted for the inner and outer parts of rockfill zone, respectively. These thin layers are compacted to a denser state with heavy vibrating rollers to minimize settlement.

5.2 Stability analysis

"Design Criteria for Dams" established by JANCOLD gives the general and fundamental rules in designing dams over 15 m in height. These criteria had been revised several times since 1957 and the latest revision was made in 1978.

In calculation of safety factor of fill dams, the slip circle method is generally used because it gives the lowest safety factor. As for the criteria of earthquake load, the seismic body force must be taken as the value of weight of dambody multiplied by a seismicity coefficient of dambody, and should be treated to act horizontally. The seismicity coefficient of dambody shall be determined within the range shown in Table 8, taking into consideration the ground seismicity coefficient indicated in Table 9, type of dam, height of dam, characteristics of dynamic response of dam and other conditions.

Table 8. Design seismicity coefficient for dambody

| Type of dam | Strong zone | Weak zone |
|---|---|---|
| Gravity and hollow gravity dams | 0.12 – 0.20 | 0.10 – 0.15 |
| Arch dams | 0.24 – 0.40 | 0.20 – 0.30 |
| Fill dams | | |
|   Earth type and facing type | 0.15 – 0.25 | 0.12 – 0.20 |
|   Zone type | 0.12 – 0.20 | 0.10 – 0.15 |

Table 9. Ground seismicity coefficient classified by region

| | Strong zone | Weak zone |
|---|---|---|
| Seismicity coefficient | 0.12 – 0.20 | 0.10 – 0.15 |

Computer programs for the slip circle method have been developed and the most critical circle (or line) is now easily found for any combination of materials, cross sections and other conditions. Besides, dynamic analysis of fill dams have remarkably progressed owing to the development of calculation techniques using the finite element method and test methods for dynamic properties of fill materials. Therefore, for important dams, it has become a common practice to carry out dynamic analysis in addition to static analysis. Detail of earthquake-resistant design features of rockfill dams are presented in the paper by Dr. H. Watanabe of Central Research Institute of Electric Power Industry (CRIEPI).

The following are brief explanations of the recent study activities on deformation characteristic of rockfill dams, etc.

a) Deformation properties of fill materials during embankment

At the stage of embankment, confined stress $\sigma_3$ for a given element increases with the progress of embankment, in other words, with increase of vertical stress $\sigma_1$. According to Mr. Matsui of CRIEPI, the modulus of deformation under such condition should not be decided directly from stress-strain curve under a given value of confined stress $\sigma_3$ obtained by triaxial tests but envelope of tangent at each starting point of the curves as shown in Fig. 17 should be used as stress-strain curve for incremental analysis. This fact was confirmed by actual measurement made at Niikappu Dam of Hokkaido Electric Co., as shown in Fig. 18. At present, actual deformation during dam embankment is being analyzed taking into consideration these physical properties.

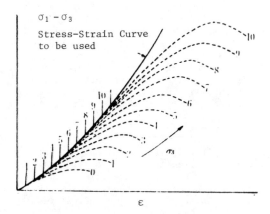

Fig.17. Stress-strain curve during embankment.

b) Settlement characteristics of fill materials during water impoundment

Depending upon the quality of rock, shrinkage of embankment is caused by deterioration, lubrication, etc., due to immersion in water. Techniques for behavior analysis and material tests during water impoundment have been developed at CRIEPI. The idea that settlement results from decrease of effective stress due to buoyancy which is an amplification of the idea of modulus of deformation during embankment mentioned in a) and the idea of strain due to immersion proposed by Dr. Hayashi were introduced in the process of numerical analysis, and settlement trends of upstream rockfill and effects on stabilities of impervious core are being studied.

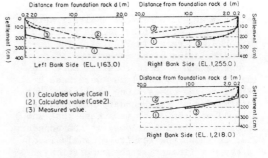

(1) Calculated value (Case 1).
(2) Calculated value (Case 2).
(3) Measured value.

Fig.19. Settlement at the abutment.

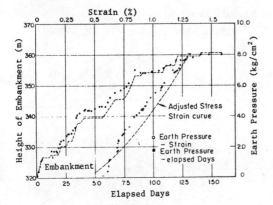

Fig.18. Measured earth pressure and strain.

c) Large shearing deformation during embankment

Impervious core in contact with the abutment is subjected to shearing deformation during embankment, water impoundment and earthquakes, which is assumed to be the major cause of failures due to seepage along contact planes. Against such unfavorable phenomena, selected fine core materials of 20 to 50 cm in width have been used in contact with the abutment and compacted at considerably wet side for all large rockfill dams in our country.

At Takase Dam, 176 m in height with steep abutments, contact clay having high consistency was placed on the foundation rock. Numerical analysis on shearing deformation in contact clay was made and compared with measured values. The results are shown in Fig. 19.

In this figure, case 1 shows calculated settlement using modulus of deformation $E = 100$ t/m$^2$ which corresponds to the average value of the contact clay actually used and case 2 is settlement using $E = 1,000$ t/m$^2$ for comparison. Settlement in case 1 have fairly good coincidence at the right abutment. 50 - 70% of total shearing deformation occurred in the contact clay of only 20 cm in width. Such deformation improves unbalanced stress distribution and removes tensile stress in the vicinity of the abutment. From this figure it is confirmed that contact clay is very effective. Fig. 20 is the distribution of local safety coefficient S.C. (ratio of resistant stress at a given point of acting stress at the same point) analyzed by Mr. Matsui and the values of S.C. are 1.2 in minimum in case of Takase dam. Furthermore, contact clay material was tested by torsional shear and permeability test apparatus and it was also confirmed that enough residual strength and low permeability are maintained even if larger shear deformation occurs in the material.

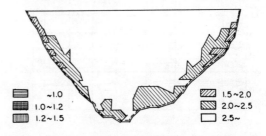

Fig.20. Distribution of local safety coefficient in the core zone.

Another study is on shearing deformation resulting from difference in rigidities of embankment materials. At Niikappu dam, measurement and numerical analysis were conducted. Deformation values actually measured are shown in Fig. 21. The observed relative displacement was 30 cm in core zone and 6 cm in rockfill zone from the filter zone. Analyses were made by CRIEPI on two cases. One is a case that deformation at the boundary of zones is allowed and the other is a case that any deformation is not allowed at the boundary. Fig. 22 shows the results of analysis of safety coefficient at the completion of embankment and at full water level of reservoir. It was found that when deformation is allowed the safety coefficient is much higher and distribution of S.C. is improved compared with the case where no deformation is allowed at the boundary. From the above, it is found that there occurs shearing deformation in the vicinity of boundary of zones and such deformation improves the stability of dam body.

a) Upon completion

CASE-1  Deformation not allowed

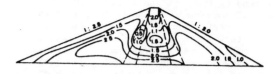

CASE-2  Deformation allowed

b) Full reservoir

CASE-1  Deformation not allowed

CASE-2  Deformation allowed

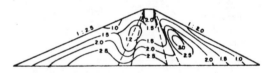

Fig.22. Distribution of safety coefficient upon completion and full reservoir.

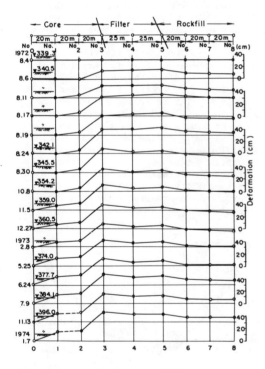

Fig.21. Deformation in zones during embankment at Niikappu dam.

d) Unsteady seepage flow

The water level of a reservoir of a large-scale pumped storage power station recedes about 5 m/hr. by rapid drawdown. In such case, large hydraulic gradients and pore pressures are produced in the upstream shell, and will cause instability of slopes, especially during earthquakes. The stability against this kind of unsteady seepage flow is studied by finite element analysis.

e) Hydraulic fracturing of impervious core

In order to study the stability of impervious core against erosion, experiments on piping resistance of materials at rock contact, core materials, and on the effect of filter have been carried out. On the other hand, research on evaluation of seepage flow velocities by numerical analysis has been brought into practical use. Faults and sheared zones exposed at contact plane are of importance in actual works. Since they would have the effects to reduce effective stress due to arching and resistance to hydraulic fracturing, special attention has been paid to the treatment of such zones in design and construction.

5.3 Dynamic model tests

In the history of rockfill dams which is less than 30 years in Japan, only several dams including Miboro dam were subjected to the earthquake. Earthquake accelerations observed or estimated at these dam sites were less than the design values and the dams suffered no damage. What would happen to a rockfill dam subjected to a catastrophic earthquake? Dynamic model test is considered to be one of the valuable approaches to answer this question, although there is a problem in regard to similitude between model and prototype. Many tests using shaking tables as shown in Table 10 have been performed so far.

Table 10. Large shaking tables in Japan

| Owner | Capacity (ton) | Size of table (m) |
|---|---|---|
| National Research Center for Disaster* | 450 | 16 x 16 |
| Central Research Institute of Electric Power Industry* | 120 | 6 x 6.5 |
| Ministry of Construction* | 40 | 4 x 4 |
| Institute of Industrial Science of University of Tokyo* | 170 | 2 x 10 |
| Electric Power Development Co.** | 22 | 5 x 5 |

\* Electro hydraulic vibration
\*\* Mechanical vibration

a) Failure of dam

Following is the results of model test conducted by University of Tokyo for Takase Dam, as an example.
The model (H = 1.4 m) was vibrated sinusoidally at a constant period of 0.22 sec and the amplitude was gradually increased until the dam failed. At first, slight crest settlement was recognized at acceleration of around 140 gal. As the 2nd stage, very small displacement developed in the central part and then in the downstream rockfill zone with increase in acceleration, but these individual displacement showed practically no further development. At around 350 gal as the 3rd stage, sliding of the surface layer parallel to the slope was recognized and this develops rapidly so that the entire slope surfaced collapsed.

Fig.23. Progression at damaged zones in model.

In case of the core with the same rigidity as the shell, the core settled at the crest and the upper half shifted towards the downstream side. Such deformation of the core was limited to the crest portion and did not extend to the lower parts. In case the rigidity of the core was higher than that of the shell, displacement of the core to the downstream could not be recognized but cracks developed at the upper part of the core. (Fig. 23)

Phenomena in the 1st and 2nd stages are not considered serious. Slight crest settlement might contribute to make dam embankment more stable and cohesionless rockfill material would settle down again when the earthquake is over. Relative displacement in the core might have the possibility to introduce leakage through the core. Therefore, provision of the filter zones at both sides of the core having a self-healing function are very important for an earthquake-resistant dam. The 3rd stage is considered as substantial collapse of dambody and the phenomenon coincides with the fact that plane slip along the dam slope is the critical failure surface by the conventional analysis method for cohesionless body like rockfill dam.

Therefore, slope failure becomes an important subject and many slope failure tests have been conducted so far.

b) Slope failure

Relation between the gradient of model slope and the critical acceleration necessary to cause a slope collapse tested by EPDC is shown in Fig. 24. Fig. 25 shows the relation between size of rock material and acceleration at slope failure by experiment conducted by Prof. Okamoto of University of Tokyo. From these results, existing large rockfill dams in our country would be safe enough against even very strong earthquakes.

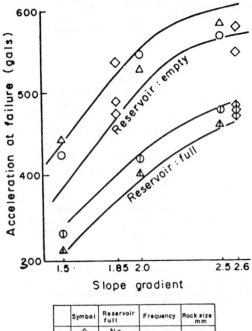

Fig.24. Acceleration at slope failure.

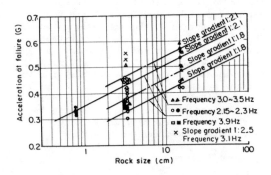

Fig.25. Acceleration at failure for rock sizes.

## 5.4 Noteworthy individual dams

### a) Vertical core dams

Coarse and well graded soil material is of low compressibility and high resistance against piping when compacted. In case that such a material is available for an impervious core of the dam, not only differential settlement and incidental cracking will be minimum but also the range and amount of tensile stresses in the core and the upstream rockfill zone are small. The zoning with a center core is then considered to be the best for an earthquake-resistant dam and is adopted for most of the large rockfill dams in Japan.

Takase Dam (H = 176 m, 1978):- This dam of Tokyo Electric Power Co. is the highest rockfill dam in Japan, and is the upper reservoir dam for pumped power generation of maximum output 1,280 MW. Total dam volume is 11,600,000 m$^3$. The foundation rock is granite and the depth of sand-gravel deposits on the riverbed is 30 to 40 m. The core material is re-sediment from weathered granite and is rather coarse, namely (Fig. 26)

| | |
|---|---|
| -200# | 7 to 15% |
| Optimum moisture content | 7.5% |
| Dry density | 2.11 t/m$^3$ |
| Permeability coefficient | 2 x 10$^{-6}$ cm/sec |

Therefore, the width of core was designed at the maximum taking into consideration the available quantity of these materials. Stockpiling was performed to adjust the gradation and to homogenize the different properties of core materials after removing oversize larger than 20 cm by grizzly. Most of the embankment materials were obtained from thick river deposits within the reservoir area near the dam site and were placed at filter and shell zones depending on the size of materials.

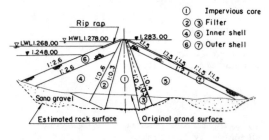

Fig.26. Takase dam.

As a result of dynamic analysis by FEM and model tests, it was confirmed that the maximum horizontal acceleration near the dam crest is 2 to 4 times of that at the foundation depending on the characteristics of earthquake, and that failure of dam takes the pattern of sliding of the surface of slope first and followed by development of serious deformation or cracks in the core. Therefore, larger and angular rocks with high shearing strength were placed in the outer shell and wide filter zones were provided.

Prior to the stability analyses of the dam, maximum acceleration was decided based on the expected earthquakes, which were derived from the statistical investigations on the records of earthquake damages and the results of seismological observations. In case of Takase dam site, the maximum acceleration on the foundation was estimated to be 200 gal. The seismic coefficient used in slip circle method was 0.15 g, taking into consideration the response characteristics of the dambody. Furthermore, the magnification of acceleration near the dam crest was taken into account. Thus, seismic load conditions used in stability analyses were the following two cases; (1) seismicity coefficient of 0.15 uniformly distributed throughout the dam, (2) seismicity coefficient linearly increases from 0.15 at the 3/4 in height of the dam to 0.30 at its crest. The minimum safety factors obtained were 1.2 in the case (1) and 1.0 in the case (2), respectively. Sixteen seismographs were installed inside the dam and foundation but remarkable earthquakes have not been recorded up to date.

Tedorigawa Dam (H = 153 m, 10,000,000 $m^3$, 1979):- This dam of EPDC is a multipurpose dam including hydropower (250 MW), flood control and water supply for city and industrial uses. The foundation rock mainly consists of gneiss and conglomerate.(Fig. 27)

The seismic coefficient of the dam foundation was determined to be 0.15 on the basis of the design criteria of JANCOLD as well as of the estimated values obtained through probability calculation of the seismic records around this area. In addition to slip circle method, dynamic analyses were also conducted using the dynamic physical properties of the dam materials obtained by dynamic triaxial compression tests. Moreover, dynamic model tests were carried out in order to verify the results of the analyses as well as to make sure the effects of earthquake vibrations on the stability of the slopes of dam. The tests and analyses made it clear that the location of the impervious core would be better to be in the center of the dam and that the seismic vibration is amplified around the top of the dam. Therefore, the dam was designed to be of center core type and special measures were taken to compact the part around the top (zone 5a) of the dam.

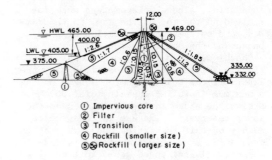

① Impervious core
② Filter
③ Transition
④ Rockfill (smaller size)
⑤⑥ Rockfill (larger size)

Fig.27. Tedorigawa dam.

Impervious core material of about 1,000,000 $m^3$ is talus deposits and weathered rock. The material was excavated from several borrow areas near the dam site. By means of stockpiling, homogeneous material was obtained as shown in Fig. 28 even though materials of different properties were used. The shape of impervious core zone was designed rather thin taking the quality and volume of available materials into consideration, but fillets were constructed at the base of core to lengthen percolation pass at the foundation rock. Because of lack of suitable river deposits near this dam site, partially weathered rock (gneiss) from quarry was used for filter.

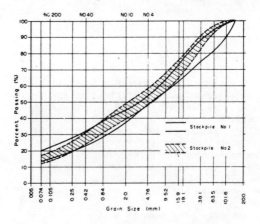

Fig.28. Gradation distribution of core material of Tedorigawa dam.

Therefore, wide transition zones of finer rock materials were provided to serve as coarse filter zones.

Since the completion of the dam, earthquake observations by using 27 seismometers have been carried out for the purpose of obtaining informations about displacement and acceleration of the dambody and the foundation. These seismometers were installed on the top of the dam, in the dambody and on the rock foundation of both banks and the bottom of the dam.

Ohuchi Dam (H = 102 m, under construction):- This dam of EPDC provides the upper reservoir for Shimogo pumped scheme having a maximum installed capacity of 1,000 MW. Ohkawa Dam shown in Fig. 3 is used as the lower reservoir. The dam is constructed mainly on tuff foundation and a part of the foundation at the left abutment is mud-flow deposit. Taking these geological conditions and the embankment materials into account, the dam has gentle slopes with weighting zones as shown in Fig. 29. Impervious core was designed wider than the other dams. Core material consists of residual tuff and talus deposit of chert distributed near the dam site and these two materials are blended at a mix proportion of around 1:1 at stockpile. Chert is used for pervious zones, namely, weathered finer portion for filter and larger more fresh and hard rock for rockfill zone.

flood control and a pump-storage scheme with maximum capacity of 220 MW. The dam foundation consists of conglomerate, sandstone, slate and shale. Conglomerate at the riverbed is sound but rock at the higher elevation at both abutments is weathered. There exists many small faults tightly filled with clay. One of the characteristic of the dam site is high annual precipitation and available core material (talus deposit) is rather wet.

In case that cohesive soil of high water content is used for the center core, there is the danger of cracking in the core due to large differential settlement caused by extreme difference in settlement with the filter zone. Further, if a thin core is selected to avoid difficulty in construction work and development of high pore pressure, there will be more risk of cracking in the core due to arching action. Particularly, during earthquakes, unconsolidated portion below arching subsides rapidly due to vibration, and it may be readily imagined that the cracking will introduce a concentrated leakage.

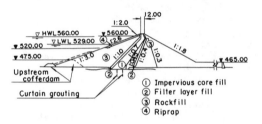

Fig.30. Kuzuryu dam.

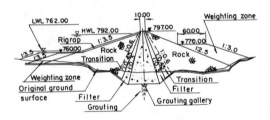

Fig.29. Ohuchi dam.

Because of the unfavorable foundation condition, seismic coefficient of the foundation was decided to be 0.15 for stability analysis by the slip circle method. Stability against seismic coefficient of 0.18 was also checked. Deformation analysis of dam embankment including foundation was made by the F.E.M. and dynamic analysis was also conducted. Sufficient number of seismometers are planned to be installed.

b) Inclined core dams

Kuzuryu Dam (H = 128 m, 1968):- This dam of EPDC is a multipurpose dam includes

Therefore, inclined core type was adopted and transition zone of excavated material from spillway was provided at the central part of the dam in order to support the core as shown in Fig. 30. The advantageous points of this type are, to reduce undesirable deformation of the core by the well-compacted transition zone and to expect the upstream rockfill zone as weighting for the impervious core. Filter material is river deposits except zone 2 between the core and downstream filter where finer excavated materials from spillway having -200# mesh portion of 5 to 10%.

Miho Dam (H = 95 m, 1978):- This dam of Kanagawa Prefecture is a multipurpose dam for water supply, flood control and power generation. Since this dam is located in an area where earthquake of the strongest intensities in Japan may happen, the design seismic coefficient employed was 0.2, the highest value for a rockfill dam in Japan.

Therefore, the dambody was designed with flatter slopes. The impervious core material consists of scoriaceous loam deposited at the left bank and muck of weathered rock from the quarry and these materials were blended by stockpiling at a ratio of 1:4.

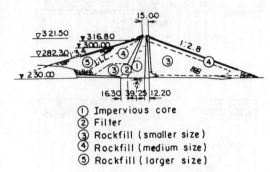

① Impervious core
② Filter
③ Rockfill (smaller size)
④ Rockfill (medium size)
⑤ Rockfill (larger size)

Fig.31. Miho dam.

c) Dams with asphaltic concrete membrane

Ohtsumata Dam (H = 52 m, 1968):- This, belonging to EPDC, is the first dam of asphaltic concrete faced structure in Japan. Granite, rockfill material was compacted in 1 m layer employing 11 ton vibrating roller. Settlement observed after completion was very small. Thereafter, dams with large daily fluctuation in water level for pumped storage project such as Numappara (H = 38 m, 1973) and Tataragi (H = 65 m, 1974) are of this type. The membrane is of multi-layers including an intermediate drainage layer through which seeped water is drained into a gallery in the cutoff concrete. No leakage has been observed at Ohtsumata and Numappara Dam so far. Following are other examples.

Miyama Dam (H = 75.5 m, 1974):- This dam is for irrigation, municipal water supply and power generation (lower reservoir dam for Numappara pumped storage project). The foundation rock consists of rhyolite and andesitic tuff breccia, and there is a deposit of sand-gravel of 10 to 15 m on the riverbed. Judging from the shape and geological conditions of the valley, a fill dam was a suitable type of structure but soil material available for an impervious core was not found in the vicinity and taking large fluctuation in water level into consideration, a dam with asphaltic concrete membrane was finally selected. This dam is one of the highest dams of this type in the world, and in its design and construction various field tests,

laboratory model tests and stability analysis including FEM were carried out.

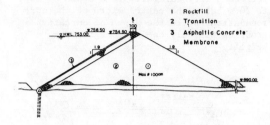

Fig.32. Miyama dam.

Muri Dam (H = 16 m, under construction):- This dam is for power generation constructed by Hokkaido Electric Power Co. As shown in Fig. 33, a concrete cutoff wall was constructed in the riverbed deposit, and on which an asphaltic concrete impervious center core fill dam was constructed. Mastic asphalt of superior deformation characteristics was used at locations where the impervious asphaltic core comes into contact with concrete of the spillway and the cutoff wall.

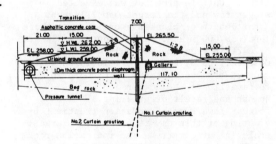

Fig. 33. Muri dam.

5.5 Foundation treatment

As Japan is located in the circum-Pacific orogenic belt, there are distributed many kinds of rocks and numerous tectonic lines throughout the country and the geological structures are complicated. Therefore, geological conditions of dam foundations are not favorable and foundation treatment is of great importance in the design and construction of dams in Japan.

Takase Dam:- The site is an example of fairly good foundation. It consists chiefly of granite. There are no large faults, but the surface of the bedrock is very cracky, and permeability is fairly high to a depth of 30 to 40 m, and decreases rapidly below that point. Blanket grouting was performed very carefully to a depth of 10 to 20 m below the entire width of the core foundation, while the depth of curtain grouting was only one half of the dam height.

was 63,000 m, and total length grouted was 200,000 m. (Fig. 36)

### a) Blanket grouting

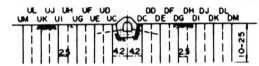

### b) Curtain grouting

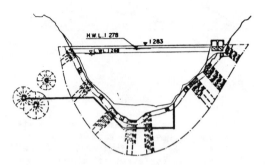

Fig.34. Curtain grouting at Takase dam.

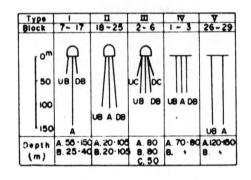

Fig.35. Grouting pattern at Tedorigawa dam.

Target of permeability for blanket grouting was not more than 5 lugeons for 18 m width at the central portion of the core and not more than 10 lugeons at other portions. Curtain grouting was executed in single row and the target was 2 lugeons. Total length of blanket grouting was, therefore, 58,000 m and, on the contrary, length of curtain grouting was only 40,000 m. Observations by bore hole television were made at 8 holes, a total length of 770 m, to investigate the number and widths of joints and result of grouting.(Fig. 34)

Tedorigawa Dam:- Foundation treatment by grouting of this dam is one of the largest scale in our country. The foundation rock mainly consists of hard gneiss and conglomerate. There exist many open cracks in limy gneiss at the riverbed and conglomerate at the right abutment. Besides, a large fault zone of 25 m in width exists parallel to the river channel at the left abutment. Clayey material in the fault is well compacted but crushed zone has rather high permeability.

Blanket grouting 10 to 25 m in depth were performed for the core base and total length of blanket holes were 62,000 m. The target of improvement was the same as Takase Dam. Total length of curtain grouting in two or three rows was estimated to be 75,000 m and depth of holes were designed to be 70 to 120 m. However, holes of 150 m in depth were necessary for several portions and many additional holes were also required at crushed zone of the large fault, cavities and open cracks in limy gneiss and conglomerate. Thus, length of additional holes

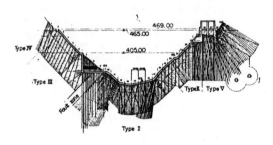

Fig.36. Curtain grouting at Tedorigawa dam.

Kassa Dam:- The geological features of the foundation at this dam (H = 90, 1977) of EPDC is complicated as shown in Fig. 37. It consists of deposits of uncemented volcanic mud flows, thick silt and sand layer, andesite lava and dacite. The surface portion of dacite was cracked and andesite lava had many structural joints. Therefore, shallow hole grouting was performed for all the area. About 950 tons of cement and 1,500 $m^3$ of mortar were injected in shallow holes having a total length of 2,800 m. The surface portion of volcanic mud flow was excavated to a depth of 2 to

3 m after grouting.

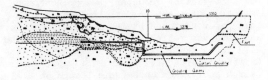

Fig.37. Geology at Kassa dam site.

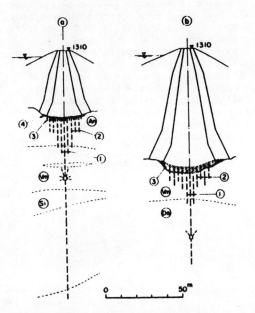

(a) Vertical cross section on Andesitic lava
(b) Vertical cross section on volcanic mud flow

Vm: Volcanic mud flow
Da: Dacite
An: Andesite
Si: Sand

Fig.38. Foundation treatment at Kassa dam.

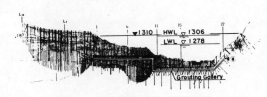

Fig.39. Curtain grouting at Kassa dam.

Total length of holes for blanket and curtain grouting was about 70,000 m and 7,700 tons of cement, 4,900 m$^3$ of mortar and 70 m$^3$ of chemical grout were injected. Chemical grouting was performed in silt-sand layer where permeability was $10^{-3}$ to $10^{-4}$ cm/sec. Maximum pressure applied was 25 kg/cm$^2$, 20 kg/cm$^2$ for rock foundation and mud flows, respectively. (Fig. 38 & 39)

Funagira Dam (H = 25 m, 1977):- This dam of EPDC is a combination of a concrete structure with a large capacity spillway and a fill dam with 430 m in length and 20 m in height. The dam lies on alluvial deposit of maximum thickness of 60 m. Foundation treatment of the sand-gravel layer was not effective with cement grouting, and, therefore, clay-cement was grouted. Drilling was performed employing overburden drills from a platform on the impervious core as shown in Fig. 40. Grouting was performed employing sleeve-pipe method or the so-called "manset-tube method".

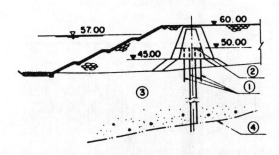

(1) Curtain grouting
(2) Platform for grout injection
(3) Alluvium, sand-gravel
(4) Rock foundation

Fig.40. Impervious work of Funagira dam.

Total length drilled was 3,060 m in the impervious core and 14,300 m in alluvium. Coarse and fine grained clay-cement grout was made and total injected volume was 52,600 m$^3$ (coarse clay: 37,100 m$^3$, fine clay 15,500 m$^3$) in an area of 14,200 m$^2$. The mix proportions are listed in Table 11. The coefficient of permeability of the sand-gravel layer was improved from $10^{-1}$ - $10^{-2}$ cm/sec to $10^{-4}$ - $10^{-5}$ cm/sec as shown in Fig. 41.

Table 11. Grouting material of Funagira Dam

| Kind of clay | | Mix. proportions | |
|---|---|---|---|
| | | C+S/W | C/S |
| Coarse-grained grout | Coarse clay | 70/100 | 30/100 |
| Fine-grained grout | Fine clay | 20/100 | 5/100 |

| Mix. proportions | | Qualities | | |
|---|---|---|---|---|
| Admixtures | | Specific gravity ($t/m^3$) | Breezing (%) | Prepact flow (sec) |
| $Na_2SiO_3/S$ | NaOH/S | | | |
| – | – | 1.363 ±0.015 | <5 | 11 ± 2 |
| 0.75/100 | 0.5/100 | 1.120 ±0.006 | <5 | 9 ± 2 |

Note) C: Cement by weight, S: Clay by weight, W: Water by weight

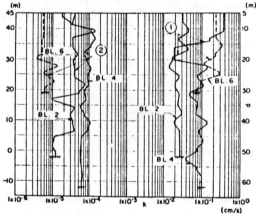

(1) Coefficient of permeability before grouting
(2) Coefficient of permeability after grouting
d: Depth

Fig.41. Results of permeability test.

Acknowledgment

The Japanese National Committee on Large Dams present a General Report on dam engineering activities for a period covering three years to the International Congress on Large Dams which is held every three years. At the 13th Congress of the International Commission on Large Dams held in New Delhi in 1979, a paper titled "Recent Dam Construction in Japan" was presented by JANCOLD in the name of its Chairman. In this paper prepared by me, much information and data was relied on the report of JANCOLD WITH THEIR PERMISSION. Dr. H. Watanabe of CRIEPI has presented a separate paper "Research Activities on Seismic Design on Fill Dams in Japan" to amplify and elaborate this country report. Mr. I. Matsui of CRIEPI made available many valuable data and kind advice on the subject concerning deformation characteristic of rockfill dams. The author takes this opportunity to express his sincere gratitude to JANCOLD and CRIEPI for their kind assistance.

A paper "Rockfill Dams in Japan" was prepared by me describing the activities of rockfill dam construction in our country for presentation at the seminar sponsored by AIT in August, 1979. Consequently, in this paper there are some parts which are a repetition of the subjects described in my paper stated above.

# The problems and practice of dam engineering in India

YUDHBIR
*Asian Institute of Technology, Bangkok, Thailand*

SYNOPSIS    Data on large dams in India is reviewed. General trends in design construction and damage frequency of dams in India are brought out and compared with worldwide trends. 90% of total dams built or under construction today are earth dams with current trends towards earth and earth rockfill dams. Physiography and Geology of India is briefly reviewed. Problems encountered at the Beas Dam at Pong and technical problems of design and construction of dams in Himalayas and its foot-hills are described.

## INTRODUCTION

An earth dam about 27 m high was built in during India the year 1000-1100 A.D. Typical cross-sections of ancient and some modern earth dams are given in figs. 1 and 2. In 1979, at the Golden Jubilee congress of ICOLD in New Delhi, the Indian National Committee on Large Dams issued a National Register of Large Dams, giving technical details of over 1550 dams in India. In this paper the data, contained in the National Register for Large Dams, has been used to examine the trends in dam construction in India. Data on types of incidents on various dams in India is compiled and the relative damage frequency is compared with incidents to dams in the world. From this survey of dams, already completed and under construction, it is brought out that 90% of dams in India today are earth dams and also only less than 2% of total dams have been so far built in the young and highly unstable rock formations in the Himalayas and its foot-hills. The majority of dams have been built in the older Deccan Plateau and the Peninsular rocks. A brief account of physiographic and geologic setting of India is also given. In order to high light the complex geotechnical problems confronting dam designers in the Himalayan region a brief description of Beas dam at Pong built in Siwalik formations, is given. Also discussed are the problems related to complex geologic conditions at sites of highest dams (Kishau dam for example) and other river valley projects in the Himalayas.

## STATISTICAL COMPILATION OF DAMS IN INDIA

### Dams Already Completed

Table 1 shows a compilation of dams already built under the categories of, (a) type, (b) purpose, and (c) height.

Table 2 presents the same details for dams under construction.

For high dams in India the details are given in table 3 and their location is shown in fig. 3. The progressive increase in dam heights built in India is brought out in fig. 4. The trend in India is compared with the data compiled by Rouve (1977) for dams built since 3000 B.C. in many parts of the world. Clearly from fig. 4 one can draw a conclusion with Rouve (1977) that "the improved design methods and the use of new construction equipments allow construction of higher dams with larger reservoir capacities "... and also" with increasing height of storage for dam sites located in the lower portion of a river system the risk factor increases exponentially because many uncertainity factors increase linearly or more than linearly."

In addition to the trends in dam heights, it is interesting to observe (fig. 5) that since 1900 the proportion of earth dams, as % age of total dams built has remained more or less fixed at 75% (gravity dams = 25%) upto 1950 and then there has been a marked increase in earth dams built in India with the current proportion around 90%. From 1950 onwards the number of gravity type dams has decreased to around 10%. Today in India majority of dams being built are of earth or earth rockfill type.

In the publication, "Lesson from Dam incidents USA" by ASCE/USCOLD in 1975, the available data from incidents/accidents to large dams in USA has been analysed. The findings about US dams have been compared with those in other parts of the world. Data on incidents to dams in India has been taken from this report and two failures reported by Rao (1961) have been included in table 4.

Fig. 6 shows evaluation of damages to dams in the world according to operating time or beginning of operations (Rouve, 1977). It is evident that the frequency of damage to dams decreases rather than increases with age until the maximum design life is reached. Furthermore the dams built in recent times have become safer (Rouve, 1977, Londe, 1980). Even though the data on damages to dams in India shown in table 4 is limited (with some cases still not reported officially) the Indian experience appears to follow rather well the general trends shown in fig. 6. Relative damage frequency has in general decreased from 0.04 in 1920-1939 to 0.015 during 1940-59 and 0.008 during 1960-1979. Except for one case (after 28 years) during 1920-39 period, maximum period of operation before the incident was 15 years. Data for incidents before 1900 is too limited to be commented upon here. The overall conclusions based on the data available are, however, both comforting and re-assuring as far as the current practice of design and construction of dams in India is concerned.

OUTLINE OF PHYSIOGRAPHIC AND GEOLOGIC SETTING

It is neither intended nor possible to go into detailed discussion of geological conditions at various dam sites in India. For details reference is made to Krishnan (1968) and Auden volume (1972) for a comprehensive account of geology of India and Geologic Settings at various dam sites in India, respectively. For the purposes of this report, it is sufficient to briefly review the physiography and major geologic formations in India so as to get an overall view of geologic conditions in different parts of India where dams have been or are being built.

India may be broadly subdivided, for this purpose, into; Peninsular India and Extra Peninsular India. The chief mountain ranges of Peninsular India are the Western and Eastern Ghats, Vindhyas, Satpuras, the Aravalis and those forming the Assam Plateau. Majority of dams have been built in Peninsular India.

The Western Ghats, 1000-3000 m in elevation (average) and 1600 km. long, extend from Tapi valley down to Cape Comorin. The northern part consists of Deccan Traps upto Dharwar & Ratanagiri in Bombay while the southern portion consists of Archaean gneisses, schists and charnockites. All the important Peninsular rivers and their valleys rise on the Western Ghats and flow eastwards into the Bay of Bengal.

On the east coast the Eastern Ghats are a series of rather detached hill ranges stretching from northern border of Orissa to Nilgiris in the Western part of Tamil Nadu. They are composed of garnetiferous gneisses.

The Vindhya mountains which separate Southern from Northern India are a fairly continuous group of hills, lying north of Narmada river and extending from Gujarat in the West to Bihar in the East. The majority of these ranges are composed of Sandstones and Quartzites of the Vindhyan System. Together with Satpuras, Vindhyan mountains form the Watershed of Central India from which rise the Narmada, Chambal Betwa, Tons, Ken, Son and other streams. The Satpuras separate the valleys of Normada and Tapi. They are composed of Gondwanas and Archaean gneisses.

The Aravalli Ranges are now remanants of once great mountain ranges of tectonic origin. They cross Rajasthan from South-West in Gujarat to North-East near Delhi and are composed of rocks of Aravalli, Delhi and Vindhyan systems. They form the major water shed of Northern India separating the drainage of the Ganges River System from that of the Indus.

The Extra Peninsular Ranges comprise of the Himalayan arc of immense radius, extending from Kashmir to Assam for a length of about 2,400 km. For our purposes the Himalayan may be subdivided as:

1. The Siwalik foot-hills, 10 to 50 km wide with a maximum attitude of 900 m. These are composed of sedimentary rocks.

2. The Central Himalayas, 140 km from the edge of plains consist of sedimentary and old metamorphic rocks with intrusions of large masses of granite.

A rough outline of geology of India may be useful for a proper overview. More than half of the Peninsular India is occupied by gneissic and Schistose rocks of the Archaean and Proterozoic ages. The Cuddapahs, Vindhyans, the Gondwanas and the Deccan Traps occupy the rest. In Extrapeninsular India, marine sedimentary systems predominate, though parts of the Sub-Himalayas and the main axis of the Himalayas are occupied by ancient metamorphic rocks and intrusive igneous rocks.

GEOTECHNICAL PROBLEMS OF DESIGN AND CONSTRUCTION OF DAMS IN EXTRA PENINSULAR INDIA

The Ganges with its numerous tributaries, constitutes the largest river system in India. It is formed in the Himalayas by the confluence of two important rivers - the Alaknanda and the Bhagirathi - at Devaprayag, about 70 km upstream of Rishikesh where it enters the plains. Inspite of the vast power-potential of Alaknanda and Bhagirathi rivers, there are hardly any power projects of significance utilising this potential in the Himalayas. As indicated earlier hardly two dozen dams have been or are being built in the Hima-Layas. In addition to few major projects like Tehri dam, a number of smaller run-of-the river schemes, are planned to be built. Many potential sites pose complex technical problems of design and construction in the upper Himalayan reaches.

Geological investigation for the intake-dams proposed under some of the run-of-the-river schemes in Uttarakhand Division of Uttar Pradesh have brought to light the existence of abondoned river courses of the present-day tributaries of the Ganges (Pant, 1972). The present-day rivers occupy newly carved out gorges, known as epigenetic gorges because of their later formation in the development of the river system. Such epigenetic gorges, being relatively young, have steep walls of hard gneisses and quartzites and narrow section without significant rock alternation due to weathering. Thus the epigenetic gorges offer suitable dam sites from topographical, geological and engineering consideration. However, the existence of older buried river Gorges in the vicinity of proposed project sites; e.g. Maneri-Bhali hydel project, Tapoban-Gulabkoti hydel scheme, etc., raise numerous problems, the most important being the leakage of water through the buried valley. Depending on the nature, thickness and extent of buried valley materials in the reservoir behind the dam, the water losses may be considerable to the extent of negating the very purpose of the planned project.

The processes responsible for formation of the present-day Gorges will control to a large degree the nature of materials occupying the burried channels as also the effects of impounding water on the behaviour of reservoir rock formations. Pant (1972) has discussed in detail the problems of planning dams under these circumstances. According to Pant (1972) these epigenetic Gorges are generally formed by the superimposition of a glacial valley on a fluvial valley or vice-versa and, by the obstruction caused by large scale landslides in the river valley. Yudhbir (1980) has discussed the geotechnical problems related to massive rockslides involving large mobile rock masses in the Garhwal Himalayas. Yudhbir (1980) has presented some information indicating that majority of such rockslides are perhaps a direct consequence of gradual build up of strain energy in the earth's crust and are fore-runners to major earthquakes in this geologically active region. Such major rockslides are currently active and bring about immense human suffering downstream due to flooding caused by sudden collapse of dams created by the slides. In addition to the problems of water leakage related to old buried gorges in the vicinity of dam sites, very serious problems are posed by these rockslides both during and after construction of dams in this region. The volumes of mobile rock mass, for example, in case of Gohana slide was 26 billion cubic meters and it dammed up the Virhi Ganga causing a rise of water level by 300 m in the lake thus formed (Yudhbir 1980). Consequences of such an occurrence in the dam reservoir were tragically demonstrated at the Vjont dam disaster in Italy.

Such clossal geological problems are fortunately not prevalent over large areas of the Himalayas being tapped for hydroelectric power in India. Nevertheless challenging geotechnical problems are posed, for example, in the Yamuna Hydel Scheme Projects. Krishnaswamy et al (1972) present an excellent account of major geological considerations pertaining to the choice of site and the type of dam for the proposed 253 meter high Kishau dam across river Tons, a major tributary of the river Yamuna. At this site the abutments are made up of intensely overfolded Paleozoic, Bansa limestones, sliced by five, minor, saucer shaped thrusts. The foundations consist of carbonaceous slates, separated by suspected major thrusts from the underlying slates and quartzite formations. The beds strike across the river and dip at low angles downstream. The left abutment shows deep weathering, critical stratifications, deep-set of relief fissures and joints that are closely spaced. The most crucial aspect of left abutment seems to be its limited width and the abutment support decreases as the height of the structure increases (hardly 2.9 times the height of the dam at about 200 m). In addition the modulus of elasticity of limestone comprising the left abutment is as low as $(0.1-0.3) \times 10^5$ kg/cm$^2$ compared with a value of $5 \times 10^5$ kg/cm$^2$ recorded for solid cores of the same rock. Inspite of the fact that a deep gorge with steep abutments was an ideal site for an arch dam at the site, for various geologic considerations, especially the inadequacy of the rock support in the left abutment, the dam type of dam was change to an earth rockfill type. Krishnaswamy et al (1972) give detailed considerations involved in deciding the type of dam suited to local geologic conditions at this site.

BEAS DAM AT PONG

In order to illustrate the nature of geologic factors and their impact on the construction and design of dams in Siwalik sedimentary rock formations, a brief account is given here of Beas dam at Pong in North India. Table 3 gives basic statistics and fig. 7 shows regional and reservoir geology and the layout and longitudinal section along the dam axis. Maximum cross-section of the dam is shown in fig. 2. Datta (1979) gives a detailed description of history of the project, geology, foundation treatment, design and construction of embankment and spillway, instrumentation, and history of reservoir filling and operation. The dam is founded on folded, stratified, sedimentary Siwalik formations. The Siwalik rocks in this region have a rather complicated geologic history. In the process of folding where the shear stresses have exceeded the shear strength, shear zones have developed in the weaker clay shale members. Henkel & Yudhbir (1966) have demonstrated that the topography at the site is controlled by the low shear strength of these zones and that the field shearing resistance in the shear zones is at residual state. The influence of shear zones on the design of excavation slopes in the intake area has been brought out by Henkel (1965). Yudhbir and Valsangkar (1973) have described the influence of jointing & shear zones on the stability of left abutment. Yudhbir and Varadarajan (1975) have demonstrated that the shear zones in the dam foundation and the complex geologic structure at the site control the mechanism of stability of dam foundations. Using a 3-dimensional stability analysis, Yudhbir and Varadarajan (1975) designed the upstream and down stream toe weights thus affecting about 30% saving in earth work for corrective measures needed to improve the stability of dam foundations. The nature of shear zone materials and their residual strength parameters at different project sites in the Siwaliks and extra-peninsular India have been reported by Yudhbir (1978).

CONCLUSION

Over the years more and more higher and larger dams, have been successfully designed and built to suit individual site conditions, its geology, hydrology and the availability of suitable construction materials. Compared to gravity type dams, earth dams form a major proportion of the total number of dams built or under construction in India. The relative damage frequency of dams in India, in relation to years of operation, follows world wide trends, i.e. the frequency of relative damage decreases with age and the recent dams, though much higher and with large reservoirs, are much more safer. This clearly is both a comforting thought and a complement to the Indian engineers and geologists involved in design and construction of dams. Majority of dams built or under construction in India are located in Peninsular rocks with only a few in the relatively younger and unstable extra Peninsular rocks. Some of the difficulties of designing and building dams in the Himalayas and its foot-hills have been briefly reviewed. Presence of shear zone and complex geologic and tectonic history of Himalayan rocks poses many difficult problems of selection of type of dam compatible with local geologic condition and also design for seepage control, slope and foundation stability and nature of tunnel supports. Clearly many challenging problems await to be tackled by engineering geologists and geotechnical engineers involved in the design and construction of dams in extra Peninsular India.

ACKNOWLEDGEMENT

Much of the data on dams in India has been taken from the special publications issued by the Indian National Committee on Large Dams on the occaision of the Golden Jubilee congress of the ICOLD, held in India during October - November, 1979.

REFERENCES

ASCE/USCOLD (1975). Lessons from Dam Incidents, U.S.A. Joint Publication of ASCE/USCOLD, New York.

Datta, O.P. (1979), Beas Dam at Pong. Design and Construction Features of Selected Dams in India, Central Board of Irrigation & Power, Publication, New Delhi.

Henkel, D.J. (1965). The Stability Problems of the Excavated Slopes in the Itake Area of the Beas Dam. Report to Bhakra and Beas Design Organisation, Punjab, India.

Henkel, D.J. & Yudhbir (1966). The Stability of Slopes in the Siwalik Rocks in India. Proc. Ist International Rock Mechanics Congress, Vol. II, Lisbon.

Krishnan, M.S. (1968). Geology of India and Burma, Higginbothams (P) Limited, Madras, India.

Krishnaswamy, V.S., Mehta, P.N. and Shome, S.K. (1972), Geological Considerations Pertaining to the Choice of the Site of the Dam for the Kishau Project, Yamuna Hydel Scheme, Stage II, Uttar Pradesh, Auden Volume, Indian Society of Engineering Geology, Calcutta, India.

Londe, P. (1980), Lessons from Earth Dam Failure, Symposium on Problems and Practice of Dam Engineering, AIT, Bangkok, Thailand.

Pant, G. (1972), The Significance of the Epigenetic Gorges of River Valley Projects in the Uttarkhand Division, Uttar Pradesh, Auden Volume, Indian Society of Engineering Geology, Calcutta, India.

Rao, K.L. (1961), Stability of Slopes in Earth Dams and Foundation Excavations, Proc. 5th Internpational Conference on Soil Mechanics and Foundation Engg. Vol. II, Paris.

Rouve, G.(1977), Review of Cases of Damage Related to Dams, Failures of Dams - Reasons and Remedial Measures Jan 6-7, Aachen, Editor, W. Wittke.

Yudhbir and Valsangkar, A.J. (1973), In-Situ Engineering Characteristics of Rock Masses-A Review. Proc. Indian Rock Mechanics Symposium, Kurukshetra, Haryana, India.

Yudhbir and Varadarajan, A. (1975), Influence of Shear Zones on the Mechanism of Stability of the Foundations of the Beas Dam, India. Engineering Geology, No. 9, Elsevier Scientific Publishing Company, Amsterdam.

Yudhbir (1978), Engineering Geologic Aspects of Earth Structures, State-of-the-Art Report, Proc. I.G.S. Conference on Geotechnical Engineering, Dec. 20-22, New Delhi.

Yudhbir (1980), Landslides in the Himalayas, Proc. International Conference on Engineering for Protection from Natural Disasters, AIT, Bangkok.

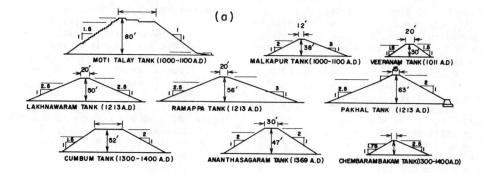

Cross Sections of Ancient Earth Dams during the Period 1000-1400 AD

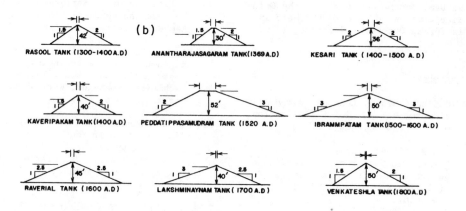

Cross Sections of Ancient Earth Dams during the period 1400-1800 AD

Fig. 1  Ancient Earth Dams in India

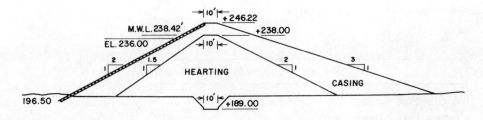

Willingdon Dam 1924 (Rao, 1961)

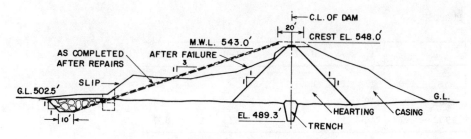

Palakmati Dam 1938 (Rao, 1961)

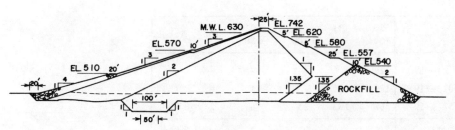

Hirakud Dam 1957 (Rao, 1961)

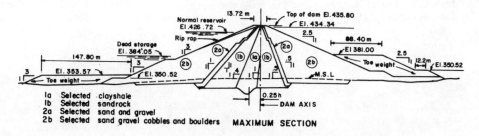

Beas Dam at Pong 1974 (Datta, 1979)

Fig. 2  Developments in Dam Cross Sections

Table 1  Completed dams (upto 1979)

a) TYPE

|  | TE | ER | PG | VA | CB | Misc. | Total |
|---|---|---|---|---|---|---|---|
| - 1900 | 32 | - | 10 | - | - | - | 42 |
| 1901 - 1951 | 110 | 1 | 41 | - | - | 12 | 164 |
| 1952 - 1979 | 716 | 5 | 128 | 1 | - | 19 | 869 |
| TOTAL | 858 | 6 | 179 | 1 | - | 31 | 1075 |

b) PURPOSE

|  | I | S | H | I-S | I-C | I-H | Misc. | Total |
|---|---|---|---|---|---|---|---|---|
| - 1900 | 27 | 12 | 1 | 2 | - | - | - | 42 |
| 1901 - 1951 | 140 | 7 | 8 | 5 |  | 4 |  | 164 |
| 1952 - 1979 | 746 | 17 | 61 | 7 | 5 | 32 | 1 | 869 |
| TOTAL | 913 | 36 | 70 | 14 | 5 | 36 | 1 | 1075 |

c) HEIGHT (m)

|  | 15 -30 | 31 -60 | 61 -100 | 101 -150 | >150 | Total |
|---|---|---|---|---|---|---|
| - 1900 | 29 | 6 | - | - | - | 35 |
| 1901 - 1951 | 126 | 21 | 5 | 1 | - | 153 |
| 1952 - 1979 | 562 | 133 | 25 | 6 | 2 | 728 |
| TOTAL | 717 | 160 | 30 | 7 | 2 | 916* |

* Difference less than 1075, as some of the dames are less than 15m in height, and for some the height is not reported.

Table 2  Dams under construction.

a) TYPE

| TE | ER | PG | VA | CB | Misc. | Total |
|----|----|----|----|----|-------|-------|
| 429 | 2 | 43 | 1 | - | 4 | 479 |

b) PURPOSE

| I | S | H | I-S | I-C | I-H | Misc. | Total |
|---|---|---|-----|-----|-----|-------|-------|
| 408 | 11 | 13 | 7 | - | 20 | 20 | 479 |

c) HEIGHT (m)

| 15-30 | 31-60 | 61-100 | 101-150 | >150 | TOTAL |
|-------|-------|--------|---------|------|-------|
| 296 | 80 | 19 | 4 | 3 | 402** |

** Less than 479, as some of the dams are less than 15m in height.

- TE : Earth dam
- ER : Rockfill dam
- PG : Concrete gravity dam
- VA : Concrete arch dam
- CB : Buttress dam including hollow gravity dam

- I : Irrigation
- S : Water Supply
- C : Flood control
- H : Hydro-electric
- M : Multipurpose.

Table 3  High Dams of India

| Name of dam | Year of Completion | Type | Ht. m. | Crest Length m | Volume ($10^3 m^3$) | Gross Reservoir Capacity ($10^3 m^3$) | Purpose |
|---|---|---|---|---|---|---|---|
| Tehri | U.C. | TE,ER | 261 | 570 | 22,750 | 3,539,000 | I-H |
| Kishau | U.C. | TE,ER | 253 | 360 | – | 2,400,000 | I-H |
| Bhakra | 1963 | PG | 226 | 518 | 4,130 | 9,621,000 | I-H |
| Lakhwar | U.C. | PG | 192 | 440 | 2,000 | 580,000 | I-H |
| Idukki | 1974 | VA | 169 | 366 | 460 | 1,996,000 | H-C |
| Srisailam | U.C. | PG | 143 | 512 | 1,953 | 8,722,000 | H |
| Cheruthoni | 1976 | PG | 138 | 650 | 1,700 | 1,996,000 | H-C |
| Sardar Sarovar | U.C. | PG | 137 | 1210 | 4,100 | 9,492,175 | I-H |
| Pong (Beas Project) | 1974 | TE | 133 | 1950 | 35,500 | 8,570,000 | I-H |
| Silent Valley | U.C. | VA | 131 | 430 | 425 | 317,000 | H |
| Ramganga | 1978 | TE | 128 | 743 | 11,013 | 2,442,600 | I-H |
| Nagarjunasagar | 1974 | TE,PG | 125 | 4865 | 7,960 | 11,550,000 | I-H |
| Kakki | 1966 | PG | 110 | 336 | 750 | 455,020 | H |
| Sholiyar | 1972 | TE,PG | 105 | 1282 | 2,496 | 160,660 | I-H |
| Koyna | 1961 | PG | 103 | 808 | 1,555 | 2,796,500 | H |
| Supa (Kalinadi Project) | U.C. | PG | 101 | 322 | 1,150 | 4,418,000 | H |
| Kulamava | 1977 | PG | 100 | 385 | 476 | 1,996,050 | H-C |
| Idamalayar | U.C. | PG | 100 | 373 | 856 | 1,089,800 | M |
| Karjan | U.C. | PG | 94 | 885 | 922 | 637,000 | I |
| Rihand | 1962 | PG | 93 | 934 | 1,699 | 1,060,000 | H |

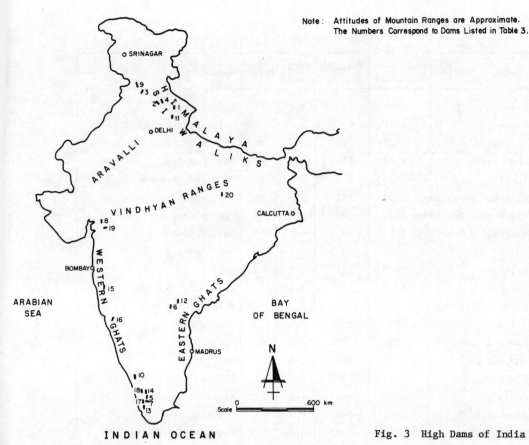

Fig. 3  High Dams of India

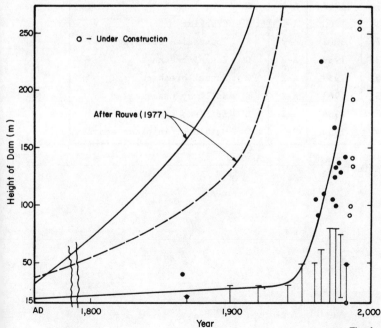

Fig. 4  Trends in Heights of Dams

Table 4  Dam Incidents

| NAME OF DAM | Year Completed | Year of incident | Incidents | Type | Remarks |
|---|---|---|---|---|---|
| Walwham | 1916 | 1932 | A-1 | PG | Dam, leakage |
| Tigra | 1917 | 1917 | F-2 | PG | Maindam, Breakage |
| Shirawata | 1920 | 1920 | A-1 | PG | Dam, leakage |
| Thokarwadi | 1922 | 1925 | A-1 | PG | Dam, leakage |
| Kundli | 1924 | 1925 | F-1 | PG | Main dam, breakage |
| Willingdon Reservoir | 1924 | 1924 | A-2 | E | Dam, slide |
| Willingdon Reservoir | 1924 | 1935 | A-2 | E | Dam, slide |
| Willingdon Reservoir | 1924 | 1936 | A-2 | E | Dam, slide |
| Mulshi | 1927 | 1955 | A-1 | PG | Dam, leakage |
| Palakmati | 1938 | 1953 | A-1 | E | Dam, slide |
| Low Khajur | 1949 | 1949 | F-2 | PG | Blow out, overtop |
| Ahraura | 1954 | - | F-2 | E | Maindam, internal erosion |
| Kaila | 1955 | 1965 | F-2 | E | Dam, slide |
| Sampana tank | 1956 | 1961 | A-1 | E | Dam, slide |
| Sampana tank | 1956 | 1964 | A-1 | E | Dam, slide |
| Kharagpur | - | 1961 | F-2 | E | Blowout, overtop |
| Kadam | 1957 | 1958 | F-2 | E | Blowout, overtop |
| Panshet | 1961 | 1961 | F-2 | E | Cracking |
| Sarda Sagar | 1961 | 1963 | A-2 | E | Foundation boiling |
| Vir | 1961 | 1962 | A-1 | E | Cracking |
| Vir | 1961 | 1963 | A-1 | E | Cracking |
| Dudnava | 1962 | 1962 | A-2 | E | Foundation boilding |
| Bhakara | 1963 | 1959 | A-2 | PG | Tunnel breakage |
| Bhakara | 1963 | 1959 | A-2 | PG | Tunnel breakage |
| Badua | 1963 | 1963 | A-2 | E | Spillway cracking |
| Badua |  | 1964 | A-2 | E | Spillway crackup |
| Kedar Nala | 1964 | 1964 | F-2 | E | Main dam, internal erosion |

\* For classification see table 5.

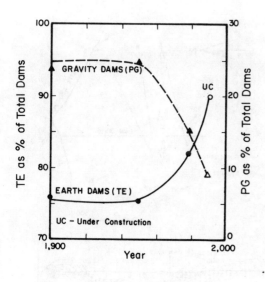

Fig. 5  Trends in Dam Types

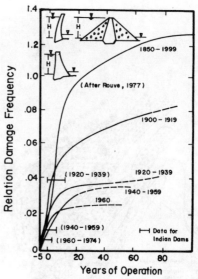

Fig. 6  Trends in Dam Safety

Table 5  Classification of types of Incidents/Accidents (from ASCE/USCOLD (1975)

| Type of Incident/Accident | Description |
| --- | --- |
| F-1 | A major failure of an operating dam which has involved complete abondonment of the dam. |
| F-2 | A failure of an operating dam which at the time may have been severe, but was of a nature and extent which permitted the damage to be successfully repaired and the dam again placed in operation |
| F-3 | An accident to a dam which had been in operation for some time, but which was prevented from becoming a failure by remedial work or operations, such as drawing down the pool. |
| F-4 | An accident to a dam, observed during initial filling of the reservoir, which caused immediate remedial measures, including such actions as drawing down the water level and making repairs before placing dam in operation. |
| F-5 | An accident to a dam before it was placed in operation and before any water was impounded. Unusual settlement of a foundation, slumping and slides of abutments, etc., after essential completion of the structure would be accidents of this type. It is not intended to include the normal run of construction problems and movements which occur during construction operations. |

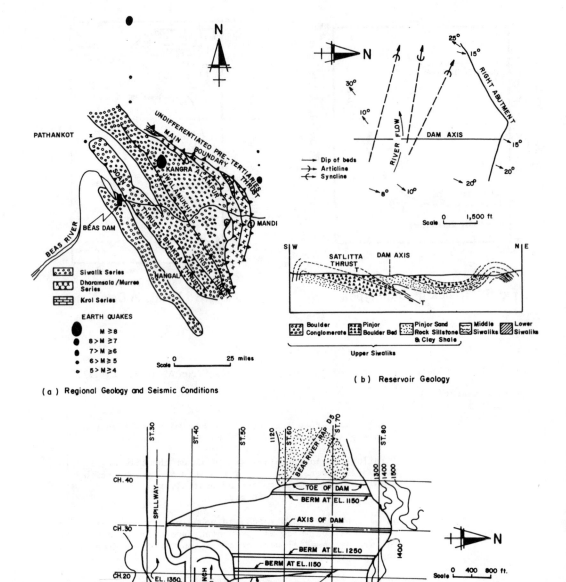

Fig. 7   Beas Dam at Pong

Symposium on Problems and Practice of Dam Engineering / Bangkok / 1-15 December 1980

# The problems and practice of dam engineering in Hong Kong

A.J.VAIL & D.J.EASTAFF
*Binnie & Partners (Consulting Engineers), Hong Kong*

INTRODUCTION

All the dams built so far in Hong Kong are for water supply or irrigation. The provision of an adequate water supply has always been a problem as the Territory has no large catchments, no natural lakes or major rivers and has no substantial underground water resources.

In 1860, 20 years after a colony was first established, the population was about 100,000 and water was already in short supply. By the outbreak of World War II the population had risen to 1 million.

After the war the population began to increase again and as the numbers and the standard of living rose, the water demand increased dramatically until today when a population of 5 million consumes over one million cu.m. per day.

Hong Kong now has three main sources of supply - natural catchments within the Territory, rivers in Guandong Province in China and a sea water desalter. Eighty percent of the water used is impounded in reservoirs formed by the construction of dams across valleys and, in more recent years, across sea inlets. There are 5 earth and composite fill dams, 2 rockfill dams, 22 concrete gravity dams and 1 concrete thrust block/rockfill dam in use and one 75 m high earthfill dam under construction.

This paper describes the geography, geology and history of the dams in Hong Kong and reviews the problems and practice of dam design and construction. Details are given of the three largest and most recent schemes, Shek Pik, Plover Cove and High Island.

GEOGRAPHY

The Territory of Hong Kong, comprising 236 islands, the Kowloon Peninsula and the New Territories (fig.1.) is 1060 sq.km. in area of which about 750 sq.km. is mountainous. Of the present population of 5 million, 1 million live on Hong Kong Island, 2 million in Kowloon and the remainder in the New Territories and the outlying islands.

The temperature in Hong Kong ranges from $33^{\circ}C$ with high humidity in the summer months of May to September to $6^{\circ}C$ with low humidity in the winter months of October to April.

The long term average rainfall amounts to 2200 mm, 80% falling in the summer months. Between the years 1963 and 1980 the total annual rainfall has varied from 900 mm to 3100 mm. The wettest year recorded in Hong Kong since 1853 when readings were first taken was 1973, the annual rainfall exceeding the long term average by 40%, the driest year was 1963, the annual rainfall falling below the long term average by 60%.

Hong Kong is a typhoon area and while typhoons can occur in any month, they normally arrive in the hot summer months from May to October, reaching a peak frequency in July and August. A major typhoon can generate gust velocities of 300 km/hr.

GEOLOGY

The Territory is composed mainly of two igneous rock types, granite intruded into older metamorphosed volcanic rhyolite lava and ashes with some metamorphosed sedimentary rocks. (fig.2.) Both rock types are tropically weathered with the formation in many areas of a thick mantle of residual soil. The granitic rocks may be completely decomposed to depths of 60 m with occasional core stones whereas the volcanic rocks are only completely weathered to depths of about 10 m and rarely contain core stones. In general, the volcanic rocks are very closely jointed while the granitic rocks have widely spaced joints; valleys tend

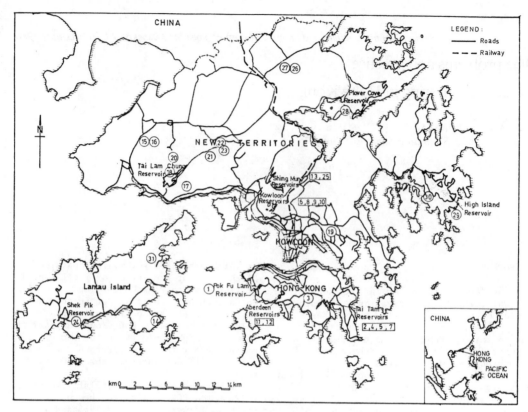

Fig. 1. TERRITORY OF HONG KONG
SHOWING LOCATION OF DAMS AND RESERVOIRS

1. Pok Fu Lam
2. Tai Tam Upper
3. Wong Nei Chung
4. Tai Tam Bywash
5. Tai Tam Intermediate
6. Kowloon
7. Tai Tam Tuk
8. Shek Lei Piu
9. Kowloon Reception
10. Kowloon Bywash
11. Aberdeen Upper
12. Aberdeen Lower
13. Jubilee (Gorge Dam)
14. Shap Long
15. Lam Tei
16. Hung Shui Hang
17. Sham Tseng
18. Tai Lam Chung
19. Ma Lau Tong
20. Wong Nai Tun
21. Ho Pui
22. Tsing Tam Lower
23. Tsing Tam Upper
24. Shek Pik
25. Lower Shing Mun
26. Hok Tau
27. Lau Shui Heung
28. Plover Cove
29. High Island East
30. High Island West
31. Discovery Bay

to form along faults and many of these have been intruded with dolerite. Thin deposits of alluvium remain in the valley bottoms overlying completely weathered to fresh rock. Colluvial deposits of boulders and cobbles, sometimes completely weathered in a matrix of clayey sandy silt form aprons along and at the toes of some of the steeper hillsides.

During the Pleistocene Period, the sea was some 100 m lower than it is today and many existing sea channels and inlets were formed by rivers which have left deposits of sand, silt and gravel. Since the inundation of the area by the sea towards the end of the Pleistocene Period, probably 8 to 10 thousand years ago, the alluvial deposits have been covered with soft marine muds.

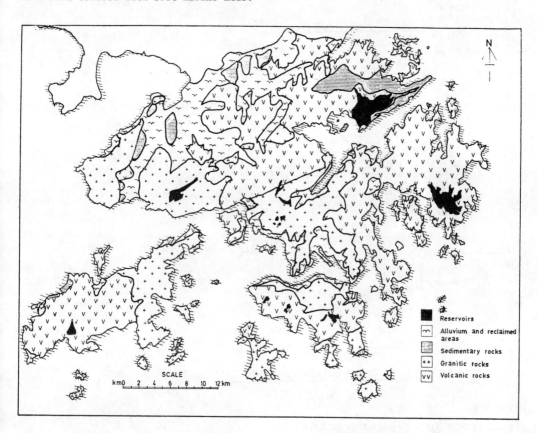

Fig. 2 GEOLOGICAL MAP OF HONG KONG

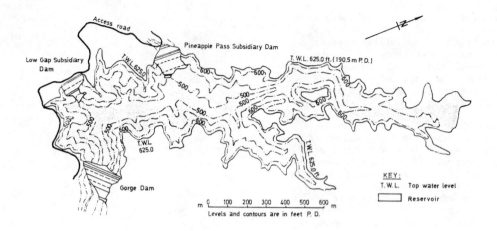

(a) RESERVOIR AND DAMS

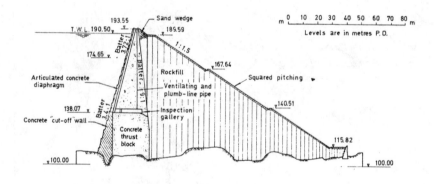

(b) CROSS SECTION OF MAIN DAM

Fig. 3:- GORGE DAM (JUBILEE RESERVOIR)

HISTORY OF DAMS

The first dam to be built in Hong Kong, a low masonry wall, was completed at Pok Fu Lam on Hong Kong Island in 1863; the impounded reservoir soon proved inadequate and the dam was replaced in 1871 by an 18 m high earthfill embankment with a core of soft grey clay and shoulders of loosely compacted residual soil derived from the local volcanic rocks. This dam, which is still in use, remained the only fill dam until 1936.

In the following forty years to 1910, four conventional concrete gravity dams were built on the Island near the major centres of demand and one concrete gravity dam was built on the mainland for the developing area of Kowloon. Between 1910 and 1935, three more concrete gravity dams were built on the Island and three on the mainland in the foothills above Kowloon. The total volume of water impounded by these six dams amounted to only 10 million cu.m.

In 1936, the Gorge dam, a composite structure consisting of a concrete semi-gravity section supported by a downstream rockfill shoulder, with two associated saddle dams, was constructed across the Shing Mun valley in the New Territories. (fig.3.) Although this dam is 83 m high, the reservoir capacity is limited by the topography of the valley to 13.7 million cu.m. which would be regarded as totally uneconomic in many other parts of the world.

The post-war expansion of Hong Kong created the need for additional storage and, by 1958 the Tai Lam Chung concrete gravity dams and two earthfill dams were completed between 1955 and 1963 including a major earthfill dam at Shek Pik on Lantau Island in 1963.

The rapid growth in the demand for water and the limited space available for reservoir storage on land, has led during the last fifteen years to the concept of offshore storage. Two marine reservoirs have been created in the remoter areas of the New Territories by the construction of dams on the seabed. The Plover Cove reservoir was completed in 1967 by building a marine barrage across a bay in Tolo Harbour and in 1972, this barrage was raised. In 1972, also, work began on the construction of two rockfill dams, 110 and 102 m high, between High Island and the mainland to form a reservoir in the Straits of Kwun Mun.

A 75 m high earth fill dam is under construction at Discovery Bay on Lantau Island to provide water for a new resort town. This dam will be completed in 1982.

PROBLEMS AND PRACTICE

The problems and practice of dam engineering in Hong Kong are affected by:-

Topographical factors

The steep dissected terrain and the limited size of the direct catchments have restricted the development of conventional reservoirs on land. The largest land reservoir at Shek Pik on Lantau Island, has a capacity of only 24.5 million cu.m. and the earthfill dam required to impound it has a height of 54 m and a volume of 4.75 million cu.m. Twenty seven of the thirty one dams in Hong Kong are higher than 15 m and may be considered as major structures. Subsidiary saddle dams have been required in four of the five largest reservoirs to allow storage up to the level of the natural topography of the reservoir rim. Two dams, Tai Tam Tuk (1917) and Shek Pik (1963), have been constructed with their foundations at or below sea level and the two most recent storage schemes, Plover Cove (1969) and High Island (1978), have involved the construction of dams on the sea bed, the formation of freshwater reservoirs in areas previously occupied by the sea and the collection of water from indirect catchments by a total of 77 km of tunnels. Two dams, Aberdeen Lower (1932) and Plover Cove (1969) have been raised to provide increased reservoir capacity where the original dams underutilised the catchments.

Geological factors

The geology of a site is an important factor in an economic appraisal to determine the type of dam to be built. The depth of weathering, the depth of alluvial and marine deposits and the effect of faulting, very common in the valley bottoms, the stability of the slopes and the availability of construction materials all play a part in the final decision.

The variation in depth of weathering and the extent and nature of the soil deposits in valleys requires extensive site investigation of foundations and potential borrow areas to avoid major design changes during construction. The first recorded site investigation was for the Gorge dam in the early 1930's. This indicated that a more favourable foundation, involving the construction of a higher dam, would be found downstream of the originally proposed site. Site investigations have been an essential feature of all the projects engineered since the second world war and, with the passage of time the quality of information obtained has improved with modifications of techniques for obtaining and recording data. In recent years geophysical methods have been used with success for investigating variation of sea bed deposits at marine dam sites and in potential marine borrow areas. Aerial photography is now an essential tool in all investigations. Particular care is required in investigating borrow areas and in interpreting results as the variable depth of weathering both in the granitic and volcanic rocks can lead to inaccurate assessments of the quantities of available borrow materials. Descriptions of rock types and the degree and zones of weathering have been standardised in Hong Kong engineering practice to provide uniform descriptions and thereby to eliminate or reduce the subjectiveness of interpretation.

Seismicity factors

Hong Kong belongs to the continental shield of eastern China. Both seismic records and historical data indicate that the shield is a stable mass with marginal tectonic fractures which are seismically active. The area is subject to frequent slight earthquake shocks but even the worst shocks have fallen below the average of the destructive earthquakes that have shaken the western provinces of China (Szechuan, Yunnan and Kansu).

The strongest earthquake likely to be felt in Hong Kong would be either a near, low magnitude shock or a distant, shallow severe shock. In both cases, the resulting ground motion in Hong Kong would have an intensity of VI + ½ on the modified Mercalli scale. A review of the effects of earthquakes on the major dams in Hong Kong has shown that earthquakes of this intensity would not have a significant effect.

The China coast is occasionally subject to seismic sea waves or seiches originating along the circumpacific belt. Seiches up to two metres high have been recorded along the coastline of Hong Kong but this is less than is already allowed in the design of marine structures for typhoon surges.

Climatic factors

Wave protection is required on the upstream slopes of all earthfill dams to resist erosion by waves generated by high wind velocities during typhoons and on the downstream slopes of dams such as Plover Cove and High Island constructed in the sea. The downstream faces of earthfill dams also require berms at frequent vertical intervals and protection with turf to prevent erosion during intense rainstorms.

The approach to dam design and the economics of dam construction in Hong Kong have changed considerably over the past twenty years. To highlight these changes these changes particular aspects of the design and construction of three recent schemes, Shek Pik, Plover Cove and High Island are described.

SHEK PIK DAM

This zoned earthfill embankment, completed in 1963 is 54 m high with a total volume of 4.75 million cu.m. It was constructed just above the foreshore across a relatively broad alluvial valley on Lantau Island. (fig. 4a). The alluvium consists of 16 m of a heterogenous mixture of boulders, gravel, sand and silt. The dam abutments are composed of residual soil formed from the deep weathering of volcanic lavas and tuffs. (fig. 4b).

The dam has a composite upstream shoulder designed to be free draining on drawdown, an impermeable central core of selected, more plastic residual soil protected by a graded intercepting filter and a downstream shoulder of residual soil with drainage blankets. (fig 4c). The upstream slope is protected from wave action by rock rip-rap and the downstream slope is bermed and turfed to protect it from erosion. The lowest point of the foundation of the dam, is level with the sea at high tide.

A cut-off was formed beneath the dam by grouting the three foundation materials - alluvium, residual soil and rock. Each type required a different technique, the work being carried out in separate operations. Tube-a-manchette grouting was used in the alluvium and residual soil and conventional cement grouting in the rock (fig. 4b).

The alluvium showed marked local variations in permeability and, to ensure concentration of the grout within the curtain zone, two lines of bored contiguous concrete piles, 550 m in diameter, were constructed to form diaphragm walls 6 m apart. The permeability of the alluvium was reduced by grouting from an average value of $5 \times 10^{-5}$ m/s to $4 \times 10^{-7}$ m/s. The underlying residual soil was also grouted by the tube-a-manchette method to open up the residual joints to allow the passage of grout. The permeability of the rock before treatment measured by well-pumping tests was of the order of $5 \times 10^{-6}$ m/s and tests after grouting showed that this value had been reduced to $5 \times 10^{-7}$ m/s.

The underlying zone of moderately weathered to fresh rock was injected with cement grout through vertical holes at 3 m centres in descending 3 m stages. The injection pressures were high ranging from 2000 to 3800 kPa (21 to 39 kg/cm$^2$).

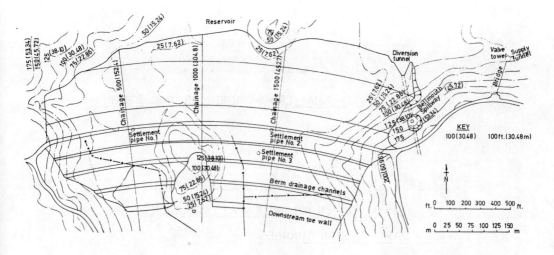

(a) PLAN OF THE DAM

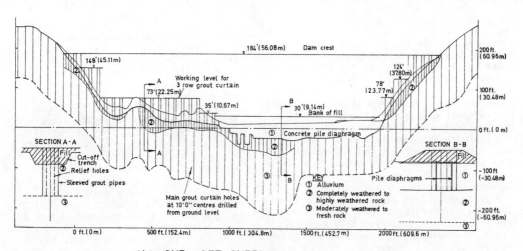

(b) CUT-OFF CURTAIN

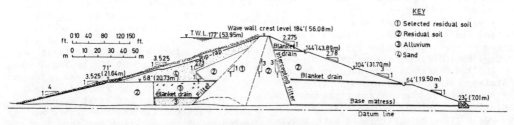

(c) DAM - CROSS SECTION AT MAXIMUM HEIGHT

Fig. 4 :- SHEK PIK DAM

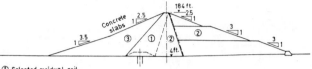

① Selected residual soil
② Residual soil
③ Alluvium
④ Sand

(a) TYPICAL CROSS SECTION ORIGINAL DESIGN

(b) TYPICAL CROSS SECTION AS CONSTRUCTED

Fig.5:- DAM — DESIGN CHANGES

The original design proposed the use of residual soil for the core and downstream shoulder and alluvium for the free draining upstream shoulder (fig. 5a.). The alluvium proved, however, to have a higher silt content and to be more variable than expected. The design was amended to incorporate only a small portion of alluvium in the shoulder adjacent to the core, to form the toe with residual soil to act as a cofferdam for diversion and to import sand which could be compacted without difficulty during the wet season to form free draining zones (fig. 5b).

The residual soil was a clayey sandy silt (fig. 6a) of medium to low plasticity. Fig.6c shows the variation in Atterburg limits and moisture content with depth in typical boreholes. Selecting the material for the core zone of the dam proved difficult as it was impracticable to select the most plastic soil by eye. The mean plasticity values for the selected core material and for the random material placed downstream of the intercepting filter are shown in fig. 6b indicating the scatter of values for the two materials as placed in the dam.

Drainage blankets were incorporated in the shoulders of the dam and piezometers installed in the fill showed that little or no construction pore pressures were generated even during periods of rapid placing of fill.

In drilling through the core to grout the residual soil foundations, a loss of water was noted in some of the holes. Exploratory trenches showed fissures in the core. Thirty one exploratory holes were drilled into the core and 10 of these showed water losses. A further 32 holes were drilled at 16 m centres, water was injected in an attempt to seal the fissures by swelling but this did not succeed. In one instance a connection was found between two holes 16 m apart and, at another location, the piezometer in the vicinity of a hole in the core responded immediately to water level changes in the hole. The spacing of the holes was reduced to 8 m and a cement/clay grout mix was used. Only sufficient pressure was applied to overcome the resistance to flow. Where grout acceptances were high the spacing of holes was further reduced to 4 m and, exceptionally, 2 m. A total of 25 cu.m of grout was injected. Pore pressures and seepage were carefully measured during filling of the reservoir and no unusual conditions were observed. The dam has performed satisfactorily for the last sixteen years.

Considerable discussion was generated at the time on the cause of the cracking which was thought to have been associated with the rigid nature of the core material and differential settlement of the foundations. Although the core material was placed at 2.5 to 3.5% wetter than Procter optimum the negligible pore pressures developed during construction and the high shear strength of the material indicate that the core was not sufficiently flexible to adapt to the differential settlement of the alluvium and completely weathered rock foundation. Settlements of up to 90 mm were recorded in the foundations during construction. The core may have cracked transversely where the alluvium and weathered rock

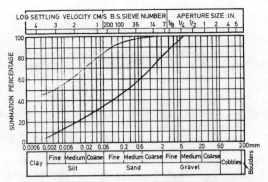

(a) ENVELOPE OF GRADING CURVES OF RESIDUAL SOIL FROM VOLCANIC ROCKS

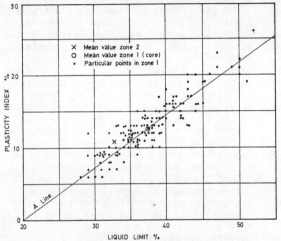

(b) PLASTICITY CHART - ZONE 1

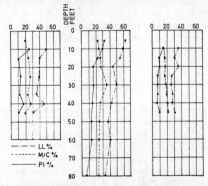

(c) LIQUID LIMIT, PLASTICITY INDEX AND MOISTURE CONTENT IN THREE TYPICAL BOREHOLES

Fig. 6 :- SHEK PIK DAM - FILL PROPERTIES

was deep and longitudinally above the more rigid piled sections of the foundations. An additional cause of differential movement was the heave of the core during grouting the underlying weathered rock. Half the length of the dam was affected by this operation and although attempts were made to limit the heave, local variations of 90 mm were observed.

PLOVER COVE DAM

The layout of the Plover Cove scheme is shown in figs 7 and 8a. The dam, a 38 m high zoned earthfill embankment, 2 km long, with a volume of 8.2 million cu.m. (fig.9) was constructed on the sea bed across Plover Cove. After closure the sea water trapped in the reservoir was pumped out and replaced with fresh water from both the direct catchment and from indirect catchments connected to the reservoir by 37 km of tunnels.

To the north the bedrock consists of deeply weathered volcanic tuffs, lavas and agglomerates and to the south, of sandstone and siltstone. Both formations are covered in the area of the sea bed by alluvial clay sand and gravel deposited during the Pleistocene period and, following subsequent inundation, by 5 to 15 m of marine muds which today lie 10 to 14 m below sea level (fig 8b)

The depth of compressible strata in the bay precluded the construction of any dam other than an earth embankment and the length of embankment required to cross the bay favoured placing under water rather than in the dry behind a coffer dam.

The dam was founded on the alluvium by first removing the overlying marine mud and a seal was effected by placing a layer of completely decomposed rock of low permeability on the surface of the alluvium and against the adjacent marine mud slopes. The design incorporated a central core of selected impermeable decomposed rock and the shoulders, of a similar but unselected material, included layers of free draining material to dissipate construction pore pressures and to accelerate consolidation.

Both shoulders of the embankment were protected from the effect of waves generated during typhoons by granite rip-rap and, in view of the large fetch involved on the seaward side and the possibility of a typhoon hovering over the Territory for a long period the designers had to provide rip-rap adequate for waves up to 2½ metres in height.

One other unusual design problem was that of producing from the reservoir a water acceptable for treatment in as short a time as possible. This involved removal of the sea water and marine life from the reservoir on closure, diffusion of salt from the mud in the reservoir floor and the biological changes inherent in converting the reservoir from salt to fresh water.

The marine mud under the dam was first removed by bucket dredge and grab exposing the underlying alluvial deposits and, in places the in-situ decomposed rock. About 5.4 million cu. m. of marine clay were dredged in depths of water up to 14 m giving a maximum dredging depth of 29m. The spoil was dumped in a small bay enclosed by islands 2 km west of the dam site.

During dredging a residue of semi-fluid mud accumulated on the bottom of the trench to a thickness in some places of more than 1½ m. Attempts to dredge this mud at first proved unsuccessful but it gradually gained strength and after a month it was removed.

The fill, obtained from deeply weathered granite and granodiorite areas consisted of silty sands with a small percentage of clay size particles. The finer core material selected with a minimum of 25% passing the No.200 BS sieve was placed by grab. Tests showed that the permeability of the core was of the order of $1 \times 10^{-6}$ m/s. Coarser material was dumped from bottom opening barges in the shoulders where increased permability caused by segregation and loss of fines was acceptable. Some of the coarser material encroached into the core zone and was removed by redredging.

Pore pressures in the embankment and foundations proved to be insignificant and imposed no restrictions on the rate of construction. An early decision was therefore taken to omit the bottom sand drains over the remainder of the dam.

A critical activity in construction was the final closure. Hydraulic model studies had confirmed the need to close on a sill of the maximum practicable length to limit the current through the gap and to keep the crest of the closure rock mound as level as possible. The closure gap was 1040 m long chosen more or less centrally in the length of the dam. In the event closure presented no problems.

The initial intention was to replace the salt water with fresh water as rapidly as possible by reducing the water level in the reservoir to -8.25 m below sea level, by diluting the remaining sea water with rain falling in the first part of the wet season and by maintaining a constant water level by pumping which would remove salt from the reservoir. The rainfall in the second part of the wet season was to be used for refilling. In the event, the first part 1967 of the wet season was exceptionally dry and little dilution was possible. As the performance of the dam on drawdown was satisfactory and the mud in the reservoir bottom

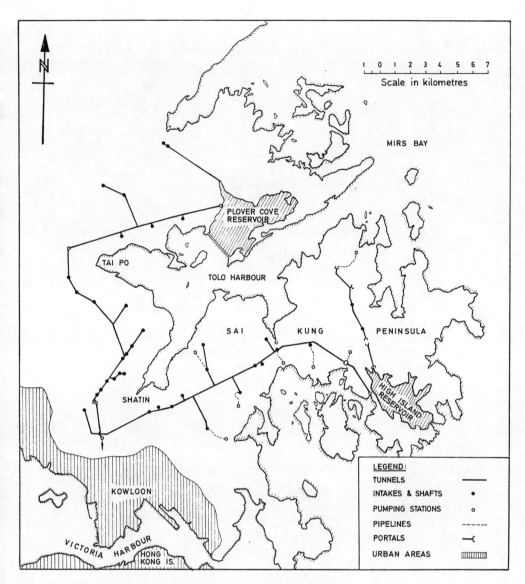

Fig. 7:- LAYOUT OF PLOVER COVE AND HIGH ISLAND SCHEME

appeared less liable to erosion than had been expected, the decision was made to pump out as much of the remaining sea water as possible. The water was lowered to -11m below sea level leaving a small pond of 650 thousand cu.m. Fortunately, the rainfall in the second half of the wet season was near average and by the end of the season, when the reservoir was about a quarter full some water was pumped to supply. At first, consumers complained of the brackish taste but inflow during the 1968 wet season reduced the salinity of water supplied to the extent that taste was no longer a problem.

HIGH ISLAND DAMS

The High Island Scheme, the largest natural water supply scheme to be constructed in Hong Kong, was completed in 1978. The scheme exploits the last untapped catchments on the mainland by the construction of intakes on the principal streams along the Sai Kung peninsula (fig.7), connecting these by branch tunnels to a

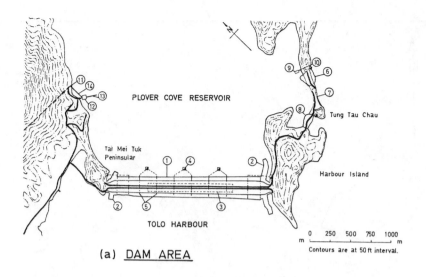

(a) DAM AREA

1. MAIN DAM
2. BEACH CUT-OFFS
3. AREA OF SILL OF CLOSURE GAP
4. GAUGE HOUSE
5. INSTRUMENTED SECTIONS
6. NORTHERN SUBSIDIARY DAM
7. SPILLWAY
8. SOUTHERN SUBSIDIARY DAM
9. TEMPORARY PUMPING STATION FOR PUMPING OUT RESERVOIR AFTER CLOSURE
10. TUNNEL FOR PUMPING OUT
11. PORTAL OF MAIN TUNNEL IN STAGE II SYSTEM
12. PERMANENT PUMPING STATION
13. INTAKE CULVERT
14. PIPELINE FOR FRESH WATER DRAW-OFF

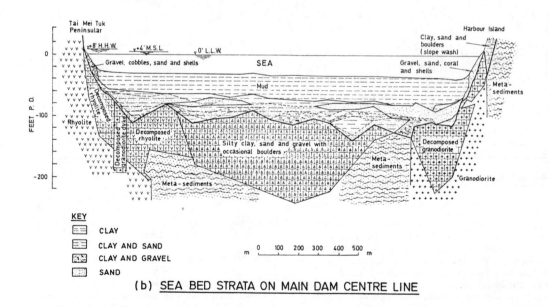

(b) SEA BED STRATA ON MAIN DAM CENTRE LINE

Fig. 8 :- PLOVER COVE DAM

main tunnel running along the spine of the peninsula to the reservoir site located in the straits of Kwun Mun separating High Island from the mainland (fig.10.)

The strait is very shallow at the western end and about 15 m deep at the eastern end. During the Pleistocene period the strait was a river valley covered with at least 12 m of alluvial deposits overlying weathered bedrock consisting of volcanic lavas and tuffs. Following the return of the sea, up to 20 m of very soft marine muds were deposited over these alluvial deposits.

Two rockfill dams founded on bedrock have been constructed to a height of 65 m above sea level across both ends of the strait to create a marine reservoir with a storage capacity of 273 million cu.m. (fig.10) The West dam is 102 m high and the East dam 110 m high. Two low points on the perimeter of the reservoir have been raised by earthfill col dams, one 30 m and one 5 m high and at a number of other points where sound bedrock is only marginally above top water level, the upper, fissured zones have been sealed with cement grout.

The designers considered earthfill, composite, concrete gravity and rockfill for the High Island dams. The mantle of decomposed rock in the area in very thin and has a high silt content. There was no nearby source of a more suitable soft fill and earth fill was therefore ruled out on grounds of cost. Composite dams were also discounted on economics and because the high silt content of the decomposed rock would create problems in providing safe filters.

Concrete gravity sections were discarded because of the fissured nature of the rhyolite in the area, the height of the dams, the possibility of differential settlement under their bases on first impounding and again, cost. Rockfill was an attractive option as it could be readily won from within the reservoir area but there remained the problem of a suitable impermeable membrane.

The reservoir is capable of being drawn down below high tide level and because of this upstream impermeable decks were ruled out as they could blow off.

Again the height of the dams and the fact that rockfill, however well compacted, will always settle more when saturated, precluded the use of a rigid mass concrete core and an articulated core would have been too expensive. The designers finally adopted asphaltic concrete cores (fig.11a, 11b) of graded rhyolite aggregate, 19 mm down to gravel size mixed with sand and cement filler preheated and mixed with about 6% of hot bitumen to form a highly impermeable but flexible membrane which could tolerate differential movement within the dams.

In each dam a secondary asphaltic concrete core was constructed to 12 m above sea level connected to an inspection gallery running the length of the dams and fitted with pipes to detect any leakage through the main cores. Cross walls of asphaltic concrete were provided between the cores every 60 m along the length of the dams to locate the source of any leak within the box formed and to enable remedial measures to be confined to that box.

The main cores are 1.2 m wide and are vertical up to the drainage gallery. Above the gallery they curve slightly seawards to elevation 30 m PD and then continue 0.8 m thick at a slope of about $75^o$ to the crest to prevent the rockfill shoulders from losing contact with the core on settlement.

Because of the higher foundation stresses of rock fill dams, the High Island dams could not be founded on alluvium as in Plover Cove but had to be founded on rock. They therefore had to be constructed in the dry between cofferdams.

The direct catchment of the reservoir is over 1600 hectares and monsoon rain with peak intensities of 150 mm an hour can produce considerable run off which would have taken weeks to pump out and would have delayed construction. The designers decided therefore to construct a pair of coffer dams round each dam site (fig.10). At the western end the cofferdams were formed of marine sand and silt dredged from the proposed dam site and placed as hydraulic fill between pairs of rockfill bunds, dumped and tipped across the straits upstream and downstream (fig.11d).

At the eastern end, the inner cofferdam was designed in this way but because of limited space and the depth of the sea at the site, the outer coffer dam was built of rock fill with a semi impervious core and blanket of sand bitumen laid on the back of the reservoir slope of the initial rock bund and extending to 30 m over the sea bed. The sand bitumen was later covered with rock fill (fig. 11c). The purpose of the sand bitumen was not to achieve watertightness but to reduce the seepage through and under the coffer dam to a volume that could be collected and pumped out during excavation of the main East dam foundations.

The water that flowed under the east sea cofferdam was collected in a row of 20 pumped drainage wells bored into a temporary toe weight constructed on the reservoir side of the cofferdam. Besides providing a platform for the wells the toe weight served the purpose of consolidating and strengthening the underlying marine muds prior to excavation for the foundation of the main dam.

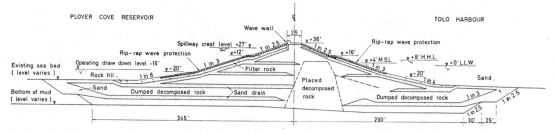

Fig. 9 : TYPICAL CROSS SECTION OF PLOVER COVE DAM

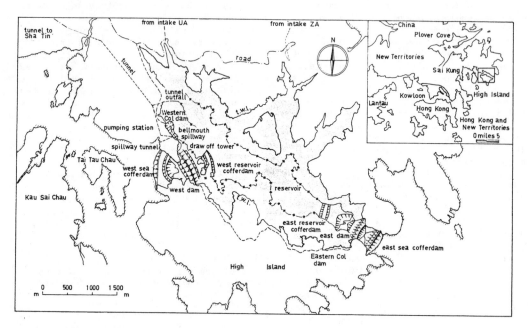

Fig. 10:- HIGH ISLAND RESERVOIR

Once the cofferdams had been compelted the water they enclosed was pumped out. The remaining marine mud and the alluvium and decomposed bedrock it overlay was then excavated by mechanical plant and, in the west dam site, by mud pumps and the surface over the area of the base of the main dams was exposed and cleared to slightly to moderately weathered rock under the shoulders and to fresh to slightly weathered rock under the core base.

The cores were designed to sit on mass concrete core bases 12 m wide and 1.5 m thick cast on the exposed fresh rock and, to complete the seal, the rock beneath the core bases was grouted through the fissured zone to a depth of 60 m. The core bases provided the weight needed to generate enough grout pressure at and near the rock surface and a convenient platform for the drill rigs.

The cores were then placed on the core bases in 200 mm horizontal layers at a temperature between 160°C and 180°C by a purpose built laying machine and were compacted with vibrating steel rollers. As the cores were so narrow extreme care was taken in quality control and supervision and core placing was not permitted at night.

The shoulders of the dam were brought up in 1 m thick horizontal layers, level with and separated from the cores by zones of transition material fine enough to contain the hot asphaltic concrete.

The other novel feature of the High Island dams was the protection of their seaward

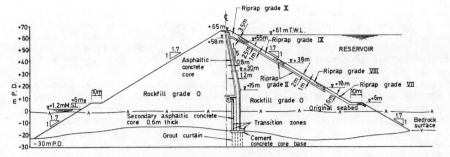

(a) TYPICAL SECTION OF WEST DAM

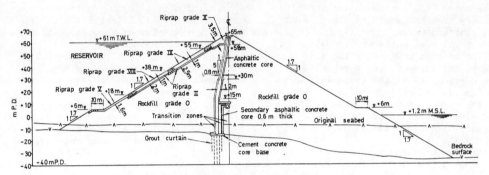

(b) TYPICAL SECTION OF EAST DAM

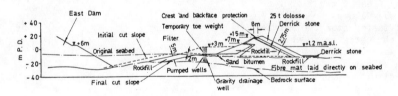

(c) EAST SEA COFFERDAM

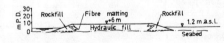

(d) RESERVOIR DAM

Fig. 11 :- HIGH ISLAND DAMS

shoulders from ocean waves. The two seaward cofferdams had, in any case, to be protected for the period of construction and it was decided to make them permanent structures incorporating the permanent sea protection. This was cheaper than providing temporary protection, removing the coffer dams at the end of the construction period and permanently protecting the main dams.

The West dam was very well sheltered by islands and the fetch was quite small. Four ton granite boulders were therefore adequate to protect the face of the West sea cofferdam.

The East dam is in a very exposed position facing the open sea and typhoons approaching the Territory from the south, as they often do, could generate waves up to 11 m in height in the 15 m of water at the site. The effect of such waves on a rockfill dam would be catastrophic and a research programme carried out at the Hydraulics Research Station at Wallingford in Britain favoured the use of 25 ton concrete 'dolosse' placed on 10 to 12 ton granite blocks graded with depth to act as a filter to the rockfill.

Seven thousand of these dolosse were used to armour the cofferdam from the crest 17 m above sea level to the full depth of water and over a length of 460 m.

Before impounding could commence the sea water in the Straits between the reservoir cofferdams had to be removed and salt water in the sea bed reduced to an acceptable concentration. As soon as the cofferdams were built the sea water was pumped out and fresh water from the direct catchment was allowed to collect until the salt diffusing from the marine sediments raised the salinity to about 4000 p.p.m. The water was then pumped out and the operation repeated a number of times until the salinity stopped rising significantly.

REFERENCES

Allen P.M. & Stephens E.A. 1971, Report on the geological survey of Hong Kong, Govt. Press, Hong Kong.

Binnie W.J.E. & Gourley H.J.F. 1939, The Gorge Dam. Jour. I.C.E. London p179.

Carlyle W.J. 1965 Shek Pik Dam. Proc. I.C.E. London vol.30 p557.

Drust T.A. and others 1964, Some aspects of Stage 1 of Plover Cove Water Scheme. Proc Hong Kong Eng. Soc.

Dunnicliffe J.D. 1968, Instrumentation of Plover Cove Main Dam. Geotechnique vol.18 p283.

Elliott S.G., Ford S.E.H. & Oules J. 1967, Construction of Plover Cove Dams ICOLD.

Fanshawe H.G. 1962, Soils in the Shek Pik Valley. Symposium on Hong Kong soils p53.

Fanshawe H.G. & Watkins M.D. 1971 A sparker survey used to select a marine dam site in Hong Kong. Quat. Jour. Eng. Geol. vol4 No.1 p25.

Guildford C.M. & Chan H.C. 1969. Some soils aspects of the Plover Cove marine dam. Proc. 7th Int. Conf. SMFE vol.2

Ho C. 1965 Lower Shing Mun main dam. Proc. Hong Kong Eng.Soc.

Jackson L. 1949 The Hong Kong Waterworks Hong Kong Eng. Soc.

La Touche M.C.D., Smith A.J.E. & Townsend M.I. 1970 Plover Cove Reservoir, Hong Kong: The transition from marine to freshwater conditions. ICOLD 10th Congress p981

Lau R. 1972 Seismicity of Hong Kong, Royal Observatory of Hong Kong.

Little A.L. & Beavan G.C.G. 1963 Discussion on "Flexibility of clay and cracking of earth dams" by Leionard and Narain. Jour. Soil Mech & Found Div. ASCE.

Little A.L. 1967 The use of tropically weathered soils in the construction of earth dams. Proc. 3rd Asian Reg. Conf. on Soil Mechanics and Foundations Eng. vol 1.

Lumb P. 1965 The residual soils of Hong Kong Geotechnique vol.15 p180

Molyneux P.S. 1977 The development of water supply in Hong Kong. Hong Kong Eng.(Oct)

Vail A.J. 1975 The High Island Water Scheme in Hong Kong. Hong Kong Eng. (Oct)

Watkins M.D. 1979 Engineering geological mapping for the High Island Water Scheme in Hong Kong. Bull IAEG. No.19

# Section B:
## Investigations, instrumentation and foundation treatment

*Symposium on Problems and Practice of Dam Engineering / Bangkok / 1-15 December 1980*

# Geotechnical investigations for dams

A.C.MEIGH
*Soil Mechanics Ltd., Bracknell, UK*

SUMMARY

The site investigation for a dam is normally carried out in stages: desk study, preliminary site reconnaissance and survey, evaluation of the foregoing, investigation at selected sites, investigation at the chosen site, and monitoring during construction and during and after commissioning.

An important component of the desk study is the initial geological assessment, which is greatly assisted by examination of aerial photographs and Landsat imagery. The preliminary field reconnaissance and survey may include a flyover, geological mapping, geophysical survey and possibly boring using relatively light and simple equipment.

The investigation at selected sites should concentrate on resolution of likely hazards and problems indicated by the evaluation of the desk study and preliminary field study. It will require such techniques as trenching, pitting, further geophysics, groundwater investigations and a limited amount of high quality boring or drilling.

The investigation at the selected site will be directed towards producing parameters for final design. The techniques to be adopted will depend on the project and the nature of the ground. It will involve boring and drilling, with particular attention to sample quality and high core recovery, careful logging of trenches, shafts and adits, in situ testing such as plate loading tests and shear tests, trial embankments, grouting trials and so on. Geophysical techniques will reappear at this stage, and groundwater evaluation will be required by means of packer tests in drillholes, installation of piezometers and pumping tests. Careful logging of rock cores and description of soils and rocks will be essential. The techniques discussed in the paper are related to various categories of ground conditions; softer cohesive soils, stiff clays and weathered rocks, pervious foundations, and rock. Case histories are presented to illustrate investigations in these categories of ground conditions.

The concluding parts of the paper discuss investigations in karstic conditions and where soluble rocks are present, materials for construction, some aspects of reservoir stability and water tightness, and seismicity.

1 PHASING AND OBJECTIVES

The investigation for a dam and its associated structures may initially be directed at chosing the more favourable among a number of possible sites. This will involve geology and geotechnics, and a variety of other aspects which will not be considered here. Whether one site only is being considered, or a number are involved, the first stage will be a desk study to acquire published and primary data, followed by, or possibly overlapping with, a field reconnaissance and survey. Evaluation of these will indicate sites for further investigation which will lead to a choice of site and a preliminary decision on dam type.

Once the site had been selected, its investigation will be designed to elucidate the geology of the site (structure, stratigraphy, faulting, foliation, jointing pattern, and so on), and to establish the ground and groundwater conditions below the dam and its abutments, and in the reservoir area, with the following objectives:

(a) To determine the engineering parameters to be used in assessing the stability of the dam foundations and in estimating the settlements which will occur.

(b) To establish the probable pattern of seepage under the dam and around its abutments, and to determine the parameters required to assess the volume of seepage, the accompanying pressure distributions, and the measures which may be required to reduce seepage, prevent piping, and modify the pattern of pressure distribution.

(c) To examine the watertightness of the reservoir area, and the stability of the surrounding slopes under submergence and drawdown.

The relative importance of these aspects of the investigation will depend on the geology of the dam site and the type of dam to be constructed.

The investigation will also be directed towards evaluating the availability, quantity and suitability of materials in the area which can be considered for use in construction of the dam, and towards determining for design purposes the engineering properties of fill materials.

The final stage of investigation will involve monitoring during construction and after completion.

## 2 THE DESK STUDY

An important component of the desk study is the geological assessment. A proper appreciation of the geology, including the hydrogeology, is essential if the ground conditions are to be adequately understood. Moreover, geology is an important component of the vehicle by which experience from one project can be used in other projects.

A simple example of failure to look at the geological situation which led to disastrous results occurred at Waltons Wood on the M6 motorway site in Northern England (Early & Skempton, 1972). This example is not from dam practice, but it illustrates the point dramatically. At the site, on side-long ground, a routine site investigation was carried out. No geologist made a desk study or visited the site, and the fact that the road crossed an old landslide was not realised. That some of the centre-lines pegs had moved downhill during the winter preceding the putting down of the boreholes was ignored. Remedial works of the failed embankments were extremely expensive, and delayed the opening of the motorway. That this kind of mistake should occur in dam practice would seem unlikely, but the author knows of three cases where unstable slopes were identified in an abutment after the dam position had been decided, and in two of these construction had started.

At the desk study stage, or later if the data has to be obtained, Landsat imagery and aerial photography will give valuable information on the geological structure, particularly in picking up major structural features, fault patterns and landslides. Increasing use is being made of colour and infra-red photography. Where geological maps or records are available, the photography will supplement these, but in many cases photography and ground mapping will be the only sources of data. The role of the geologist is of prime importance at this stage and it is essential that he should work closely with the engineer undertaking the investigation.

## 3 PRELIMINARY FIELD RECONNAISSANCE AND SURVEY

The value of flyover by helicopter or light aircraft is becoming recognised (Wulff and Perry, 1976; Bertranou et al, 1976), but it does not remove the need for a ground reconnaissance. The composition of the team to undertake the preliminary field reconnaissance and survey and the techniques they should have at their disposal are most important. This should be basically the same team as that involved in the desk study. A basic unit should consist of a project engineer, geologist and geophysicist. They should be experienced and of high calibre. The project engineer should lead.

During this stage, the preliminary geological data will be reviewed and enhanced by mapping. The geological study should extend beyond the immediate site area. Properly carried out it can lead to valuable considerations, before extensive drilling or other exploratory work is started, about the appropriate type of dam and the problems which may be encountered. It will assist in planning the exploratory programme and in choosing the exploratory methods to be adopted.

Geophysical equipment (seismic, resistivity, magnetic) is becoming less bulky and more robust so that some geophysical work can be done by the initial survey team rather than leaving it all to a separate contract at a later stage.

In many cases, it may be possible to make an approximate estimate of the thickness of superficial cover based on geological and geomorphological considerations coupled with geophysics and pitting, without recourse to boring. In some cases it may be appropriate to use relatively light and simple boring equipment during the preliminary survey. However, the objectives of a boring programme at this stage should be limited. The main boring programme should be deferred until a later phase.

## 4  EVALUATION OF DESK STUDY AND PRELIMINARY FIELD RECONNAISSANCE AND SURVEY

The evaluation of preliminary desk and field work should, besides dealing with the major potential hazards, establish at least qualitatively the likelihood of encountering any more tractable hazards. This will permit the ranking of the potential sites in order of their probable suitability.

Following the desk study and preliminary field work, there may be a requirement to extend the coverage of the seismological network, to take full advantage of the time available in order to establish a pattern and base level of seismicity for later evaluation of induced seismicity. If potentially active fault systems are identified, micro-tremor seismic arrays should be installed to monitor these. This can go some way towards assessing the need for criteria changes should seismic activity occur after the feasibility stage has been completed and the design stage is well advanced.

## 5  INVESTIGATION AT SELECTED SITES

At this stage investigation should concentrate on resolution of the likely hazards and problems indicated by the evaluation of the desk study and preliminary field survey. The effort should be directed towards establishing any features which would rule out a particular site or at least place it low on a ranking list. This will require such techniques as trenching, pits and boreholes, further geophysics, and groundwater investigations. Although the extent of detailed investigation at any one site may be limited, it is nevertheless of vital importance that the quality of investigation is adequate. For example, small diameter 'traditional' rotary coring will almost inevitably fail to identify such features as bedding plane shears or clay seams, whereas modern drilling techniques or carefully excavated and inspected shafts should succeed, even if they are few in number. It will be necessary to establish more positively the availability and quality of construction materials.

It is at this stage that the experience of the investigation team will be of importance in deciding what aspects require investigation, and how detailed that investigation should be.

## 6  INVESTIGATION AT THE CHOSEN SITE

At this stage, potential hazards and problems should already have been identified. It will nevertheless be necessary to remain alert for indications that such features may exist notwithstanding the conclusions arrived at during the earlier stages of the investigation. The main effort will be directed towards producing parameters for final design. This will involve high quality boring and drilling, with particular attention to sample quality and high core recovery, careful logging of trenches, shafts and adits, in situ testing such as plate loading tests and in situ shear tests in adits, trial embankments, grouting trials, and so on. Geophysical techniques will reappear at this stage, with the dual purpose of extending borehole information on the thickness of superficial materials and as an aid in the evaluation of engineering parameters, particularly for earthquake engineering design. Also, most importantly, groundwater evaluation will be required, by means of packer tests in drillholes, installation of piezometers and pumping tests. If deep underground chambers are involved, it will be necessary to measure in situ stresses to assist chamber design and orientation.

Methods are discussed in Section 8, under the heading Foundation Investigation.

## 7  MONITORING

Monitoring during construction will include the work of an engineering geologist on site, who will examine all excavations to see whether the expectations of the preceding investigations have been realised. The identification of exceptions may then lead to an early diagnosis and anticipation of problems.

For the post-commissioning stage, monitoring will involve regular reading of instrumentation installed to check performance against design criteria. This should serve as an 'early warning' system which will initiate a contingency programme, thus minimising the delays which would result from the development of an adverse situation.

## 8  THE FOUNDATION INVESTIGATION

As in all geotechnical investigations the techniques to be adopted will depend on the objectives of the investigation and the nature of the ground to be investigated. The discussion which follows is therefore divided into various broad categories of ground conditions. No attempt has been made to catalogue all the possible methods of exploration, but rather the intention has been to highlight more interesting or novel techniques, and to comment on the difficulties and dangers involved.

A principal feature in the investigation will be the logging of rock cores and description of soils and rocks. These are matters which concern both the geologist and the engineer. It is essential that they are done carefully and that terminology is standardised as far as possible. As there are no international codes for this purpose, national codes should be used where available. Identification and description of soils and rocks is dealt with comprehensively in BS 5930 : 1981, Code of Practice for Site Investigations. This has relied heavily on reports of a Working Party of the Geological Society covering logging of rock cores (1970), mapping (1972), and description of rock masses (1977).

## 8.1 Dams on softer cohesive soils

Large dams will not of course be built on clays which are no better than firm in strength, although such clays may be present at the site, and will be removed. However, for many small dams it would be uneconomic to remove more than a thin layer of weak clay.

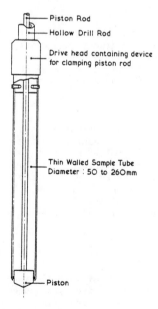

Fig 1. Thin-walled piston sampler

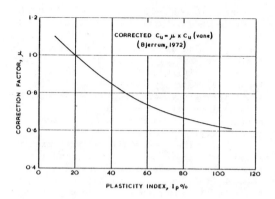

Fig 2. Correction of in situ vane test values for rate effects and anisotropy

Seepage is seldom a problem. Stability and settlement must be considered. The techniques of investigation are relatively straightforward, involving thin-walled piston sampling, (Fig 1), generally in 100 mm diameter, and laboratory testing to determine the shear strength parameters. In simple cases it will be sufficient to use undrained shear strengths in the design. These can be determined from undrained tests in the laboratory, or from in situ vane tests provided that there is not a high organic content or pockets of silt and sand which may produce erratic results. With vane tests care must be taken to correct the measured values for rate effects and anistropy, (Bjerrum, 1972), (Fig 2). Where stability presents a severe problem, drained testing will be required to determine shear strength, and it will be important to assess the rate at which consolidation will occur. This will depend on the fabric of the soil, and in particular on the presence of horizontal drainage paths provided by sand or silt layers or laminations. A general indication of this can be determined by the Delft continuous sampling technique, in which a continuous sample of either 29 mm or 66 mm diameter can be obtained down to a depth of about 18 m (Fig 3). Samples of the larger size are suitable for laboratory testing. Continuous pore-pressure probing can also be used to detect thin more or less permeable layers, (Torstensson, 1979).

Results of standard oedometer tests will not reflect drainage anisotropy and it is preferable to use the Rowe consolidation cell, (Fig 4), in which vertical and horizontal drainage can be measured. This will require large diameter samples (250 mm or 150 mm diameter) taken with a piston sampler. As an alternative to this procedure, compressibility measured in standard oedometer tests can be combined with the results of in situ permeability tests to arrive at a coefficient of consolidation.

### 8.1.1 Aughinish embankment

Within the category of weak cohesive soils occur soils which possess a collapsible skeletal structure. Such soils were encountered at Aughinish, on the Shannon estuary in Ireland where embankments are being constructed to retain red mud waste from an alumina plant. At the site recent estuarine deposits overlie glacial till.

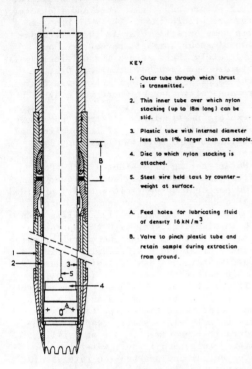

Fig 3. Delft continuous sampler

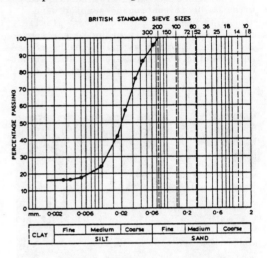

Fig 5. Aughinish; particle size distribution.

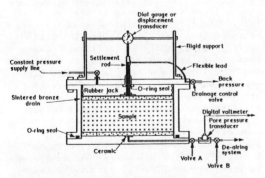

Fig 4. Rowe consolidation cell.
Note: Cell is shown set up for vertical drainage; it can also be set up for horizontal drainage.

The construction of trial banks is another useful procedure, particularly where conditions are complex. They require careful monitoring of porewater pressures, settlements, and lateral deformation, if they are to be really useful in design.

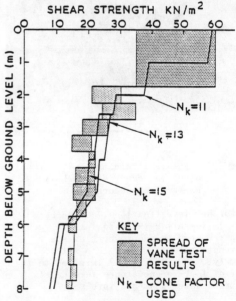

Fig 6. Aughinish; Comparison of shear strength profile obtained from core penetration tests with upper and lower strengths from vane tests.

The estuarine deposits are predominantly very loose homogeneous clayey silts, often sandy and sometimes slightly organic and shelly, with a desiccated crustal layer one or two metres thick. The grading is shown in Fig 5 and undrained shear strengths, as measured in vane tests and interpreted from static cone tests, are shown in Fig 6.

Porosities determined from samples taken with the Delft continuous sampler were generally above 50 per cent, and plasticity indices were generally below 20 per cent, indicating that the material can be regarded as having a collapsible skeletal structure. Electron micrographs confirmed this, (fig 7).

Fig 7. Aughinish; electron micrograph of clayey sandy silt, showing skeletal structure.

Further investigations, including a trial excavation taken to failure, confirmed the need to adopt very low shear strength parameters for this material, namely $c = 0$, $\emptyset' = 15°$. This led to the use of four-to-one slopes for the six metre high embankments involved.

8.2 Dams on stiff clays and weathered rocks

As for dams on weak clays, underseepage is seldom a problem for dams on stiff clay, with the possible exception of stiff clays in arid regions. Nevertheless, care must be taken to detect the presence of any more permeable layers and, if encountered, their nature and persistence and the water pressure within them. The main problem with dams on stiff clay is the determination of shear strength parameters, and immediately there arises the difficulty of obtaining undisturbed samples. Driven-tube sampling will inevitably cause disturbance, often severe where the clay is fissured or contains stones. Weathered rocks are particularly difficult to sample because of the presence of lithorelics and alternations of more weathered and less weathered material.

Rotary coring of weathered rocks can often give satisfactory core recovery (100 per cent or very close to it) provided that a core barrel of reasonably large diameter is used, generally not less than 100 mm diameter core, with the appropriate bits, and with an experienced driller using a heavy drill and a stiff drill string to avoid vibration. However, with a normal double-tube core barrel, it is sometimes difficult to get satisfactory core recovery even if these requirements are satisfied. The use of retractable-type triple-tube core barrels will usually lead to a considerable improvement. During drilling, water returns and losses, water levels, water colour and penetration rate should be carefully observed and recorded.

The triple-tube core barrel is a modification of the double swivel-tube barrel with a liner incorporated in the inner tube. The core is extracted from the core barrel within the liner, which is often split longitudinally to facilitate examination of the core. Removing core from the barrel within a liner avoids the disturbance which often occurs when core is extruded from the barrel whatever method of extrusion is adopted. The retractable triple-tube core barrel consists of a conventional outer tube with a spring loaded inner tube adapted to receive a liner, (Fig 8). This barrel was developed to core very weak rocks, especially those containing layers or lenses of stronger rock. When drilling in very weak rock the leading edge of the inner tube is held in position in advance of the core bit by the action of the spring. The core entering the barrel is completely protected from the flush water. When somewhat harder material is encountered, the spring allows the inner tube to retract into the outer tube and the barrel then operates as a conventional triple-tube core barrel.

In strong rocks the retractable barrel will not be effective.

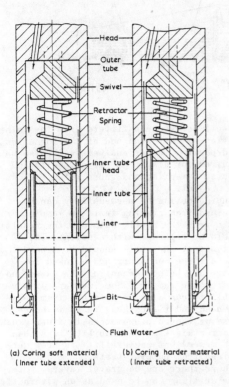

Fig 8. Diagrammatic arrangement of triple-tube retractor core barrel.

As an alternative to using special triple-tube core barrels for weak rocks, a method has been developed whereby the modified inner tube of a standard core barrel is lined with a split plastic sheath. On completion of the core run the core is withdrawn from the barrel encased in the semi-clear plastic. This greatly reduces the disturbance that often occurs at this stage and helps to preserve the natural structure of the rock.

Complete core recovery is required for full identification of the weathered rock profile. Core which is suitable for testing should be wrapped without delay to avoid loss of moisture. However, in many weathered rock profiles there will not be sufficient representative specimens of core suitable for testing, and in these circumstances in situ determination of shear strength parameters will be required. But again care must be taken to minimise disturbance of the ground to be tested. A particularly dangerous form of disturbance is that arising from seepage forces where a trial pit in which plate loading tests or in situ shear tests are to be carried out is extended below groundwater level without prior lowering of the water-table. In some cases the effects of this would be easily observed, but in others softening or loosening may significantly reduce measured strengths without this being obvious.

For in situ testing within a borehole a dilatometer, such as the Menard pressuremeter, can be used to measure strength and deformability. There are some limitations and disadvantages of this technique, but it is nevertheless useful in certain closely fissured rocks which do not provide suitable core for testing but in which the sides of a drillhole will be relatively undisturbed (Meigh and Wolski, 1981).

Where weathered bedrock is of sedimentary origin, care should be taken to identify soluble deposits such as evaporites, and unstable materials such as compaction shales and swelling rocks.

8.2.1 Empingham dam

Empingham dam in the United Kingdom is built on heavily over-consolidated Jurassic Upper Lias Clay in a valley which had been subjected to valley bulging. The dam is some 1100 m wide at the crest and 85 m high, with rolled clay shoulders and an upstream inclined core. The investigation under the direction of the consulting engineers for this project, T and C Hawksley, now Watson & Hawksley, was a comprehensive one involving driven open-tube sampling in light cable-percussion borings, deep rotary drillholes using 131 mm diameter core barrels with both water flush and mud flush, effective stress laboratory testing and in situ plate loading tests in 300 mm diameter boreholes. The design required the construction of sand drains under much of the foundation area and to aid the design an extensive programme of in situ permeability measurements was undertaken.

However, the most interesting feature of the investigation was the construction of a trial embankment, which was later incorporated into the main upstream fill, the purpose of which was to check the design assumptions concerning the Lias clay, which showed anistropy of shear strength. It was some 70 metres high with a very steep upstream face (Fig 9) and it was instrumented to measure pore pressures and displacements. Rate of construction was fast up to 60 metres height and then very slow, 250 mm per day, for the remaining 10 metres. Significant deformations did not occur, and the performance of the trial embankment

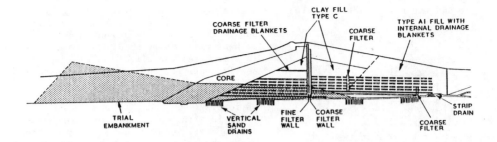

Fig 9. Cross section of Empingham Dam.

gave the designers confidence in their assumptions.

A detailed investigation was also undertaken along the route of the 14 km long water supply tunnel (Dumbleton and Toombs, 1978).

8.3 Dams on pervious foundations

For a dam built on a granular soil the problem is normally one of seepage rather than stability, unless it is located in a seismic zone. In practice, the ground conditions are usually complex. In river alluvium for example, permeable deposits (sometimes highly permeable "open work" gravels) will often be interbedded, or interdigitated, with less permeable soils and relatively impermeable soils. The same is true of glacial soils, where there is sometimes the added complication of large boulders. In these mixed conditions, cable-tool boring rather than rotary drilling is probably the best method of determining the ground and groundwater conditions in the superficial deposits. During boring the hole should be permanently filled with water to prevent piping, and shelling should be done carefully using an undersized shell to minimise suction. Standard Penetration Tests or static cone penetration tests are generally done in cohesionless soils; undisturbed sampling is extremely difficult, although where coarse particles are absent "air-bubble" samplers can be used and also the Delft continuous sampler. Thin-walled piston samples can be taken in the cohesive layers. Water levels should be measured at the start and finish of each working shift, and water inflows observed during pauses in boring. Piezometers should be installed in selected layers.

Measures to control seepage may include downstream relief wells, in which case it will be necessary to analyse the groundwater to determine its potential effect on well screens and filters from the point of view of corrosion and of the formation of deposits. Water analysis will also be required for diaphragm cut-off walls and grouted cut-offs, and where soluble rocks are present.

An indication of permeability can be obtained from grading tests on samples taken from the shell, although the results should be regarded with caution. The contents of the shell should be tipped into a large container and allowed to settle out before decanting the water and bagging the sample. Some measure of permeability can also be obtained from falling-head and rising-head tests in boreholes or though piezometers in finer grained materials (say permeability less than $10^{-6}$ m/sec) but permeability is best determined from pumping tests.

Where there are no gravel layers, rotary core drilling can be used as an alternative to cable-tool boring. Even fine sand can be successfully cored using retractable-type triple-tube core barrels with a plastic inner tube. There is frequently an advantage in using a combination of techniques to suit complex conditions. One approach is to 'overlap' the techniques in adjacent boreholes.

8.3.1 Backwater Dam

Backwater dam in Scotland, is another example of a dam built on variable deposits, in this case glacial, extending to some 50 m below valley bottom, in which underseepage was controlled by a grouted cut-off, (Fig 10). The earthfill embankment is 550 m long and 43 m high, and it has a central core, (Geddes and Pradoura, 1967).

Bedrock is overlain by some 9 m of glacial lake deposits consisting of finely laminated silts and sands. This is overlain by some 24 m of lenticular beds of sand and gravel deposited by glacial melt water which was denoted as the "sand and gravel complex". Uppermost was a thick layer of relatively impermeable mountain

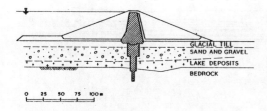

Fig 10.  Cross section of Backwater Dam.

till.  Artesian groundwater was found at lower levels.  Boring was done by cable-tool methods, starting in large diameter casings to facilitate penetration through the bouldery glacial deposits, followed by rotary coring in the bedrock.  A pumping test was carried out in the sand and gravel complex, over a period of seven days, within a 450 mm diameter borehole.  Pumping at a rate of 1.5 litres/sec produced negligible drawdown.

To determine the effectiveness of a grouted cut-off, trial grouting was undertaken around a shaft, with three rows of grout holes using a combination of clay-cement grout, bentonite-phosphate-silicate grout and silicate-aluminate grout.  The trial showed a permeability after treatment of $5 \times 10^{-7}$ m/s compared with from $5 \times 10^{-4}$ to $1 \times 10^{-6}$ for the untreated ground.

### 8.3.2 Derwent Dam

Derwent Dam in the north-east of England is on glacial deposits extending some 55 metres below valley bottom, (Ruffle 1965).  A first investigation of the site had suggested a profile consisting of glacial deposits, granular for the most part, with some silt and clay (Fig 11, Profile A).  This was a preliminary investigation and it would appear that the methods adopted were not entirely suited to the ground conditions, particularly to the high artesian water pressures.  Rotary drilling was used in a fairly small diameter and it is believed that most of the logging was based on washwater returns.  On the basis of this investigation it was proposed to construct a conventional cut-off in trench to bedrock under the protection of a deep-well dewatering system.  To obtain data for this installation a number of trial wells were put down.  These showed that ground conditions were significantly different from those indicated by the preliminary investigation.  Further investigation was then carried out, including holes put down from elevated platforms to overcome the problem of artesian water pressures.  The profile which then emerged (Fig 11, Profile B), was of a blanket of boulder clay, over the whole dam area, overlying the bedrock and forming an impermeable barrier between artesian pressure in the bedrock and the permeable glacial deposits immediately overlying the boulder clay.  These were again overlain by lake deposits and boulder clay of low permeability.  Furthermore, it was shown that observation wells sealed into bedrock showed little response to pumping from overlying aquifers and vice versa.  The further investigation also showed that the upper lake deposits and upper boulder clay extended across the whole valley profile and extended for a considerable distance upstream and downstream.  This made it possible to redesign the dam with a partial cut-off, in the form of an upstream core trench, of maximum depth 18 metres, taken some three metres into the boulder clay overlying the upper lake deposit, (Fig 12).  Downstream relief wells were installed into the upper aquifer to reduce uplift pressures at the downstream toe.

### 8.4  Dams on Rock

The requirements of a geotechnical investigation for a dam on rock will depend on whether a "fill" dam or a concrete dam is contemplated, and since in some cases the choice may depend on the findings of the investigation, it may be necessary to cover both options.  In either case however, a full understanding of the geology is paramount.

For a concrete dam consideration of the stability of the foundation will require a careful examination of the nature and orientation of discontinuities in the rock.  Rotary coring will assist in this, but its limitations must be recognised; it will only represent a minute fraction of the rock mass.  To appreciate the general rock structure and its fabric it will be necessary to examine it in natural exposures and in trial pits and adits.  Adits can sometimes be so placed as to be useful later as grouting or drainage galleries.

Strength along discontinuities is best determined by in situ testing, although laboratory testing can sometimes be useful.  It will also be necessary to determine deformation modulus; laboratory testing will be useful provided due allowance is made for the influence on deformation modulus of discontinuities within the rock mass which are not represented in the test specimen, (Hobbs, 1974).  However, laboratory testing is best supplemented by in situ testing, either with borehole dilatometers

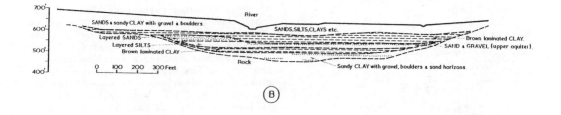

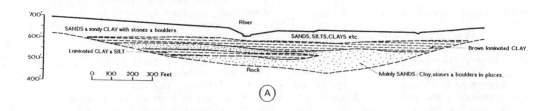

Fig 11. Derwent Dam profiles: A, from preliminary investigation; B, from later investigation.

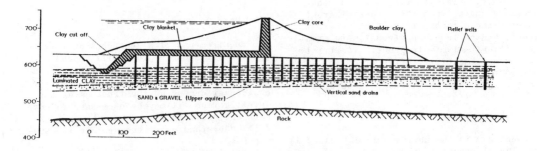

Fig 12. Cross section of Derwent Dam.

(Goodman et al, 1968), plate loading tests in pits or jacking tests in adits. The loaded area in an in situ test should be scaled to the spacing of discontinuities in the rock. It is important that excavation of pits or adits is done carefully, with hand excavation of rock in the vicinity of the test area. Inevitably there will be some disturbance at or near the surface of the rock to be tested, and it is therefore desirable to instrument the rock to measure strains within its mass. Large-scale in situ tests, such as pressure chamber tests (Monahan & Sibley, 1965) although expensive are extremely valuable.

A measure of rock deformability can be obtained from borehole seismic shooting, in which shear waves generated in one borehole are picked up in another borehole or on the surface, or by down-hole logging. Such measurements are best calibrated against more direct measurements such as plate loading tests.

Shafts should be put down into faults to determine the nature of the gouge material. Location of faults is assisted by inclined drilling.

Concrete dams on rock are often built in steep-sided valleys, and stability and strength of the abutments is an important consideration, particularly if an arch dam is to be constructed. As for the foundation, the nature and orientation of discontinuities must be established by the examination of natural exposures and of exposures in pits and adits.

To evaluate seepage flows and pressures below and around a concrete dam on rock requires permeability measurements. The usual method is to make water inflow tests

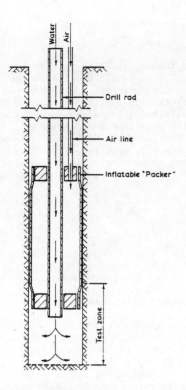

Fig 13. Drillhole packer for water inflow test.

within drillholes over lengths isolated by packers, (Fig 13). Both single and double packers are used. Double packers are sometimes unreliable due to undetected leakage under the bottom packer. Single packer testing is more expensive as it must always be done during drilling, whereas double packer tests may be done in a continuous operation after drilling. However, the single packer test is to be preferred since it is more reliable especially if a water-level dipmeter is used above the packer to monitor leakage. Considerable scatter of results frequently occurs since flow is determined by the incidence of joints and fissures within a given test length. Evaluation requires a detailed appreciation of the geology and considerable judgement.

For earth and rockfill dams on rock the seepage problem may be less critical since seepage paths will be very much longer, but is is still important, requiring careful attention to design details, and continuous inspection during construction. Stability of the foundation is often not in question, so joint strength will be relatively unimportant. Settlement of the foundation may still be a consideration particularly where an upstream impermeable facing is to be used.

### 8.4.1 Victoria Dam

Victoria Dam, in the central highlands of Sri Lanka, which will form part of the Mahaweli development, is an example of a fairly comprehensive investigation for a dam founded on rock. The double - curvature concrete arch dam will be some 110 m high, founded on rocks which are predominantly quartz biotite gneiss, together with quartzite, granulite and metamorphosed limestone. Depth of weathering is variable.

In the main investigation carried out between October 1978 and July 1979, under the direction of Sir Alexander Gibb & Partners, thirty-six boreholes were put down, including seven in the power station area, to a maximum depth of 260 m, totalling about 4000 m. This included some holes inclined at 30° and 45° from the vertical. Azimuth of the holes was monitored with an Eastman 'one-shot' device. Drilling was 'water-flush' with cutting agents, using Craelius thin-walled core barrels. Initially it was envisaged that the cutting bits would have fine set diamonds or natural impregnated diamonds, but after trials it was found that cutting speed was significantly increased by using copper-infiltrate or occasionally nickel-infiltrate matrix bits. This was important as the investigation was done under a very tight time schedule. To obtain maximum recovery in the upper weathered horizons, plastic inner liners were used inside modified SF and PF double-tube core barrels, with face-discharge surface-set bits. In areas of low water supply, along the tunnel line, when drilling in fractured limestone, polymer drilling additives were used.

Permeability testing within the boreholes was done with mechanically operated single packers, of about 0.6 m fixed length (Fig 13). Good seals were obtained in the metamorphic and igneous rocks; less satisfactory seals were obtained in the weathered rocks.

All cores were described, photographed and point-load tested on site. The investigation also included the installation of piezometers and inclinometers to monitor valley movement. Jacking tests were done in adits to determine the deformation modulus of the rock.

### 8.4.2 Calder Dam

Calder Dam to be constructed in Scotland,

is an example of a dam to be founded on volcanic lava flows, the main feature of which is weathering in the upper part of each flow. Such conditions are to be found worldwide, in South America, in Australasia, and in Asia, for example in southern Thailand.

It will be a rockfill dam some 63 m high, with a central asphaltic core. A concrete inspection gallery will run below the core, with a grout curtain below it, and an access gallery will run along the valley bottom.

Site investigation was done under the direction of Babtie Shaw and Morton, Consulting Engineers to the Strathclyde Regional Council Water Department. Drilling with 101 mm and 131 mm core barrels revealed a complex sequence of basaltic lavas belonging to the Calciferous Sandstone Series of Lower Carboniferous age, under a thin drift cover which will be removed when the dam is constructed. Similar rocks occur in the rockfill quarry area. Surface geological mapping and aerial-photo interpretation have shown a number of faults crossing the dam site.

The basalts are of two kinds, an olivine basalt of the Markle type and a trachytic basalt known as Mugearite. There has been autobrecciation and hydrothermal alteration; contemporaneous weathering can be seen at the upper surfaces of the lava flows.

Two aspects of the investigation will be commented on here, estimation of probable settlement under the dam and some properties of the rockfill.

To get a first approximation of settlement, values of the deformation modulus, E, of the intact rock were obtained by two approaches. Firstly by measurement of E in laboratory compression tests, uniaxial and triaxial, on intact specimens. Secondly from point load test results, using the relationship

$$E = 20\ I_s\ (MR).\ (CF) \quad \ldots\ldots (i)$$

where $I_s$ is the point load strength

MR is the modular ratio, $E/q_u$

$q_u$ is the unconfined compression strength

and CF is a confinement factor

MR and CI were also obtained from the laboratory compression tests. The ratio between $I_s$ and $q_u$ was determined from a number of Brazil tensile tests.

Having obtained values for the intact rock modulus it was necessary to derive values for the rock mass modulus, taking account of the discontinuities in the rock mass, using the relationship,

$$E_{mass} = j_c \cdot E \quad \ldots\ldots (ii)$$

The mass factor, j, which is the ratio of mass modulus to intact modulus varies with fracture frequency. Fig 14 shows such variation in a strong rock. (The values attributed to Deere, curve 1, were originally presented in terms of Rock Quality Designation; RQD). The corrected mass factor, $j_c$, used in equation (ii) is derived using a relationship established by Hobbs (1974):-

$$j_c = \left[1 + \frac{E}{K.f}\right]^{-1} \quad \ldots\ldots (iii)$$

Where $K = E_j/t$

E is the modulus of the intact rock

K is the joint stiffness

f is the joint spacing

$E_j$ is the joint modulus

and t is the joint thickness

This is a purely geometrical relationship, and therefore requires some assumptions about the effects of weathering on the thickness and stiffness of joints in the rock.

It was assumed that for slightly to moderately weathered rocks the width of the joints would be increased by 25% and the joint modulus reduced by 25% as a result of weathering with the intact rock remaining unchanged. For moderately to highly weathered rocks it was assumed that the joint modulus and the modulus of the intact rock would be reduced by the same factor and weathering would cause the joints to close by 50%. For highly to completely weathered rocks it was assumed that the joint modulus and the modulus of the intact rock would be reduced by the same factor and that weathering would cause the joints to close by 95%.

Fig 15 shows the relationships derived between the rock mass factor for fresh rocks, ie j (uncorrected), and rock mass factor taking into account the weathered state, ie j (corrected).

The settlement calculation involved the derivation of modulus profiles from a consideration of rock mass modulus values as described above, the division of the foundation area into a number of zones bounded by fault lines, to each of which is assigned the modulus profile corresponding to the borehole in that area, and the

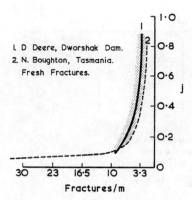

Fig 14. Variation of rock mass factor with fracture frequency, for strong rock.

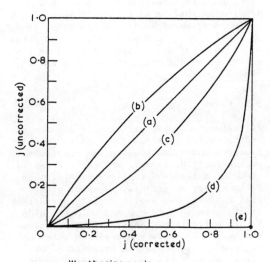

Fig 15. Calder Dam; corrected values of rock mass factor against uncorrected values.

use of a computer programme for the calculation and summation of strains. The program is based on the assumption that the stress distribution is independent of the modulus profile, and assumes an homogeneous elastic half space. Fig 16 shows a profile of calculated settlement at the rock surface under the dam crest line. Settlements were also calculated for the case of inclusion of the galleries.

The results can only be considered as representative of the order of settlement due to the assumptions made in the analysis together with such factors as the presence of a weathered margin to faults and of layers outcropping in the formation to the galleries which will influence the magnitude of settlements. However, viewed together with the conditions found in the excavations they will facilitate the location of movement joints in the galleries.

A further aspect of the investigation for Calder dam involved the provision of information on rock conditions in the proposed quarry area, and laboratory testing to determine the suitability and properties of the rock for use as rockfill. In addition to drilling in the quarry area, trial excavations were made to locate zones of weathered and altered rock for laboratory testing.

The slake durability test (Franklin and Chandra, 1972) was selected as that most likely to represent the susceptibility of the rocks to breakdown. As an extension of the slake tests, simple soaking tests were carried out on material which gave low slake durability values in order to assess the susceptibility to breakdown from contact with water, but without mechanical disturbance. Freeze-thaw tests were also made. The slake durability index values are plotted against saturation moisture content in Fig 17.

It can be seen that the rock described as unaltered and unweathered had saturation moisture contents below 6% and slake durability indices greater than 90%, and with one exception those described as altered also had indices above 90%, but with saturation moisture contents up to 10%. At the other end of the scale, the two specimens which gave very low slake durability indices would have been recognisable from visual examination in the field as potentially behaving as a soil. Of particular interest are the materials described as both altered and weathered, in that although many of these had high slake durability values, a few fell in the range of low to medium durability, and yet there was nothing in their appearance to indicate this different behaviour. Petrographic

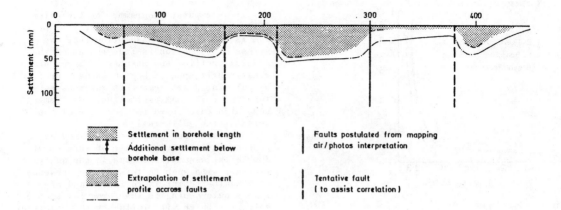

Fig 16. Calder Dam; calculated settlement profile along line of crest.

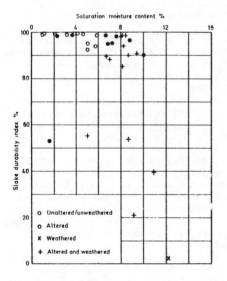

Fig 17. Calder Dam; slake durability index against saturation moisture content.

examination revealed no clear boundary between specimens of high durability and those of medium and low durability. However from X ray diffraction analysis, the specimens which gave low to medium durability were seen to be characterised by the presence of chlorite and haematite with open microfissures. It would appear that the presence of chlorite or micaceous alteration products, which were also present in the specimens exhibiting high durability, is necessary to facilitate the mechanical breakdown which may be initiated by disintegration along lines of weakness formed by the microfissures.

Space does not permit a discussion of the engineering implications of these results, nor the description of the other laboratory testing of rockfill, which included triaxial tests and oedometer tests at small scale, in Rowe cells, and at large scale in a one metre diameter floating ring oedometer at the Building Research Establishment.

## 8.5 Karst

In carbonate rocks, principally limestone and dolomitic limestone, extensive solution features may exist, making dam construction particularly difficult, and it is essential that the extent and configuration of such features are determined. A study of structural features may assist in interpretation of the karstic landform and help to guide the investigation. Shallow features usually have some surface expression which can be seen by stereographic examination of aerial photographs. Geophysical methods (resistivity, magnetic and more particularly gravity) are sometimes useful but their effectiveness diminishes with increasing depth of the solution features. Seismic survey methods have also been used, either surface or downhole, but with limited success. Undoubtedly the most positive method is a combination of adits and drilling at close centre. For economy the majority of the holes can be percussion drilled, the presence of an open or unfilled cavity being detected by increased rate of drilling. This should be supplemented by downhole television inspection, and by internal mapping of the features where they are large enough to permit access.

In many cases adits will serve to give access both for investigation and for ground treatment. At the Khao Laem dam now under construction in Thailand for example, (Anon 1980), cavernous limestone in the right abutment was investigated initially by means of drillholes from the surface and subsequently by driving six small diameter tunnels, total length 3.5 km, spaced about 14 m apart vertically. All limestone caves and cavities intersected by the tunnels and drillholes will be connected and sealed, and the intervening rock will be drilled and grouted.

## 8.6 Soluble Rocks

Even where karstic conditions have not developed there will be situations in which soluble rocks (gypsum, anhydride, calcium carbonate and halite) are present and where the increased hydraulic gradients created by a reservoir might lead to removable of material by solution with consequent settlements and increased water flows.

An important factor in determining solution is the frequency and size of fissures (James and Kirkpatrick, 1980). These parameters can be estimated from sets of borehole water-test data (Snow, 1969), but there are uncertainties involved, and it is therefore necessary to study the frequency of fissures in core from drillholes. This requires high quality coring in a sufficiently large diameter to give a reasonable certainty of 100 per cent core recovery. In good core some indication of fissure size may also be possible.

It should be noted that in practice high permeabilities are often underestimated in borehole water tests and artifical fractures in drill core may be mistaken for water-carrying fissures. Both of these errors lead to an underestimate of effective fissure width. It may be desirable therefore, although not essential in certain cases to examine the soluble material and the size and spacing of fissures in situ by means of shafts or adits, and if natural exposures are present these should be examined. It will also be necessary to study the volumetric proportion of soluble material and its distribution, and this again is most easily done in shafts or adits.

## 9 MATERIALS FOR CONSTRUCTION

Methods of investigation and testing materials for earth-fill are too well-known to require discussion here. Up-to-date information can be obtained from the Proceedings of the Conference on Clay Fills held in London in November 1978. However, there is one point to be made concerning the exploration of earth-fill borrow areas. This is frequently done by soft-ground boring methods (augering or cable-tool boring). Inspection in machine-dug pits or trenches can save much time, can be cheaper, and can give the engineer a better appreciation of the lateral variability of the deposits.

Difficulties can arise in assigning characteristics to fill soils on the basis of index properties if unusual materials are present. Where unusual soil properties are suspected, recourse can usefully be made to scanning electron microscopy and X ray diffraction analysis to examine the minerals present and their arrangement.

Investigation of rockfill sources may be straightforward where adequate sound rock is available, or complex in cases such as Calder Dam, as described above. Testing procedures of rock for concrete aggregate are wellknown, but the possibility of alkali-aggregate reaction should not be overlooked.

## 9.1 Val-de-la-Mare Dam

This 23 m high, 168 m long concrete gravity dam was completed in Jersey in 1962 on Precambrian sedimentary shales, (Coombes et al, 1975). In 1971 small upstream movements of the handrail of the crest bridge were noticed, darkening and damp patches were seen on the downstream face of the dam and parts of the surface showed random cracking of the concrete. Alkali-aggregate reactivity was eventually diagnosed as the cause of the defects, notwithstanding that the quarry from which the aggregate had been obtained had been in operation for several years and that there had been no published record of an occurence on Jersey.

In addition to a review of the sequence of construction and of the sources of the materials, and testing of local samples of aggregates and beach sands, the programme of investigation included in situ sonic measurements, core drilling in specific sections of the dam, extension testing on cores from the dam and on made up specimens and the installation of electrically-operated piezometers in certain sections of the dam.

The degree of concrete deterioration varied throughout the dam. This was borne out by the sonic test results and by visual inspection, and was consistent with the known variation of alkali content of the cement delivered and the fact that most of the reactive material was thought

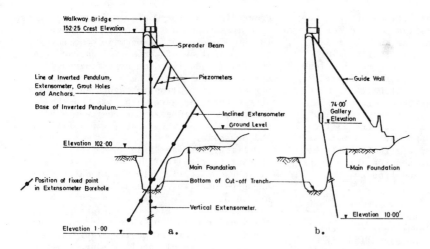

Fig 18. Val-de-la-Mare Dam; remedial measures.

    a. Anchors, grout holes and instrumentation

    b. Drainage holes

to be derived from a number of veins in the quarry and from beach pebbles containing either chalcedony or opal.

Remedial measures have been confined to one block which was the worst affected, and which was the only one where it was found that the original design uplift pressures were being exceeded. Following additional exploratory drilling below this block, remedial measures consisted of anchoring the block to the foundation rock, the installation of monitoring instrumentation, grouting using an inert polythixion grout, and the drilling of drainage relief holes in the block and a number of adjacent blocks. The instrumentation included vibratory wire load cells at the head of each anchor, vertical and inclined extensometers, electro-levels, and an inverted pendulum, electrically-operated piezometers and superficial mechanical strain gauges. (Fig 18).

## 10  THE RESERVOIR

Stability of the reservoir slopes is an aspect which cannot be neglected. Large scale failures such as that which occurred at Vaiont (Kiersch, 1964) can have catastrophic results. They can be triggered by saturation on submergence, by rapid drawdown of reservoir level, or by seismic activity. Equally important is the assessment of possible leakage from the reservoir. This requires an overall comprehension of the geology, based on existing records, mapping and aerial photography, supplemented if necessary by exploratory drilling and water pressure measurements in various locations and horizons to determine existing flow patterns.

### 10.1  Kastraki Lake

At the Kastraki dam site in Western Greece the dam is an 87 m high gravel shell embankment with a clay core. The rocks are Ionion Zone Flysch, consisting of inter-bedded conglomerates, sandstones, siltstones and mudstones with well-developed bedding. After completion of the dam doubts were raised about the stability of the right bank. Bedding planes were considered to be the principal discontinuities and in situ shear tests were done on bedding planes containing argillaceous materials, and on suspected tectonic shear horizons.

Test blocks were one metre square by 400 mm deep. The test set-up (Fig 19) used two 200 ton jacks for shearing. Normal load was applied uniformly through a pair of flat jacks using a pair of hydraulic rams each of 200 tons capacity. Electrically operated pumps were used to maintain constant normal stress and to operate the shear rams. A roller system was incorporated to ensure minimum resistance to shear displacement. The test chambers were hand excavated, without blasting, and during the final stages of excavation and preparation, specimens were protected and presented from relaxing by a system of rock bolts, beams and screw

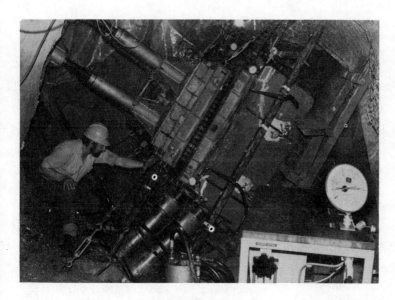

Fig 19. Kastraki Lake; set up for in situ shear test.

props.

## 11 SEISMICITY

The recent conference held in London on "Design of dams to resist earthquakes" has provided an up-to-date coverage of this topic, and investigation requirements are referred to in a number of the papers.

The seismicity of the region may already have been rated and zoned. Nevertheless for an important dam it will be desirable to make a full study based on earthquake records and the regional geology.

Must of the geological data required will be acquired during the normal foundation investigation. However, it will be necessary to pay particular attention to the identification of active faults, and this will need geological data over a wider area than for a non-seismic site.

The first step in the evaluation of the seismic risk will be to identify the main geological structures on a regional scale, in particular the major faults and to decide which faults have been active in recent geological times. At the same time the historical record of earthquakes in the region will be assembled, and an attempt made to match the apparent epicentres to the geological features. This will be followed, if the data is adequate, by the construction of a probabalistic or semi-probabalistic model from which, in combination with the attentuation law appropriate to the area, an estimate is made of the risk of exceeding a given intensity or given acceleration at the dam site.

Where seismic records are not available, it will be necessary to install monitoring devices in the locality to record small magnitude events. Seismograph networks will as far as possible be located so as to examine the seismic behaviour of the fault systems which have been identified.

For a full assessment of earthquake effect on a dam and its foundation, it is necessary to measure dynamic moduli, damping factors and liquefaction potential. The dynamic parameters of the foundation materials are best established by a combination of seismic velocity studies on site (crosshole-, downhole-, uphole-shooting, borehole logging), laboratory cyclic tests and semi-empirical procedures (Anderson, 1978).

Where liquefaction potential is an issue, valuable insight may be gained by careful geological and geomorphological studies coupled with a historical review of the occurrence of features indicating liquefaction of the foundation soils. (Youd & Perkins, 1978).

Much has been published about cyclic shear tests in the laboratory and their use in the evaluation of liquefaction potential. It must be said however that sampling difficulties in cohesionless material make such testing difficult, unless reconstituted specimens are used for testing, in which case there must be some doubt about the relevance of the test results. Furthermore there is a dearth of correlation between laboratory test predictions and field performance. The available correlations are between field performance and Standard Penetration Test values, and the SPT is therefore a first requirement where the liquefaction potential of granular soils is under investigation. Results of this test are however easily invalidated by poor site procedures, and it is considered highly desirable to carry out static sounding tests as well as SPT's so that future correlations can be with this more stable parameter.

Reliable and accurate piezometric levels are needed for study of the potential build-up in porewater pressure under earthquake loading. Artesian and sub-artesian pressures in stratified deposits with alternating aquifers and aquitards may be important in the analysis of foundation stability.

Blasting trials can also be considered for the study of liquefaction potential (Ruxton, 1980).

Attempts have been made to study the stability of a rockfill dam during earthquakes by testing a simulated scale model of small size material on a shaking table (Nose & Baba, 1980). There must be limitations in this procedure however, arising from lack of similitude between model and prototype.

For the geotechnical engineer who is not a specialist in seismicity, the literature on this topic can be confusing and overwhelming, and he is not always certain that a very detailed evaluation is justified in his particular project. It may be helpful therefore to repeat some general conclusions put forward by Seed et al (1978), which may guide him towards the appropriate investigation procedures:

1. Virtually any well-built dam on a firm foundation can withstand moderate earthquake shaking, say with a peak acceleration of about 0.2g from earthquakes with magnitudes up to about 7, with no detrimental effects.

2. Dams constructed of clay soils or rock foundations have withstood extremely strong shaking ranging from 0.35g to 0.8g from a magnitude 8-1/4 earthquake with no apparent damage.

3. For dams constructed of or on loose or medium dense saturated cohesionless soils and subjected to strong shaking, a primary cause of damage or failure is the build-up of porewater pressures in the cohesionless soil and the possible loss of most of its strength which may result from this pore pressure increase. It is not possible to predict this type of failure by pseudo-static analyses and dynamic analysis procedures are required to provide a more reliable basis for evaluating field performance for such cases.

ACKNOWLEDGEMENT

This paper was prepared while the author was Managing Director of Soil Mechanics Ltd and is based on his experience with that company. The permission of the Dirtors of Soil Mechanics Limited to publish the paper is gratefully acknowledged.

REFERENCES

Anderson D G, C Espania and V L McLamore 1978. Estimating in situ shear moduli at competent sites. Proc. ASCE speciality conference. Earthquake engineering and soil dynamics, 181-197.

Anon 1980. Khao Laem multi-purpose project. Information brochure. Electricity Generating Authority of Thailand; Snowy Mountains Engineering Corporation.

Bertranou R L, A A Pacma and J L Gonzales 1976. Essais methodologiques pour evaluation rapide de lieux d'emplacement des barrages. ICOLD 12 Q46 R45, 3, 699-708.

Bjerrum L 1972. Peformance of earth and earth supported structures. Proc Speciality Conf, Purdue Univ Lafayette, Indiana, ASCE.

BS 5930: 1981. Code of Practice for Site Investigations. British Standards Institution, London.

Coombes L H, R G Cole and R M Clarke 1975. Remedial measures to Val-de-la-Mare Dam, Jersey, Channel Isles, following alkali aggregate reactivity. BNCOLD Proc of Symp held at Newcastle upon Tyne on Inspection, Operation and Improvement of Existing Dams, Paper 3.3.

Conference on Clay Fills, 1978. Inst Civ Eng, London.

Dumbleton and Toombs 1978. Site investigation aspects of the Empingham Reservoir Tunnels. Transport and Road Research Laboratory, Report 845.

Early K R and A W Skempton 1972. Investigations of the landslide at Walton's Wood, Staffs. QJ Engng Geol Vol 5. 1 & 2, 19-41.

Franklin J A and R Chaudra 1972. The slake durability test. Int J Rock Mech Min Sci 9, 325-341.

Geddes W G N and H H M Pradoura 1967. Backwater Dam in the county of Angus, Scotland, grouted cut-off. Trans 9th Int Conf Large Dams, Istanbul. 1, 253-274.

Geological Society of London 1970. Work-Party report of the Engineering Group on the logging of rock cores for engineering purposes. QJ Eng Geol 3, 1.

Geological Society of London 1972. Workparty report of the Engineering Group on preparation of maps and plans for engineering purposes. QJ Eng Geol, 5, 4.

Geological Society of London 1977. Working party report of the Engineering Group on the description of rock masses for engineering purposes. QJ Eng Geol, 10.

Goodman R E, T K Van and P E Heuze 1968. The measurement of rock deformability in boreholes. Proc 10th Symp Rock Mechs, Texas, 19, 523-555.

Hobbs N B 1974. Factors affecting the prediction of settlement of structures on rock. Review paper, IV. Proc Conf on Settlement of Structures. BGS, Cambridge, Pentech Press, London.

James A A and I M Kirkpatrick 1980. Design and foundations of dams containing soluble rocks and soils. Q Jl. Engg Geol 13, 189-198.

Kiersch G A 1964. The Vaiont Reservoir disaster. Civ Eng, March.

Meigh A C and W Wolski 1981. Design parameters for weak rocks - Final General Report. Design parameters in geotechnical engineering. 7th Eur Conf Soil Mech Brighton, Sept 1979.

Monahan C J and E A Sibley 1965. Rock Mechanics for Dvorshak Dam, N. Idaho. Mtg Geol Soc American, Kansas City, Missouri.

Nose M and K Baba 1980. Dynamic behaviour of rockfill dams. Conf on Design of dams to resist earthquake.

Ruffle N J 1965. Derwent Reservoir. J Instn Wat Engrs, 19, 361-408.

Ruxton T D 1980. Liquefaction potential of of Wash sands. Conf on Design of dams to resist earthquake. Inst Civil Engrs. London.

Seed H B, Faiz I Makdisi and P DeAlba 1978. Performance of earth dams during earthquakes. J Geol Eng Div, ASCE, 104 GT7.

Snow D A F 1969. Rock fracture, spacings, openings, and porosities. Soil Mech Fdn Eng 73, 73-91.

Torstensson B A 1979. The pore pressure probe. Archiwum Hydrotechniki, Stockholm, 26, 2, 325-335.

Wulff J G and W Perry 1976. Efficient methods of site appraisal and determination of type of dam - a discussion of the basic philosophy and procedures in project planning and site selection. ICOLD 12, Q46, R8, 3, 135-151.

Youd T L and D M Perkins 1978. Mapping liquefaction induced ground failure potential. J Geotech. Engng Div, ASCE, 104, GT4. Proc paper 13659, 433-446.

*Symposium on Problems and Practice of Dam Engineering / Bangkok / 1-15 December 1980*

# Instrumentation requirements for earth and rockfill dams

A.D.M.PENMAN
*Building Research Station, Garston, Watford, UK*

## CONTENTS

1 INTRODUCTION
  1.1 Marte R Gomez dam
  1.2 Sasumua dam

2 MEASUREMENT OF LEAKAGE
  2.1 Svartevann dam
  2.2 Teton dam site
  2.3 Embankment construction
  2.4 Signs of failure
  2.5 Cause of failure
  2.6 Material removed by seepage

3 SETTLEMENT MEASUREMENTS
  3.1 Overtopping risk

4 DIFFERENTIAL SETTLEMENTS
  4.1 Duncan dam
  4.2 Vertical settlement gauge
  4.3 Water overflow settlement gauge

5 HORIZONTAL MOVEMENTS
  5.1 The horizontal plate gauge

6 SURVEYING METHODS
  6.1 Reference monuments
  6.2 Triangulation and trilateration

7 PRESSURE MEASUREMENTS
  7.1 Pore pressures
  7.2 Two tube hydraulic piezometer
  7.3 Special tests
  7.4 Other types of piezometer
  7.5 Total pressres

8 EARTHQUAKE

9 OLD DAMS

10 DAM SAFETY

11 REFERENCES

## SUMMARY

The choice of type and the positions for instruments in embankment dams must ultimately be the responsibility of the engineer who designs and supervises the construction of the dam.

The accurate assessment of seepage flow and its dissolved mineral and solids contents is of paramount importance and derives from continuous measurement with some of the simplest of the instrumentation used to monitor the performance of embankment dams.

Strains within the body of the fill due to differential movements can cause large redistributions of stress and lead to cracking or hydraulic fracture of an earth fill core. Predictions can be made by analyses using finite element techniques and movements measured with water overflow settlement gauges, horizontal plate gauges etc. Overall horizontal and vertical movements can be accurately measured from stable reference monuments by triangulation and/or trilateration. Reference monuments should be established prior to dam construction and can be very useful for setting out.

Stresses within an impervious core may be measured with total pressure cells and piezometers. Fine-pored intake filters enable the initial pore water suctions in partly saturated fill to be measured. Subsequent changes indicate total stress conditions as well as rates of drainage. It can be argued that pore pressures at the end of construction should exceed the reservoir water pressures that will act on the core during first filling as an indication of sufficient total stress to avoid hydraulic fracture.

In dams not fitted with total pressure cells an assessment of existing total stresses can be made with hydraulic piezometers by carrying out critical pressure tests. In very old dams, both pore pressures and total stresses can be measured by installing piezometers and spade-shaped total pressure cells through boreholes. Such a permanent installation will record future changes, but current pressures can also be measured with self-boring instruments.

The magnitude of strains developed in a dam by earthquake may depend on the natural frequency of vibration of the dam. This can be measured by use of artificial exciters attached to the dam - an approach which may also reveal in-situ properties of fill and foundation material.

Dam safety is of fundamental importance and can be assured when the behaviour of the dam is fully understood and is constantly monitored by suitable instrumentation.

# 1 INTRODUCTION

The choice of type and the positions for instruments in embankment dams must ultimately be the responsibility of the engineer who designs and supervises the construction of the dam. He knows the difficulties of the particular site for the dam and those likely to arise through the use of the particular fill associated with the site.

## 1.1 Marte R Gomez dam

As an example, the 49 m high Marte R Gomez dam, built in Mexico in 1946, was founded partly on a terrace of compressible sand and silt with void ratios in the range 0.9 to 0.6. This terrace extended for a considerable distance on the left of the river and it was clearly not economic to excavate it all under the foundation area for the dam. In order to cause some preconsolidation, the terrace was flooded and during dam construction settlements were not unusual. During first filling of the reservoir, when the water came over the terrace, a sudden collapse occurred and caused a settlement of 1 m at the dam crest for a length of 1 km. The differential movement between the parts of the dam to the left and right of the old river channel caused longitudinal and transverse fractures visible at the upstream face. Measured leakage increased to about 180 l/sec. This behaviour was not unexpected and after trial pits had shown that the transverse cracks disappeared at a depth of 8 m, all fractures were filled and sealed with a clay grout suspension at low pressure.

The designer expected that collapse settlement might occur when the loose terrace material was flooded and tried to minimise the settlement under the dam by pre-soaking. He was careful to measure settlement and seepage during first filling of the reservoir and in this way observed the collapse settlement when is occurred. After repair, settlement and seepage observations have continued and shown a satisfactory behaviour of the dam.

## 1.2 Sasumua dam

The borrow area at the site of Sasumua dam in Kenya provided a clay fill material with unusual characteristics. It was quite abnormal in its properties, so far as ordinary experience with earth dams at the time was concerned. As more laboratory tests were carried out, however, it became apparent that, from the engineering point of view, it was a very satisfactory material. Its behaviour has been described as being like that of an ordinary clay in all respects except that the water contents were consistently about 25 per cent higher than one would expect. It seemed probable that part of the water was in some way bound in or on the particles. Some potential contractors felt that the handling and placement of such material was a task they would prefer not to attempt.

Mineralogical analyses revealed that the clay, which was of volcanic origin, consisted predominantly of halloysite of a relatively low state of hydration with about 5 per cent each of gibbsite, geothite and rather coarse quartz. This caused it to have, as it were, two water contents; one in the voids between the particles, which controlled the behaviour of the clay, and the other within the particles themselves that played no part in controlling behaviour but, unfortunately, could not help but be measured as part of the water content of the clay.

Proctor optimum compaction tests were made, following the standard of the time, on oven-dried material and these gave values of optimum water content much lower than the value that would apply during construction with clay fresh from the borrow pits. Because of this, the material was dried as much as possible by harrowing and turning under the equatorial sun, until it was placed, in general, slightly above the laboratory optimum value, but in fact well below what would have been a true optimum for the natural clay not previously dried in a laboratory oven at $105^\circ C$.

Stability calculations based on pore pressure predictions for clay placed at or slightly above optimum water content gave factors of safety of 1.5 to 1.8. In order to check on the actual pore pressures developed, a large number of piezometers were placed in the fill during construction. These were to satisfy the dam designer that the assumed pore pressures were not exceeded and that the factor of safety against rotational slips would not fall below 1.5.

Because the material had, in fact, been placed at water contents well below the true optimum value, measured construction pore pressures were always small and in several positions, below atmospheric pressure. These measured values indicated a value for the factor of safety approaching 2.8.

These two example illustrate the importance of instrumentation in obsering the behaviour of embankment dams.

## 2 MEASUREMENT OF LEAKAGE

Knowledge of water seeping through a dam is essential in assessing its behaviour. First indications of dangerous deterioration are often given by an observed change of seepage rate. Equipment to measure seepage flows formed some of the earliest instrumentation installed at dam sites. Where under-drains or relief wells are connected to manholes, flows are often measured with Vee notch weirs or, when rates are low, by diverting the flow temporarily into containers of known volume. More sophisticated equipment includes remote reading level sensors to record water levels and hence flows over weirs and flow meters that can be either read directly or arranged with electrical connections so that they can be read remotely or recorded.

### 2.1 Svartevann dam

At Svartevann dam, constructed on granitic gneiss bedrock in Norway, a small bund was built across the valley floor, just downstream of the core as shown in Fig 1, to collect seepage water. The pond area between the bund and the core was divided into right and left halves by a central concrete wall, and pipes were laid in the foundation for the dam to lead the water to measuring chambers in the instrument house at the downstream toe of the dam. The chambers contained Vee notch weirs, as indicated by Fig 2, and the level of the water behind the weirs is measured by a float attached to an instrumented cantilever. Vibrating wire strain gauges respond to variations of the bouyant weight of the float and allow the chamber water level, ie seepage flow, to be recorded continuously at a remote station. Water and outside air temperatures are monitored as well.

A record of seepage measurements, given by DiBiagio et al (1982), is shown by Fig 3. The effects of rainfall, snowmelt, frozen chambers, etc, are clearly shown by the early part of the record. About 1050 days after the beginning of construction, while the reservoir was filling fairly rapidly, the recorded flow rate rapidly increased from 1.2 to nearly 3 l/s. At a later stage, during a period of only small change in reservoir level, the recorded flow dropped to about 1.4 l/s. As shown by the curve, Fig 3, it later increased rapidly to 3.1 l/s and then fell again to 1.5 l/s without corresponding changes of reservoir level. Total seepage was small for a dam of this size and these fluctuations were of no cause for concern, but they illustrate changes that commonly occur in measured seepage not caused by leakage through or past the core of a dam.

The measuring arrangements at Svartevann are unique in having a collecting area within the body of the dam so close to the core to reduce the ingress of surplus water entering the downstream fill of the dam from the valley sides. Usually measurements are made at the discharge of under-drains and relief wells that collect a great deal of water entering the site from the valley sides. Prior to construction of a dam, the phreatic surface in the ground on either side of the river often slopes gently towards the river and at some distance under the valley sides, is still not much above river level. After the reservoir has been impounded, however, the phreatic surface just upstream of the core slopes gently towards reservoir level, whereas that downstream of the dam is still at river level. It is impossible to have a sudden discontinuity in the phreatic surface of the ground going from upstream to downstream of the dam and just downstream of the core the phreatic level in the ground forming the valley side, maintained by the rainfall and snowmelt on adjacent land, will be close to reservoir level. At increasing distance from the core in a downstream direction, the phreatic surface can be expected to fall until it rejoins its old position near river level. This effect will cause an increase of water discharge from the valley sides into the downstream fill of the dam and has been known to cause landslips in the valley sides just downstream of a dam.

The bedrock at the formation level of

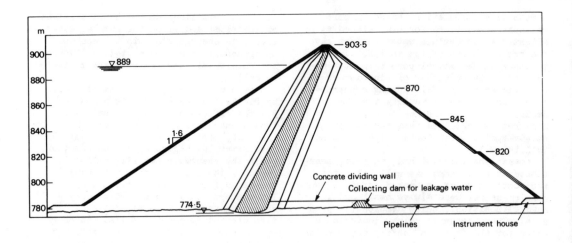

Fig 1   Cross-section of Svartevann dam showing seepage collecting barrier

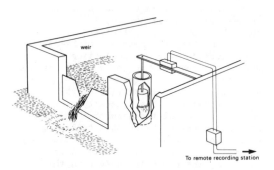

Fig 2   Equipment to measure leakage

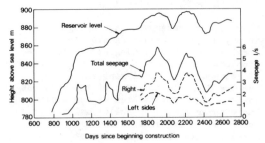

Fig 3   Record of seepage at Svartevann dam

Svartevann dam was relatively impervious, so that almost all the water just downstream of the core was collected for measurement. At other dam sites, however, the permeability of the ground may be so high that water leaking past the core may flow undetected below ground surface. This can lead to a false sense of security as exemplified at Teton.

2.2   Teton dam site

The site chosen for the dam was in a steep-walled canyon incised by the Teton River into the Rexburg Bench, a volcanic plateau draining into the Snake River Plain in the state of Idaho. Valleys and lakes formed during the volcanic period contained alluvium that was buried by later molten flows so that the rock mass contains large zones of sands and gravels. The whole depth of the canyon was in volcanic rocks and on the high lands flanking the canyon it was covered by aeolian sediments up to 15 m thick. The canyon was about 93 m deep to the river and it contained about 30 m depth of river alluvium below river level. Deep boreholes showed that there were lake and stream deposits below the valley, in the volcanic rock. The rhyolite has been described as a welded ash-flow tuff and the openness of its joint system can be judged by the high overall permeability of the rock. A well only about 2 km from the site produced 140 l/s with a drawdown of only 0.6 m. Well drillers in the area found that the rock yielded much more water than the included pockets of alluvium and they have been known to abandon a well when it was found that it was going into lake deposit.

The regional water table was about 54 m

below river level and the Teton loses water as it passes through the lower part of the gorge. Measurements of river flow showed losses of 0.7 to 1.4 m$^3$/s downstream of the dam site, although upstream the losses were negligible due to a perched water table. Drill holes made during the site investigation lost their drilling water and one NX size borehole in the right abutment had 30,000 m$^3$ of water injected during 15 days without filling it. The water level in other boreholes in the right abutment rose during injection and dropped abruptly when it was stopped. A borehole camera was used to examine some of the exploratory holes. It showed many cracks and joints of apparently random orientation: the widest crack measured was 43 mm - most were 2 to 12 mm wide.

A trial grouting programme was carried out on the left abutment during 1969. There were large grout takes in several holes and the take in just two of the exploratory drill holes exceeded the estimated quantities for the entire programme. Further boreholes were put down in 1970 to check the effectiveness of the grouting and after a further review of the site investigation it was decided that the rock forming the sides of the gorge was too open for efficient grouting and that it would be better to replace the upper part of the grout curtain by a key trench 21 m deep, to be filled with core material.

The rock in the valley floor below the alluvium and up the sides to a level of about 15 m above river level was satisfactory for grouting. The key trench was to be used above this level, up to dam crest level at the top of the gorge.

During trench excavation, large openings were uncovered. Two large open fissures about 25 m apart crossed the trench nearly 200 m from the spillway. One was about 1.2 m wide and was explored by a geologist who was able to get a distance of about 30 m upstream, downstream and downwards below the level of the trench floor. The second large fissure was not quite wide enough for a man to get into. Both were filled, in the zone around the trench, with high slump concrete. The section of the dam that failed was about 100 m to the left of the spillway.

The prevalence of large fissures is shown by the fact that there are several exposed in the right wall of the canyon in the reservoir area 200 to 400 m upstream of the dam. Also, subterranean caverns are often discovered by well drillers. At one well to the north-west of the right abutment, the drill bit dropped 2 m and at another the bailer was lost in a cavity. The hole had to be abandoned and a replacement started nearby. It too ran into the cavity, where the lost bailer was fished out!

The cause of these fissures and caverns may be in the cooling stresses in the rhyolite, combined perhaps with steam generated when the molten mass spread over the lakes and rivers trapping alluvial deposits such as those revealed by boreholes below the present canyon floor.

Whatever the reason, it is clear that the rock mass is well broken by cracks and fissures and as a whole, is quite pervious to water.

To limit the flow around the dam, the design called for the grout curtain to extend about 300 m into the right abutment and about 150 m into the left abutment. It was formed by a single row of holes at about 3 m spacing made through the 0.9 m wide by 0.9 m deep grout cap which extended continuously along the whole length of the dam. On the right abutment it was cast into a groove of the correct size that had been made in the key trench floor using careful pre-splitting methods to avoid damage to the in-situ rock. To contain the grout from the central holes, two outer rows of holes at 6 m spacing were drilled 3 m either side of the central row. Thick grout mixes, laced with liberal amounts of calcium chloride were put into these outer holes so that a more complete curtain could be formed from the central row of holes.

During the construction of the curtain below the right key trench, there were some grout leaks around some of the large rock blocks. While the men were caulking these leaks they saw numerous bats flying out of the cracks.

The total grouting programme entailed drilling 36 km of grout holes and injecting 17,000 m$^3$ of dry material at a contract cost of 3.8 million dollars. The actual quantities of cement injected were more than twice the bid quantities.

2.3 Embankment construction

Fill placement began on 25th April 1972. The main body of the embankment was constructed in four placing seasons extending from about May to the end of October of the years 1972 to 1975. The silt was placed in 15 cm layers, compacted by 12 passes of a 6 t/m sheepsfoot roller at a water content of 0.5 to 1.5 below optimum. This produced a very strong and brittle fill.

The surface of the bedrock was carefully prepared to take the core material, which was made slightly wetter (0.5 per cent below optimum) for the layer in contact with the rock. Preparation of the rock surface was carried out 1 to 3 m ahead of the advancing fill. Any soil remaining attached to the rock was removed by handwork and high pressure air jet, and the accumulated material taken away by rubber-tyred backhoe. The rock surface was watered just prior to fill placement. No metal roller or tracked vehicle was allowed to contact the rock surface and compaction was by hand-held power rammers or, if the slope was suitable, by rubber-tyred equipment. A minimum of eight passes were made by a loaded Euclid 74-TD end dumper or other approved machine forcing the core material into the wetted cracks in the rock abutment.

In the key trench, compacting conditions were more difficult. Surface preparation was carried out by labourers using hand shovels and crowbars to remove any loose rock or soil. An air jet was then used to clean any remaining finer material down to a clean rock condition. Any grout that had been spilled in the key trench was loosened by paving breaker and cleaned by air jet. Any open joints were filled with concrete dumped from mixing trucks. Sometimes funnels were formed with core material to guide the concrete into the cracks.

The specification laid down stringent conditions for the first layer of core material. Moisture was controlled by mixing wet and dry materials until the specified water content was obtained. It was compacted with petrol and compressed air tampers in $7\frac{1}{2}$ cm lifts in irregular areas and by rubber-tyred plant in 15 cm lifts where there was room. If laboratory tests revealed that the moisture limits had been exceeded the material was removed, reworked and then replaced. If the tests had shown that the specified density had not been achieved, the area was recompacted. In the length of key trench from 270 m on the right to 220 m on the left of the spillway, a total of 425 density tests were made. The difficulty of the placing conditions was reflected in the fact that 114 of these tests failed and required reworking of the material to bring it within the specification. Difficulty must also have been experienced due to variations of the optimum water content for the material which varied from 16 to 24 per cent. The mean for the material being placed in the key trench was 19.1 per cent and the aim was to place it at 0.6 per cent dry of optimum. The fact that so many areas were reworked shows the conscientious approach to this important task, even though the significance of the rock fissures was being overlooked.

This special contact layer of core material was specified to be at least 30 cms thick, or 60 cms measured horizontally from steep rock slopes. Once clear of this zone, the drier material was placed and compacted by the heavy machinery as has been described.

When the 2 m diameter tunnel for irrigation water was completed, the river was passed through it so that the main tunnel to the power station could be completed. Its limited size meant in effect that impounding started in October 1975. Its design discharge rate under full reservoir head was 24 m$^3$/s. It was the intention to raise the reservoir level at no more than 30 cms a day, but in March 1976 a request was made to head office for permission to use a higher rate and 60 cms/day was agreed to. Because the main tunnel could not be used, even this rate was too slow and it was exceeded for the whole period 11th May to 5th June 1976 when, with the reservoir level just 1 m below sill level the dam failed.

2.4 Signs of failure

While the reservoir was filling, water levels were monitored in a number of observation wells in the vicinity of the dam. They all responded to the rise of reservoir level, although they all showed water levels considerably lower than reservoir level. The rate of rise of water level in the wells increased as the reservoir level rose and during the 5 weeks prior to failure, it was increasing at a greater rate than the rise of reservoir water. This was attributed to water flowing more readily from the reservoir through dominant horizontal joints that exist in the upper part of the rock mass.

A particularly rapid rise occurred in a borehole about 75 m downstream of the grout curtain in the right abutment when the reservoir level reached 17 m below sill level, about 18 days before failure. Even though the level in the hole rose rapidly, it was still 31.8 m below reservoir level when the failure occurred.

No special provision was made for measuring leakage, but as the reservoir neared top water level, inspection was made to see if any springs developed, as might be expected through the general raising of the regional groundwater table due to the presence of the reservoir. The first seeps at Teton did not come until two days before the failure when the reservoir was only about 3 m below sill level. They were well away from the dam and the issuing water was clear. Their appearance had been expected and it was regarded as normal behaviour. Their late arrival no doubt reflects the very low initial regional water table and it is quite possible for some leakage to have been occurring through the dam at that time.

Next day the flow from the seeps did not appear to have increased and it was still clear water. Some further small seeps had developed, closer to the dam, but there was absolutely nothing to indicate impending failure. Only on the day of failure was water found coming from the dam itself and at 8 o'clock in the morning it was still a small discharge of clear water. It was probably not evident that failure would occur until about 10 o'clock and the dam was breached by 11.50. The small loss of life can be attributed to some very quick telephoning and the fact that the flood came during daylight.

2.5 Cause of failure

Investigations have been made by the Independent Panel and by a Review Group established by the Department of the Interior. Both are in agreement that the cause of failure was centred in the key trench of the right abutment. On the section where failure occured, the grout cap was about 30 m below reservoir water level and the hydraulic gradient across it, assuming that reservoir water pressure could reach its upstream side through the fissures, was about 33 : a value high enough to cause some trouble in the silt if there was the slightest imperfection in the core.

The total pressure in core material must exceed the pressure of the reservoir water to avoid hydraulic facture. The bulk density of the impervious type of fill used for cores is commonly more than twice the density of water, so that at any depth below dam crest, the overburden pressure can be expected to be considerably in excess of reservoir water pressure. Unfortunately, small differential settlements can cause a great deal of core weight to be transferred by the shear strength of the core, to the dam shoulders. This causes reduction of the vertical total stress in the core to values lower than the theoretical overburden pressure. In a direction along the dam axis the total stress (resisting the opening of vertical cracks through the core) is further reduced by the shear strength of the core material. It is therefore desirable to have a weak, flexible core rather than the strong, rigid material used at Teton, to avoid hydraulic fracture.

An analysis using finite element techniques to assess the total stress conditions in the core material in the key trench was commissioned by the Independent Panel. This analysis, made for several sections through the dam near the failed section (see Fig 4), showed convincingly that hydraulic fracture could have occurred through the core material in the key trench when the reservoir water reached the almost full level that it had at the time of failure.

This example of Teton site illustrates the difficulty of collecting leakage water that can occur with very permeable foundation strata. In these cases changes of groundwater levels have sometimes to be used to estimate flows. This aspect will be considered with the discussion of piezometers below.

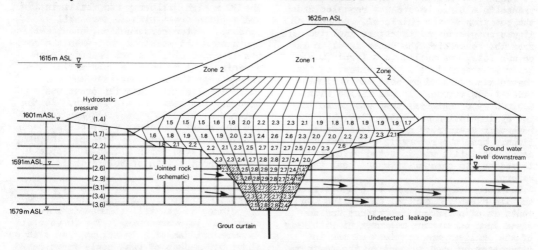

Fig 4 Analysed section of Teton dam. Figures show computed values of normal stress on transverse section and hydrostatic pressure of reservoir water $kN/m^2 \times 10^2$. Shaded zones where hydraulic fracturing could occur.

## 2.6 Material removed by seepage

From time to time the dissolved mineral and solids contents of seepage water should be measured so that estimates may be made of any loss of material occurring from the dam or its foundation. Examples have been quoted of measured concentrations of calcium in the seepage water passing through the foundation of a dam in a limestone region. The amount represented an increase from that of the reservoir water and could account for the removal of 450 tonnes of material a year.

Cracks in a clay core can allow the slow erosion of clay flocs, which may pass a traditonally designed filter and lead eventually to piping failure. Although design should provide a weak and flexible core material to avoid cracking, it is advisable to provide filters that will retain clay flocs. A design for suitable filters has recently been given by Vaughan and Soares (1981).

Rock may be dissolved and removed by groundwater flow over a long period, unconnected with dam construction or reservoir impoundment. At Keban dam in Turkey, large caverns were found in the limestone foundation rock during excavation work and these were filled with concrete. During first filling of the reservoir, springs appeared in a creek $2\frac{1}{2}$ km downstream from the dam and their discharge increased roughly in proportion to the further increase in reservoir level. As the reservoir became almost full, a large whirlpool developed about 200 m upstream of the dam near the left side. Measures were immediately taken to begin lowering the reservoir and the entrance to a large cavern was revealed under the position of the whirlpool. As soon as it was uncovered and therefore isolated from the reservoir, the water level in the cavern fell, eventually to a level 70 m below reservoir level. The extent of the cavern was explored by boat and estimates made of its volume. It was filled with 600,000 $m^3$ of material that was consolidated by inundation and water flow when the reservoir again rose under flood conditions. When the reservoir was again drawn down, the infill was injected with grout to reduce its permeability and seal the cavern.

The accurate assessment of seepage flow and its dissolved mineral and solids contents is of paramount importance and derives from continuous measurement with some of the earliest and simplest of the instrumentation used to monitor the performance of an embankment dam.

## 3 SETTLEMENT MEASUREMENTS

Since the advent of the surveyors' optical level it has been normal practice to measure crest settlements by periodic level surveys along its length. The main purpose of these observations was to ensure that adequate freeboard remained to avoid overtopping during flood conditions. In the past, errors have been caused through lack of reference points on the dam and lack of stable datum positions. There were many examples of measurements made with staff and surveyors' level on to kerb stones forming the sides of a crest roadway. Excessive apparent settlements have been caused by vehicles mounting the kerb and other strange records are found after repair work to the road. Despite these difficulties, observed settlements have shown that in general crest settlement is proportional to the height of fill below the crest and has led to design rules such as the requirement that the crest shall be built to level above the nominal crest level of 1.5 per cent of fill height. This leads to a crest highest over the old river channel and usually the extra height requirement is adjusted to give a smooth camber over the dam length, although there have been cases where the rule was applied so rigorously that undulations were formed in the crest reflecting undulations in the valley shape along the centre line of the dam. As with a bridge, a slight upward camber is aesthetically pleasing and imparts a feeling of stability and security.

### 3.1 Overtopping risk

The 30 m high Billberry dam, built in 1840, had a plain crest with no wave wall or roadway. After construction, crest settlements were not measured, but when the dam was 12 years old, a storm raised the reservoir level to cause a substantial discharge over the spillway wier and the water level showed that the crest was dangerously low near its mid length. As the water continued to rise, it began to overtop the crest and after a few hours, began erosion which eventually breached the dam, causing a flood which killed 81 people in and around the town of Holmfirth in England.

It was sometimes the practice to reinforce the crest of earthfill dams by adding stones to the fill to make it more resistant to overtopping. The 80 year old Cwmwernderi dam of 20 m height, has a layer about 0.6 m deep of what could amost be described as rockfill, blinded with earth, covering the crest. It is not certain whether this was a feature of the original design, or if the layer was added to compensate for settlements and increase freeboard.

Settlement observations can be made with reference points of the type described by Cheney (1974). These consist of stainless steel sockets, 65 mm long and 22 mm diameter, which can be set into holes drilled in masonry or concrete wave walls, valve towers, etc. A spherical ended plug accurately fits the sockets and is used to support the end of the levelling staff. Reference can be made to stable positions established on the valley sides at places where movement is unlikely to occur.

Nowadays it is usual to measure horizontal as well as vertical movements of not only the crest, but also other parts of a dam. This aspect will be considered in further detail under the heading 'Surveying methods' below.

## 4 DIFFERENTIAL SETTLEMENTS

Strains within the body of the fill due to differential settlements can cause large redistributions of stress and lead to cracking or hydraulic fracture. Settlements of the foundation caused by the weight of the dam and hence greatest under the highest parts, can induce considerable undesirable strains in the dam fill.

### 4.1 Duncan dam

The ability of an embankment dam to tolerate large differential settlements was illustrated by the 41 m high Duncan dam in British Columbia, completed in 1967, on a foundation unsuitable for any other type of dam. The foundation consisted of a 380 m deep canyon in the bedrock that had been infilled with sedimentary material as indicated by Fig 5. The lower infill was a dense gravel which became more sandy with depth and was overconsolidated. It was overlain by a lacustrine deposit about 90 m thick of silt and fine sand interspersed with sand and gravel lenses. It was covered by layers of highly pervious unsorted sands and gravels 6 to 24 m thick. An extensive field investigation and detailed laboratory testing programme led to estimates of settlement, shown by fig 5, in the range 3 to 4.3 m with the maximum occurring near mid-length of the dam.

To counter the effects of the expected, saucer-shaped settlement, the dam was designed with a sloping core to be placed 1 to 2 per cent wet of Proctor optimum so that construction pore pressures might keep

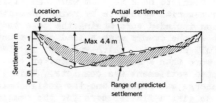

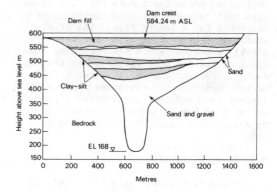

Fig 5 Section through Duncan dam and foundation showing settlements. East to left - West to right

the core sufficiently flexible to distort with the differential settlement rather than crack. The upper 20 m of core for distances of 35 m from each abutment was not to be placed until the last moment. The available core material was a glacial till with a plasticity index of only 4.3 per cent. When placement was stopped over the 35 m lengths at each abutment, regular inspections were made to detect any cracks that might occur as the rest of the dam was brought up to full height. Settlement gauges showed that the maximum settlement was developing nearer to the east abutment than had been predicted and this was producing the most severe differential settlements adjacent to the east abutment. Unfortunately pore pressures dissipated rapidly from the core and cracks were found over a very restricted length of only 25 m along the dam axis, close to the east abutment. Rate of crack opening was measured by installing temporary reference points on either side of them and it was found that the ratio of crack opening to settlement was about 1.6 to 1.

Trial shafts were excavated to follow the cracks down into the core and they were found to be 2 to 8 cm wide and filled with material washed in by rain. Remedial action included surcharging the core to the west of the cracks with a load equivalent to full reservoir, which was left in place for more than a month until most of the induced settlement was complete. This surcharge was removed and the design amended to give a flatter core for the remaining, upper 20 m height and flatter downstream slopes to the dam. The plastic index of the core material was raised to 20 per cent by the addition of 6 per cent bentonite, producing a more flexible material that retained its construction pore pressures for a longer time. The trial shafts were backfilled with this more plastic material before the remaining sections of the dam were completed to the new design. Fortunately the alluvial infill of the buried canyon responded rapidly to applied load and settlement rates reduced markedly on completion of construction. On reservoir filling, no leakage was detected.

4.2 Vertical settlement gauges

Foundation settlements have been measured by a variety of vertical gauges, such as the well-known cross-arm gauge developed by the United States Bureau of Reclamation and described in their Earth Manual (1974) under Appendix E-29. A vertical gauge developed at the Building Research Station and described by Marsland and Quarterman (1974) uses magnet markers placed round a central access tube. To measure settlements at different depths in the foundation, the magnets are attached to expanding units that can be placed in boreholes. When the units have been lowered to the required positions, they are expanded to anchor themselves to the surrounding soil. During dam construction, the access tube can be extended upwards through the fill and marker plates containing magnets can be added at various fill heights. A reed switch sensing unit is used to detect the magnets and it is lowered into the access tube at the end of a measuring tape so that the positions and hence the settlements of the markers can be measured.

These vertical devices have the serious disadvantage that they restrict the movement of earth placing and compacting machinery. They also suffer uncontrolled horizontal movements due to local fill displacements caused by the machines. In general it is better to keep access tubes and connecting leads for instruments on a plane parallel to the surface of the fill so that they can be placed in trenches during construction and buried for ever.

4.3 Water overflow settlement gauge

The water overflow gauge has been used extensively to measure settlements of particular positions in the foundations and fills of dams. The gauge consists simply of an overflow unit buried in the fill and connected by small flexible tubes to a vertical standpipe in an instrument chamber.

A simple version of this gauge was used by Spangler (1933) in North America and a complex version, described by Mallet and Pacquant (1951), was used in the 22 m high Sarno dam in Algeria.

The type currently in use by British engineers is illustrated diagrammatically by Fig 6. The buried container is connected to an instrument house by three 6 mm bore tubes made from nylon 11, one of them coated with polythene to prevent any water loss by evaporation from the walls of the tube. All the tubes are connected to the bottom of the container, the polythene coated one being used to connect the water overflow tube (see Fig 6) to the Perspex standpipe in the instrument house. One of the other tubes is used as an air tube to ventilate the buried container and ensure that it is kept at atmospheric pressure. The other is used as a drain to return water to the instrument house.

De-aired water at ambient temperature is usually obtained from de-airing apparatus provided for hydraulic piezometers. It is fed through a block of 3 sleeve packed cock valves to the coated tube. A volume equal to about 1.5 V, where V = volume of whole length of coated tube, is allowed to pass to ensure removal of all air from the tube and to fill it with water of uniform temperature. The buried container should have sufficient capacity to accept this water without submerging the end of the water overflow tube. As the last of the measured volume of water is flowing, the cock valve in the block of three, connected to the Perspex standpipe, is opened slightly to fill the standpipe before the cock valve controlling the supply of de-aired water is closed. The water level in the Perspex standpipe should not be allowed to fall too rapidly to avoid leaving drops of water in the pipe, but as the rate of fall slows, the valve may be opened more, until it is fully open. The stationary level of the water will indicate the level of the top of the overflow tube in the buried container, but may be in error due to over-run. The valve controlling the supply of de-aired water

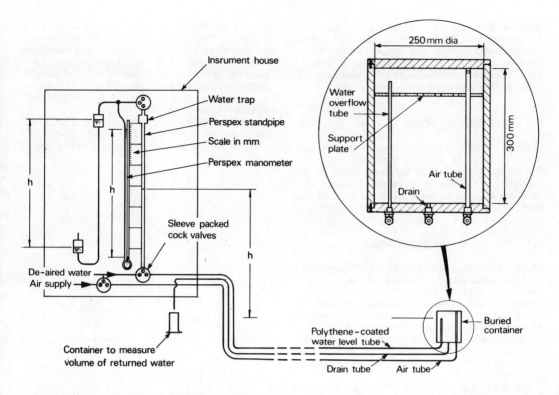

Fig 6    Water overflow settlement gauge

should be opened slightly to raise the water level in the standpipe by about 2 cm and then closed.  When equilibrium has been reached, this process should be repeated until at least three identical level values are obtained, reading the scale to an accuracy of 0.1 mm.  Poiseuille's equation for laminar flow in pipes shows that the time required for a given degree of equalisation varies inversely as the fourth power of the pipe diameter.  By using a 6 mm bore tube, the 99.9 per cent equilibrium time for the lengths of tube normally required in an embankment dam is reduced to a few minutes.

Errors will be introduced if there are any water drops in the air tube or if the air tube is blocked in any other way, and if the water overflow tube becomes submerged.  Water from the buried container should be returned to the instrument house for measurement then to waste, by closing the valves in the three valve block and applying air pressure to the air tube.  Only moderate pressure should be used to avoid overstraining the buried container (if that becomes fractured, errors may occur) and a careful check should be kept of amounts of water used and returned to indicate any leaks that may develop in the system and to ensure that the buried container has been emptied.

To enable measurements to be made at positions in a dam above or below the level of an instrument house, a back pressure or suction can be applied to the top of the Perspex standpipe by a head of water, as indicated by Fig 6.  Accuracy is then dependent additionally on a correct knowledge of the applied head.  This will vary with use of apparatus and must be measured at each set of readings.  To improve accuracy, a water manometer is connected to the back pressure unit so that water levels can be read against a scale as with the Perspex standpipe itself.

The buried container must be cast into a concrete cube of about 1 m side to protect it from both the external pressure of the surrounding fill and the internal air pressure used to return the used water.  The connecting tubes are noramlly placed in a trench about 1 m deep cut in the fill, and protected by sand, sieved fill or clay depending on the type of fill.  It is usual to carry them through earthfill cores so that settlements of the upstream fill can be measured at instrument houses near the downstream slope of the dam. The trench through the core must be backfilled with core material and compacted to the same extent as the surrounding material.

## 5 HORIZONTAL MOVEMENTS

Strains in the fill are usually manifested as horizontal as well as vertical movements and to obtain a fuller understanding of the behaviour of a dam and its foundations it is desirable to measure both horizontal and vertical movements of chosen points within the fill.

### 5.1 The horizontal plate gauge

The horizontal plate gauge, developed at BRS and described by Penman, Burland and Charles (1971) consists of a horizontal column of plastic tubes passing through steel marker plates. The horizontal positions of these plates are measured by induction coils and their vertical positions by water overflows. Accuracies of ± 2 mm are obtained for both vertical and horizontal movements.

The positions of three gauges in the 90 m high Llyn Brianne dam are shown by Fig 7. The plastic pipes made of rigid pvc, were 66 mm inside diameter with 5 mm thick walls and were fitted with watertight telescopic joints at every 3 m length. They were taken 6 m into the clay core and fitted with a watertight stop end. Steel plates 300 mm square and 5 mm thick with a central clearance hole were fitted over the pipes to be placed in the clay core, on either side of the fine filter and at about 15 m spacing in the rockfill of the downstream shoulder. They were laid on a bed of pea-gravel (a rounded gravel 3 to 10 mm size) and buried under about 15 cm of pea-gravel in trenches about 1 m deep. These were backfilled with selected excavated fill.

Instrument houses, as shown by Fig 8, were pre-cast off site, and when the fill had reached a level that would bring the

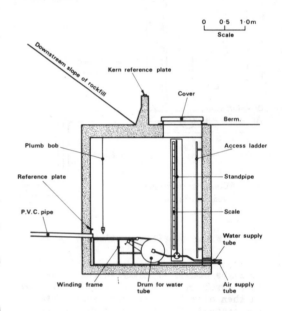

Fig 8 Section through an instrument house

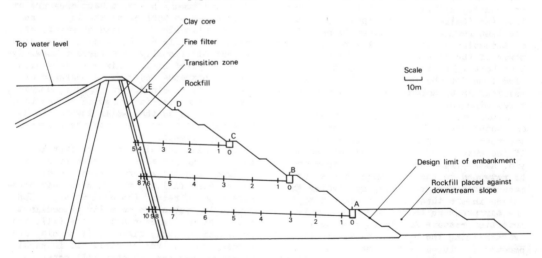

Fig 7 Horizontal plate gauges in Llyn Brianne dam

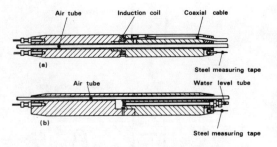

Fig 9 Cross-sections of the two measuring probes

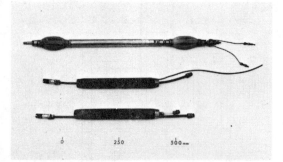

Fig 10 Duct motor and measuring probes

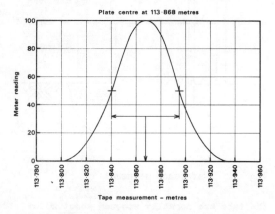

Fig 11 Meter reading v probe movement in pipe

top of a house to berm level, it was towed across the fill to a levelled area prepared for it and secured with grout injected underneath. The fill, which was being placed with an outward fall of about 1 in 40, was brought up a further 2 m before the trench was excavated.

The positions of all the plates were measured by level and distance from the instrument house to be sure that they were where required and for subsequent location. As soon as the trench had been backfilled their zero positions were measured accurately from within the pipe with the induction coil and water overflow measuring probes shown by Figs 9 and 10. They were pulled into the pipes by a duct motor, shown in Fig 10. It is driven by compressed air and consists of a double acting ram with plastic bags at each end. In operation, the rear bag inflates to grip inside the pipe and the ram extends. The front bag then inflates, the rear one deflates and the ram contracts. In this way the motor 'walks' along the pipe, taking 0.2 m steps while compressed air is supplied to it. The motor drags a probe, trailing a coaxial cable (or water level tube), steel tape and a tube for the compressed air. It is normally allowed to run to the end of the pipe and readings are taken at each plate on the way back.

The steel tape passes under a cursor in the instrument house so that length measurements can be read to 0.1 mm. Accuracy depends on tape tension and this is controlled by the sliding resistance of the deflated bags inside the tube. It has been found that the pipe stays clean and repeat readings on a stationary plate give an accuracy of $\pm$ 1 mm, indicating little variation of sliding resistance at a particular plate position.

Before passing into the pipe, the induction coil is balanced against an induction bridge. The sensitivity of the bridge is adjusted while the probe is pushed through the reference plate of the instrument house (see Fig 8) so that the indicating meter moves from 0 to 100 as the balanced induction is upset by the steel plate. Length readings at each plate are subsequently taken as the meter reaches 50 while increasing and again when it reaches 50 as the coil comes out of the plate. The centre of the plate is taken as the mean of the two readings, as indicated by Fig 11.

The level at each plate is measured by use of the water overflow probe, connected to a Perspex standpipe in the instrument house and operated in the same way as the water overflow settlement gauges described above.

These initial readings are used as the zero positions of the plates. They are more accurate and usually slightly different from the positions measured in the trench before backfilling. Subsequent sets of readings measure the relative movements of the plates in relation to the reference plate, attached to the rear wall of the instrument house. To measure the overall movements of the plates, the movements of the instrument houses have to be measured. This is done, together with the measurement of other positions on the dam, by precise surveying.

The horizontal plate gauge has been installed in Scammonden (70 m), Llyn Brianne (90 m), Winscar (50 m) and Megget (56 m) dams in Britain. For this last dam the apparatus was modified by using a separate induction coil probe at each plate to reduce the time required to take readings. The probes were permanently connected together by hollow metal rods and to take readings, it is only necessary to move the string of rods in and out by small amounts to produce the type of curve shown by Fig 11 and to position each plate. A second line of pipes was laid beside the first during installation, to house a string of water overflow probes, similarly arranged with a probe at the position of each plate.

# 6 SURVEYING METHODS

Movements of a dam are best measured from stable reference monuments built outside the construction area on bedrock or in positions where movement will not occur. If carefully positioned, these monuments can be of considerable value in setting out the dam work, as well as for measuring subsequent movements. They should be built at the very beginning of construction so that they can be of greatest value to the whole project.

## 6.1 Reference monuments

An example of the fairly complex reference monuments used at Scammonden and Winscar is shown by Fig 12. They were founded in bedrock and protected from seasonal movements of the overlying soil by an outer casing. Both the outer casing and the lower part of the monument were formed by pre-cast commercial manhole rings. A special cover was pre-cast for the outer casing and the upper part of the monument constructed with a special steel tube with a protection disc welded to it. This was arranged to protect the gap between the tube and the outer casing cover so that stones would not jam in the gap (thereby transferring ground surface movement to the pillar) and to prevent the loss of small objects that might be dropped while carrying out a survey.

When bedrock is close to the surface, the steel tube for the pillar can be founded directly into a hole chopped into the rock. The reinforcing cage and concrete inside

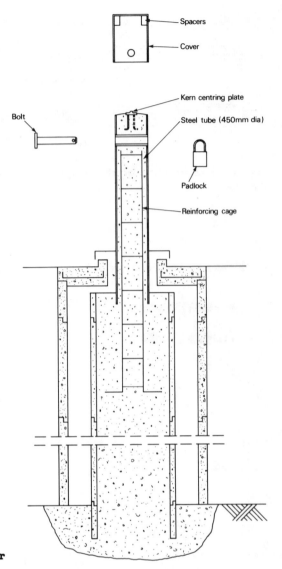

Fig 12  Reference monument

the tube are normally stopped about 20 cm below the top of the tube during construction. A Kern centring instrument plate, fitted with three screwed steel rods about 12 cm long and bent over at their ends, is then positioned in the tube by a simple steel frame that rests on the end of the tube. The frame is fitted with adjusting screws so that the Kern plate may be accurately levelled before concrete is placed to almost fill the tube and anchor the three rods. Nuts securing the rods to the plate are left exposed, and after a few days when the concrete has cured, final adjustments are made to level the Kern plate and the remaining space under and adjacent to it filled with mortar, finished to an outward fall to shed rainwater. Various types of cover have been used to protect the Kern plate and prevent unauthorised interference. One consists of a heavy steel 'hat', filled with spacers so that it does not rest on the Kern plate, and secured by a 25 mm diameter steel bolt passing through a hole in the pillar, fitted with a padlock. Another is made flush on the outside held down by a heavy bayonet type catch and secured by a secret bolt, so that there is no obvious way of getting it off.

Instruments and targets have projecting, accurately machined pins which locate in a central recess in the Kern plate to ensure that they are always precisely re-positioned. The reference point on the instrument chamber (see Fig 8) was constructed as the top of a monument and some reference points on berms and dam crest were constructed as pillars founded in blocks of concrete cast into small excavations in the fill. The simple, stainless steel sockets described by Cheney (1974) have also been used for reference points on concrete or masonry units such as wave walls, instrument houses etc when the sight distances from the monuments did not require use of the large Kern targets. A small bullseye target and a small cube corner reflecting target have been designed at BRS for use with these sockets.

### 6.2 Triangulation and trilateration

In general the monument positions and the initial positions of instrument houses and reference points on the dam were established by both triangulation and trilateration. The Kern DKM2A theodolite used could measure both horizontal and vertical angles, to an accuracy of 1 second of arc. The Tellurometer MA100, an electronic distance measuring instrument with an accuracy of 1.5 mm ± 2 part per million and the Mekometer, a slightly more accurate device, were used for the distance measurements.

The approximate positions of monuments and reference points at Llyn Brianne dam are shown by Fig 13. Dam movements were usually measured as angular changes, although from time to time distance measurements were made. Results were analysed by a 3-dimensional computer program which calculated the coordinates of each reference point from an input of measured angles, as read from the theodolite, without modification.

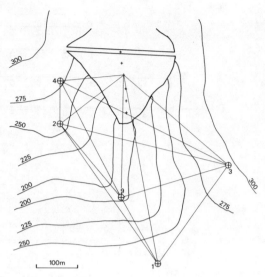

Fig 13 Approximate positions of monuments and reference points at Llyn Brianne dam

At first a precise surveyors level was also used to check vertical positions, but it was found that comparable accuracy and more consistant results were obtained from the vertical angles measured by theodolite.

These measurements showed the movements of the instrument houses at the ends of the horizontal plate gauges, which also housed the standpipes for the overflow settlement gauges, so that the movements of all the marker points within the body of the dam were related to the stable reference monuments. To analyse movements finite elements can be drawn on the dam sections with node points at plate positions, as illustrated by Fig 14.

## 7 PRESSURE MEASUREMENTS

Knowledge of the effective stresses in the fill is necessary not only to determine the shear strength and deformability but also to assess the risk of hydraulic fracture in the impervious zone of a dam.

### 7.1 Pore pressures

It has been normal practice to measure pore pressures in the fill of dams built during the past 40 years and a wide variety of piezometers are commercially available. Any piezometer consists of three parts:

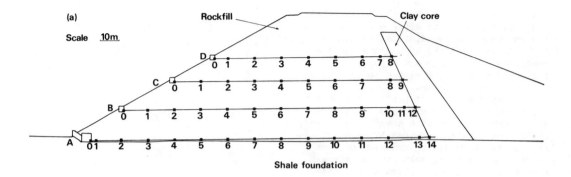

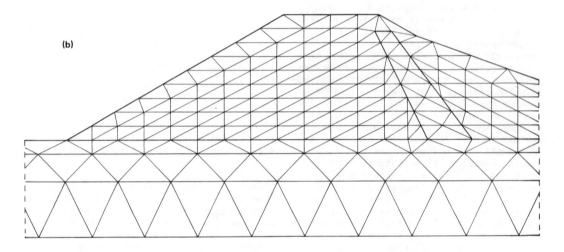

Fig 14 Simplified cross-section of Scammonden dam showing:
(a) position of four horizontal plate gauges
(b) mesh of finite elements used in analysis

(1) the intake filter; (2) a pressure measuring device; and (3) a connection between the two. The intake filter must be strong enough to withstand the maximum total pressure that will act on it without appreciable deformation; it must be large enough to give the required response time to the piezometer and it must have suitable pores to allow correct contact with the pore water of the soil. In partially saturated fill this is a particularly important aspect.

The simplest type of piezometer and the first used was a standpipe made from steel tubes with some perforations at the lower end to allow the ingress of pore water from the soil. Such a device was used in Waghad dam in 1907 (Nagarkar et al, 1981). The head of water in the standpipe opposed the pore pressure and when equilibrium had been reached, the level of water surface in the standpipe was a measure of the pore water pressure. This simple apparatus was improved by Casagrande (1949) for use at Logan airport to measure pore pressures in the silty clay foundations. He used plastic standpipes and ceramic filter candles that were placed in long sand pockets formed in the installation boreholes to improve the intake area from the silty clay and so improve the response time of the piezometers.

The idea of using a sand pocket round the intake filters of piezometers has persisted, even though it may not be necessary with some instruments and it should not be used with filters in partly saturated fill. In order to measure pore water pressures in these fills, which are likely initially to be below atmospheric pressure, a fine pored filter must be used. The pores should be of uniform size and small enough so that curvature of the water-air meniscus across them can withstand more than atmospheric pressure. The problem is well known and intake filters for piezometers can be obtained commercially that will withstand almost two atmospheres air pressure before water in their pores is blown out. When a saturated filter of this type is pressed into contact with a partly saturated fill, it will contact the pore water in the fill and enable its suction to be measured if the piezometer has a suitable pressure measuring device.

A standpipe can only measure suctions if the absolute air pressure above the water surface in it is reduced to a value below that of the absolute pressure to be measured. This is not a practical approach for field measurements and standpipe piezometers are normally only used where their coarse pored filters can be placed (with advantage in a sand pocket to improve response time) below the lowest expected phreatic level. They are thus commonly used in dam foundations.

It is usually of value to install standpipe piezometers in the exploratory boreholes made during the investigation of a dam site, rather than simply backfilling the holes. Regular readings can reveal much about ground water conditions on the site before dam construction. Chosen piezometers can later be converted to be read remotely to form part of the permanent piezometer layout of the dam.

This can be done by taking a small tube to the bottom of the standpipe and measuring the head of water by the air pressure required in the small tube to blow a few bubbles from the end. In practice a trench is dug from the standpipe to an instrument house and the standpipe cut-off below ground level. A special head, carrying 3 small tubes as shown by Fig 15 is attached to the standpipe so that the central tube reaches the filter. The three tubes are laid to the instrument house and pore pressure at the filter measured as the air pressure required to cause a small flow through the central pipe, air returning through the other two. If, at a later stage during dam construction or reservoir impounding, the water level in the standpipe rises to the head so that water comes through the return tubes, these can be connected to a measuring system and the air pipe closed, to make the standpipe into a two-tube piezometer. The hydraulic gradient through the foundation, revealed by the piezometers can be used to calculate flow that is occurring below formation level and does not pass through the seepage measuring equipment.

7.2 Two-tube hydraulic piezometer

This type of piezometer was developed by the U.S. Bureau of Reclamation in 1939 and has been described in their Earth Manual (1974) under E-27. Apparatus developed and manufactured at BRS was described by Penman (1956). This apparatus has since been improved in the following ways:

(a) Intake filter: Coarse filter replaced by fine pored conical shape with a blow through air pressure of about 2 atmospheres. This filter was described by Bishop et al (1960) and is shown in Fig 16. It is installed in a saturated condition into a prepared tapered hole in the fill and pressed into place with the aid of an extension handle that is later removed.

(b) Connecting tube: This used to be 3 mm bore x 1 mm wall polythene: a material slightly pervious to air. It was later replaced by nylon 11: a material that will absorb water causing loss by evaporation from the tube. These problems were overcome by combining the two materials and the tube that has been used for nearly 20 years is polythene coated nylon 11 of 3 to 6 mm bore, depending on the distance to the buried filter.

(c) Pressure measuring device: Corrosion and calibration troubles with Bourdon gauges led to the use of mercury manometers. Inexperienced operators could easily blow these out and it was not uncommon to find that mercury had entered the connecting tube and sometimes entered the filter. To assist with automatic recording, electrical transducers are often used. Their zero readings and calibration must be checked at least once a year against known pressures and they must be protected from severe temperature changes. At one installation all the transducers were arranged to be in an insulated box kept at a constant temperature slightly higher than the maximum experienced in the instrument house, by a thermostatically controlled heater. Pressure transducers are usually so stiff that for manual reading all piezometers can be read with one unit, connected in turn through leak-proof sleeve-packed cock valves.

The apparatus may be supplied with air free water by a reciprocating pump driven by compressed air. The air and water cylinders of the pump are connected by a piston rod which operates valve gear to

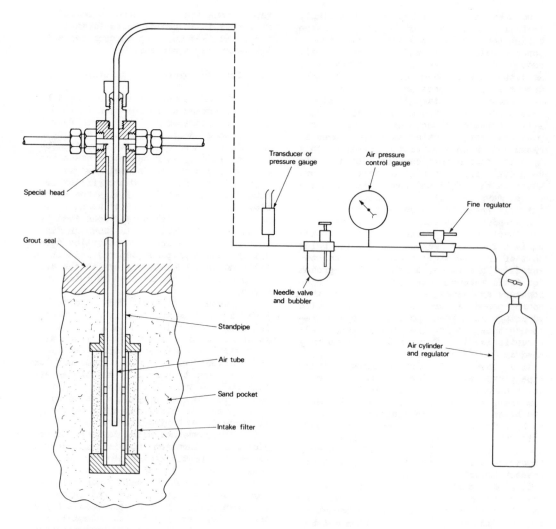

Fig 15  Standpipe piezometer with special head and arrangement for remote reading

give continuous pumping and maximum delivered water pressure can be pre-set by an air pressure regulator. Pumped volume can be measured by the number of strokes and pre-set so that the pump stops when the required volume has been delivered.

When a filter unit is to be installed it is first saturated with air-free water, preferably allowing water to pass slowly inwards over a period of about 24 hours. This can be done by attaching temporary short lengths of connecting tube to let the air out, and standing the unit upright in air-free water. When the connecting tubes have been laid out in a trench to the required position, they are filled with air-free water by the pump. They are cut to length and connected to the filter under water in a container. The altitude of the water surface in the container is measured by site level and staff and a check made to see that the pressure transducer records a pressure corresponding to the head of water. A taper hole is then formed in the fill and the saturated filter pressed into place. Continuous readings should then be taken with the transducer to show the minimum pore pressure (maximum suction) achieved to use as the zero value for that piezometer. Subsequent dam construction may cause pore pressure increase, but the magnitude of this increase cannot be found unless the zero value is known. It can be argued that end of construction

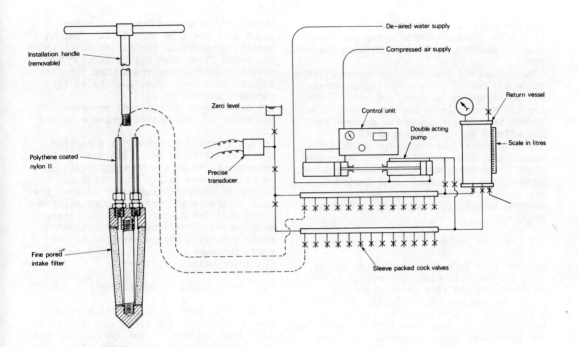

Fig 16  Two tube hydraulic piezometer

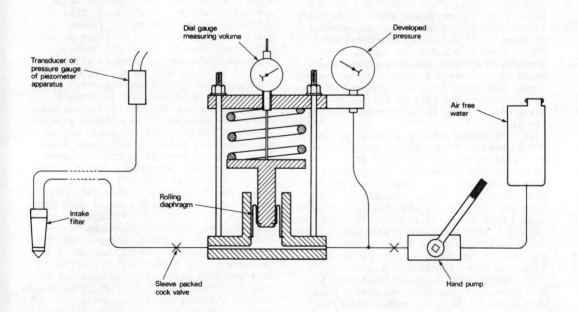

Fig 17  Critical pressure apparatus

pore pressures in an earthfill core should be greater than reservoir water pressure after impounding to avoid hydraulic fracture.

Use of coated tube and fine-pored filters keeps air out of the system and it should not be necessary to circulate water through the two tubes unless they have been subjected to very low pressures or some fault has allowed air into the system.

A weakness of the apparatus is that the connecting tubes should not rise more than about 7 m above measured piezometric level to avoid low pressures. With instrument houses for horizontal plate gauges at several levels, this is seldom a serious problem.

7.3 Special tests

The two tube hydraulic piezometer can be used for the following purposes in addition to measuring the pore water pressure:

(a) To measure permeability. Constant head inflow or outflow tests can be made by supplying suitable constant pressures at the instrument house to one of the connecting tubes. Connection is made through a flow meter which is often the enclosed burette type. This consists of an accurately calibrated glass burette inside a Perspex cylinder containing coloured paraffin over water. Cock valves are arranged so that water flow causes paraffin to enter the burette and volume change is measured by movement of the paraffin/water interface. When this nears the bottom of the burette, the valves are re-arranged to cause the interface to rise with increasing flow. The actual pressure achieved at the filter is measured through the second tube which has no flow, by the pressure transducer.

(b) To assess total pressure. If the water pressure in the filter is increased, flow will occur into the pores of the soil at a rate corresponding to the pressure difference and permeability of the soil. When the water pressure exceeds the total pressure of the soil, however, soil will be pushed away and the flow rate will increase rapidly. When there are pronounced differences in the values of the principal total stresses in the soil, a fracture may develop on the plane of minor principal stress.

In clayey soils, such as those used in dam cores, this approach can give a valuable guide to the total pressures in the core and is of particular interest in relation to risk of hydraulic fracture that might be caused by pressure from the reservoir water.

Bjerrum and Andersen (1972) found that in normally overconsolidated soft clay, the lateral pressure and hence the value of $K_o$ could be found from this test. They increased pressure in the filter until there was rapid flow, then slowly reduced the pressure to find the value when the flow reduced to a rate corresponding to the permeability of the soil.

This close-up or critical pressure gave the value of the lateral pressure in the clay.

Penman (1975) designed the apparatus shown by Fig 17 to carry out these critical pressure tests on two tube hydraulic piezometers in dam fill. He carried out a number of tests on piezometer filters that were in groups of earth pressure cells and found that the measured critical pressures were generally somewhat greater than the minor principal stress measured by the earth pressure cells, but much less than the major principal stress.

7.4 Other types of piezometer

In general other types of piezometer have diaphragms which isolate their intake filters from the rest of the system and make them unsuitable for the special tests described above. They fall into two groups:

(a) Electrical piezometers: these use the diaphragm as the pressure measuring device. It is instrumented with vibrating wire or resistance strain gauges which measure diaphragm deflection. This is caused by the difference of pressure on its two sides and if the pressure on one side remains constant, it can be used as a pressure transducer to measure the pressure of water in the filter. Usually the strain-gauged side of the diaphragm faces a leakproof container that is factory sealed during manufacture at a given pressure, sometimes below atmospheric pressure. Electrical connections are taken out of this container through fused glass seals. Provided that there is no creep between strain-gauge and diaphragm, that the elastic properties and edge fixing of the diaphragm remain unchanged, that the pressure in the sealed container remains unchanged and that the strain gauges continue to give the same electrical changes in response to a given strain change, then the piezometer should correctly measure the pressure in the filter unit. Provided that gas from the soil pores does not diffuse into the filter, the piezometer should measure the pore water pressure in the soil; however, if gas does enter the filter it may begin to measure only the gas pressure in the soil pores.

When an instrument is to be buried and become inaccessible for ever, there is always concern about its zero readings

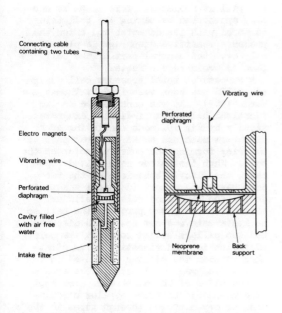

Fig 18  Vibrating wire piezometer

and changes that may occur in its calibration. DiBiagio (1974) described a development he made to his vibrating wire piezometers so that both zero reading and calibration could be checked. A line diagram of his piezometer is shown in Fig 18. The BRS vibrating wire piezometer, described by Penman (1960) used an air pipe to house its electrical connecting cable and this ensured that the instrumented side of the diaphragm was kept at atmospheric pressure. By applying air pressure through the pipe, the calibration of the instrument could be checked, but the zero reading remained unchecked. DiBiagio made small perforations in his metal diaphragm and placed a separate flexible neophrene membrane over it. In normal use, water pressure in the filter presses the membrane against the diaphragm. To check calibration, air pressure behind the diaphragm can be increased in steps and held until the diaphragm has reached its new equilibrium position. When the applied air pressure slightly exceeds the pressure in the filter, the membrane can lift off the diaphragm, leaving it with equal pressure on both sides in its zero position. A backing support prevents excessive deflection of the membrane and prevents damage if the applied air pressure is slightly excessive. Use of two tubes, built into the connecting cable, enables dry air or inert nitrogen to be circulated through the strain gauge part of the piezometer to keep it dry and avoid oxidation.

A similar arrangement to allow both calibration and zero checks to be made was described by J P Collins in a paper to be submitted to a speciality session of the 7th International Conference on Soil Mechanic and Foundation Engineering in Mexico, 1969. Instead of using a perforated metal diaphragm, one of the tubes in the connecting cable was attached to an annular space at the edge of the diaphragm so that air pressure from it would separate the membrane from the metal diaphragm. The other tube led to the space above the diaphragm so that by applying the same air pressure through both tubes, in excess of the pore pressure, the diaphragm became subject to the same air pressure on both sides.

Electrical storms have been known to produce surge voltages at a dam site which have damaged electrical piezometers and made them inoperative. Special earthing arrangements should be made to protect the strain gauge equipment in the piezometer.

(b) Pneumatic piezometers: these use the diaphragm to operate a control valve that indicates when applied air pressure is equal to the water pressure in the filter. As with the hydraulic piezometer, two tubes are used to connect it to an instrument house or manhole, etc. Air pressure applied through one tube acts on the flexible diaphragm and when it exceeds the pressure of water in the filter, movement of the diaphragm opens a valve to allow escape of air through the other tube. This is used to indicate that the water pressure has been exceeded and no further air is passed into the first tube, which is then connected only to a pressure gauge. Air continues to escape from the second tube until pressure behind the diaphragm falls to that of the water pressure in the filter, when the diaphragm begins to close the valve, at a rate dependent on the rate at which pore water can enter the filter. Eventually the valve closes, locking a pressure in the system, shown by the pressure gauge and usually taken as a measure of the pore pressure.

Leakage of the valve, due to dirt or wear, makes the cut off point ill defined and this type of faulty piezometer can sometimes be read by plotting the rate of pressure fall. Projected straight lines from the faster and slower parts of the plot may reveal the pressure at which the valve tries to shut and so indicate pore pressure. As with the electrical type of piezometer, gas that diffuses into the filter unit cannot be flushed out and the piezometer may indicate pore air pressure in a partly saturated fill.

Advantages claimed for the pneumatic piezometer are that the readout station can be at any height above the measured piezometric level, it suffers no freezing problems and the terminal station can be very simple: there is no need of special instrument houses for it.

## 7.5 Total pressures

Strains caused by differential movements can seriously alter the total stresses in the various zones of a dam. This may be of particular importance in the impervious zone where information about any reductions in the total stresses from those imposed by the overburden pressure is essential.

If, under the pressure of the reservoir water, the effective stress in the impervious zone falls to zero, hydraulic fracture can occur and subsequent erosion may lead to piping. It is therefore a design requirement that all total stresses in the impervious element are greater, by a suitable factor of safety, than the maximum pressure that can be exerted by the reservoir water.

Many different types of earth pressure cell have been used in attempts to measure total stress conditions in dam fill. In general, the more flexible the fill, such as clayey material placed wet in a dam core, the more nearly the total pressures approach a fluid pressure and the easier it is for a rigid cell to measure total pressure correctly. An ideal cell should have the same deformation properties as the fill it replaces. This ideal is usually approached by making the cell thin compared with its diameter and often the pressure sensitive diaphragm is placed only over the central part of the cell face to avoid edge effects.

A successful total pressure cell designed at BRS has been described by Thomas and Ward (1969). It was constructed as two identical halves, as indicated diagrammatically by Fig 19, each consisting of a diaphragm machined as an integral part of the ring forming the outer diameter of the cell. When the halves were bolted together, the elastic behaviour of the two diaphragms was almost identical and their deflection caused no relative sliding movement between any parts of the cell so no hysteresis was set up. Each diaphragm had two pillars welded to it at its points of inflection and vibrating wires stretched between the pillars were used to indicate diaphragm deflection. The strain-gauged sides of the diaphragms were kept at atmospheric pressure by slow circulation of dry nitrogen through tubes in the connection cable.

A thinner cell, sensitive over almost its whole surface, can be formed from two metal plates joined at their edges with the space between them filled with oil. Earth pressure, transferred to the oil, can be measured as oil pressure by a pressure transducer that can be placed a little distance from the cell so as not to influence its stress field. This design

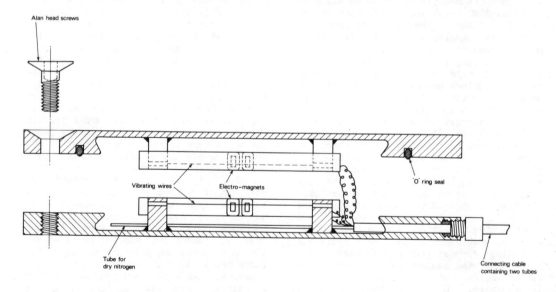

Fig 19    Vibrating wire total pressure cell

has been developed by Glötzl AB in Germany who uses a diaphragm pressure transducer similar to that of the pneumatic piezometer described above. Glötzl originally used thin oil in the connecting tubes to operate the transducer but he has also used air. He manufactures a wide range of earth pressure cells of several sizes and capacities. His basic cell can also be used with an electrical transducer to simplify data collection, although risk of damage by electrical storms must not be overlooked.

Total pressure cells are usually installed in groups to measure pressures acting in different directions. When used, eg near the centre-line of a dam, where the major principal stress is expected to act in a vertical direction, a minimum number of 3 cells should be used. They should be installed with their faces normal to the direction of the major principal stress (expected to act vertically), the direction along the dam axis and the direction across the width of the dam to measure what may be the intermediate and minor principal stresses. In other parts of the dam, or where more detailed information is required, additional cells should be included in the group, placed to measure stresses acting at $45°$ to those measured by the other cells.

An important development described by DiBiagio et al (1982) is the use of remote reading inclinometers attached to total pressure cells. These enable the attitude of the cells to be measured and possibly corrected during installation. At a later stage, the actual measured attitude of the cell can be used when interpretation is made of the measured total stresses.

The cells should never be placed in pits dug into the fill because of the danger that fill arching over the pit will prevent the true total pressure reaching the cell. It is better to arrange the group of cells and piezometers (piezometers must be placed in the group so that effective stresses can be calculated) in a shallow depression scraped out of the fill. The cells should be kept about two diameters apart and the surface of the fill sculptured to form flat surfaces for the cells placed horizontally and on planes at $45°$ to the horizon. Vertical cells can be placed in shallow slots, wide enough to allow excavated material to be backfilled and compacted by hand held rammers to the density of the adjoining fill. The area around the cells should be built up gradually using hand compaction. Frequent readings should be taken to establish zero conditions, to check correct functioning and cell attitude as backfilling proceeds. When the cells have been covered, handheld pneumatic tampers may be used to consolidate the fill, but a constant check must be made on the attitude shown by the inclinometers to ensure that compaction does not move the cells too far from their intended attitude. When the cells have been covered to a depth of about one diameter, light plant may be used, but at least three diameters height of fill should be over the cells before normal fill placing continues above them.

# 8 EARTHQUAKE

Design calculations for dams in seismic areas include allowance for the maximum expected earthquake. Predictions may be based on the seismic history of the area, but proposals are being made to install sensitive instruments at the feasibility or site investigation stage so that some records may be accumulated before dam construction. Such instrumentation would be maintained to record any disturbance caused by filling the reservoir, and to keep a record of the earthquake forces that act on the dam during its life.

Because the inertia forces developed by earthquake shock increase with dam height, it might be argued that the more modern, higher dams will be more liable to damage. Fortunately improvements in design and construction methods have more than counteracted this effect and it is the older, poorer built dams that suffer. Seed (1979) described the effects of the 1971 San Fernando Valley earthquake on the Lower San Fernando hydraulic fill embankment dam. It suffered an upstream rotational slip which removed the crest. At the time the reservoir was about 10.5 m below crest level so overtopping did not occur, but 80,000 people living downstream were immediately ordered to evacuate and the reservoir was further lowered.

Records of the motions caused by the earthquake were given by two seismoscopes installed, one on the rock abutment and the other on the crest of the dam. This one was found in the slide debris with its record undamaged. These two records showed that the slide movements involving the dam crest, did not occur during the main earthquake, but about half a minute after the main shaking stopped. This indicates that the slip was not due to the inertia forces under the shaking action but to reduction of shear strength caused by high pore pressures induced in the hydraulic fill of the upstream shoulder.

A pseudo-static analysis that used a horizontal force of 0.15 g to represent earthquake forces predicted a factor of safety of 1.3, indicating the necessity to use a dynamic analysis to assess dam stability under earthquake conditions.

This principle was further illustrated by a re-analysis of the behaviour of the Sheffield dam during the Santa Barbara earthquake of 1925 due to liquifaction of a layer of loose sand near the base of the embankment. A tailings dam on the island of Oshima, Japan, failed during an earthquake in January 1978, releasing about $1 \times 10^5$ m$^3$ of tailings contaminated with sodium cyanide. A pseudo-static analysis using a horizontal force of 0.2 g predicted a factor of safety of about 1.3.

These failures involved saturated sands and silts, but a detailed study described by Seed (1979) revealed that no failures have been reported in dams built of clayey soils even under the strongest earthquake shaking conditions. Dams of clay fills on clay or rock foundations have withstood extremely strong shaking, producing horizontal forces of 0.35 to 0.8 g caused by an earthquake of Richter magnitude 8.2, with no apparent damage.

The magnifying effects of height and shape were shown by the measurements made at Long Valley dam in California during an earthquake of magnitude 6.1 that occurred in May 1980. The 38 m high embankment rests on firm rock and the base experienced a maximum measured acceleration of 0.2 g, while at the crest the measured value was 0.5 g. A ledge of rock above the left abutment suffered a maximum acceleration of 0.99 g, showing that it was ringing like a tuning fork. The dam suffered no discernible damage.

In order to determine the response of a dam to earthquake shocks, it is becoming the practice to artificially excite the dam with large vibrating exciters whose frequency and phase can be altered to detect natural modes of vibration of the dam body.

Tests were made on an earthfill dam by Petrovski et al (1974) and what was claimed to be the first field vibration test by vibration machine on a rockfill dam as described by Nose et al (1976). Development of powerful, eccentric mass vibrators jointly by the Building Research Station and Bristol University led to their use to excite the rockfill Llyn Brianne dam. Four vibrators were deployed along the crest of the dam and their effect monitored by servo-accelerometer. All vibrators were initially set to run in-phase and their frequency increased from 2.6 to 2.9 Hz. The maximum response, indicating the first symmetrical mode of vibration occurred at 2.79 Hz. Further tests revealed antisymmetrical modes and behaviour was compared with the results of theoretical studies that used finite element techniques in both two and three dimensional dynamic analyses. This work has been described by Severn et al (1979, 1980).

In the past there has been perhaps an over-emphasis on the horizontal component, to the extent that the vertical acceleration caused by an earthquake has tended to be overlooked. Records from a local network can reveal valuable information on actual wave forms and indicate the directions and magnitudes of accelerations to be expected during a major earthquake along and normal to a given plane.

The instrumentation of a major dam in a seismic zone cannot be complete without strong-motion seismographs to record motion and acceleration in three-dimensions at sensitive positions on the crest and other parts of the dam. They will record induced seismicity, explain irregular values recorded by the other instrumentation and build up a body of experience on dam behaviour under earthquake conditions.

9  OLD DAMS

All necessary instrumentation can be installed in new dams, but there are many old dams with little instrumentation. Monuments and reference points can easily be built to enable any movements to be measured. Stress conditions in their cores are more difficult to assess.

At Cwmwernderi dam, a 20 m high, 80 year old dam in Wales, standpipe piezometers and oil-filled total pressure cells were installed in the central clay core in boreholes made from the crest. The piezometers measured the existing pore pressure distribution and enabled critical pressure tests to be made to assess total pressures. The total pressure cells were of the type described by Massarsch (1974) and consisted of a thin spade-shaped steel cell filled with oil connected to a transducer. They were pushed into the clay at the bottom of the boreholes for about twice their length by chain blocks on a crosshead over the boring rods, with reaction from screw pickets.

The equipment used by Massarsch was manufactured by Glötzl, but the equipment used initially at Cwmwernderi, made by BRS workshops, used a Glötzl steel cell 0.2 m long, 0.1 m wide and 5 mm thick, as the blade of the spade, attached to an

Fig 20  The BRS push-in spade-shaped total pressure cell of 2 mm thickness

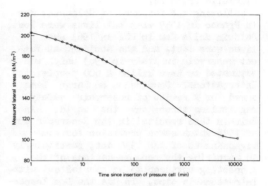

Fig 21  Dissipation of pressure after insertion of 2 mm thick spade-shaped cell

electrical transducer housed inside the boring rods. This arrangement enabled initial installation pressure and its subsequent dissipation to be measured in detail. It was found that it took 3 weeks for complete dissipation. To reduce the initial installation pressure, a 2 mm thick cell, shown by Fig 20, was made at BRS. It also used an electrical transducer and gave the pressure dissipation curve shown by Fig 21.
The cells were used at several depths in the dam core as each borehole progressed, where they were left in place only overnight. At these positions the values of the lateral total pressure after complete dissipation was extrapolated from the pressure v time plot by assuming a pressure dissipation curve geometrically similar to the one obtained with that particular cell when it was left in place for several weeks. The cells were orientated to measure the total pressures acting on vertical planes either across the core or on its centre line. At a later stage, Glötzl cells of the type described by Massarsch were put in boreholes that were backfilled, as part of permanent instrumentation for this old dam.
It is surprising that total pressures developed by pushing the spade cells into the soil ever dissipate completely. It could be expected that they would always record a pressure slightly in excess of the true lateral pressure. To check this aspect, a comparison was made by measuring total pressures also by critical pressure tests, with the Camkometer self-boring load cell and with the Camkometer self-boring pressuremeter. In the direction upstream-downstream (ie pressures on a vertical plane along the dam axis) the lateral total pressures measured by the Camkometer were, in general, slightly higher than those measured by the spade cells, while the critical pressure tests gave lower values. This work has been described by Penman and Charles (1981).

## 10  DAM SAFETY

Britain was one of the first countries in the world to have governmental regulations concerning dam safety. Under the Reservoirs (Safety Provisions) Act of 1930, a large dam may only be designed and its construction supervised by a competent civil engineer, defined as such by being a member of a special panel of reservoir engineers. Existing dams must be inspected at least every 10 years, when the inspecting engineer is able to see records of reservoir operation and measured leakage as well as records of instrumentation. With the older dams, not initially fitted with instruments to measure internal pressures, piezometers are often installed in sealed boreholes and, as mentioned above, total pressure cells are now being used to check that sufficient lateral pressure exists in the core to avoid risk of hydraulic fracture.

Following the failures of Malpasset dam in France in 1959 when 421 lives were lost, Baldwin Hells dam in USA in 1963 when 5 lives were lost, and the accident at Vaiont reservoir in Italy in 1963 which was estimated to have killed 2,600 people, the International Commission on Large Dams asked for a review of reservoir safety regulations throughout the world. In Britain this resulted in the Reservoir Act of 1975 which makes provision for some tighening up of the 1930 Act, particularly in requiring the recommendations of the inspecting engineer to be carried out within reasonable time. In USA the Dam Inspection Act of 1972 authorized the Secretary of the Army, acting through the Chief of Engineers, to undertake a national programme of inspection of all dams. An inventory drawn up by the Corps of Engineers, identified approximately 49,300 dams within the United States of which about 18 per cent had never been inspected under existing State or Federal authority. A manual was published in 1980 by the US Bureau of Reclamation as a guide for professional personnel making safety evaluations of existing dams. Entitled Safety Evaluation of Existing Dams, it is known as the SEED Manual. A background has been given by Jansen (1980).

In order to assess the problem of deterioration of dams, ICOLD set up a technical committee to collect information on type and extent of deterioration experienced in dams throughout the world. Its preliminary report in 1980 contains more than 1,100 cases of deterioration, 400 of which were associated with embankment dams and 330 with appurtenant works. Concrete dams provided 234 cases. Although first reservoir filling showed up many faults and the first five years could be regarded as the most dangerous time for a dam, the majority of cases of deterioration occurred in dams more than five years old. A full assessment will require not only consideration of the total numbers of each type of dams in existance but also consideration of design practices at the time when the dam was built.

It could be argued that a dam which survives reservoir operation for a few years could be kept in a safe condition for ever by adequate maintenance, provided that it does not suffer severe damage by fault movements, earthquake or military or other action by man intended to distroy the dam. Maintenance costs, particularly for a dam retaining a small reservoir, may eventually prove uneconomic and make it desirable to abandon the dam. British regulations provide for this and require an abandoned dam to be made incapable of retaining water so that it cannot constitute a safety hazard.

Effective maintenance can only be designed from a detailed knowledge of dam behaviour, much of which is provided by suitable instrumentation. During normal reservoir operation, sufficient information may be obtained from many of the instruments by taking readings infrequently, eg at intervals of a month or more These readings are often best taken manually so that instrument condition, calibration etc can be checked. Processing and recording the readings, however, particularly when a dam contains a lot of instruments, can often be most effectively carried out by electronic equipment. Excellent equipment has been developed to handle instrument readings from the Guri dam in Venezuela. All existing data can be retrieved from storage rapidly and the apparatus will plot any required values.

Electricite de France have regional observation offices responsible for collecting readings from, typically 40 local dams and sending the material in standard format to a processing centre. Here the readings are translated into engineering values by computer, stored on discs or magnetic tape and presented in tabular or graphic form, often with limit lines shown to indicate if the measured values are approaching design limits. Irregularities can be quickly identified and requests for repeat readings are sent to dam sites rapidly.

Similar arrangements are in operation in North America (Corps of Engineers), Spain, Portugal and Italy. A report on automated observation for the safety control of dams has been prepared by a task group of the committee on deterioration of dams and is expected to be published by ICOLD.

## 11 REFERENCES

Bishop, A W, Kennard, M F and Penman, A D M (1960): Pore pressure observations at Selset dam. Pore Pressure and Suction in Soils. Butterworths, pp 91-102.

Bjerrum, L and Andersen, K H (1972): Insitu measurement of lateral pressures in clay. Proc 5th European Regional Conf ISSMFE Madrid, vol 1, pp 11-20.

Casagrande, A (1949): Soil mechanics and the design and construction of Logan International Airport. Jnl Boston Soc Civ Engrs, vol 36, no 2

Cheney, J E (1974): Techniques and equipment using the surveyors' level for accurate measurement of building movement.

Field Instrumentation in Geotechnical Engineering. Butterworths, pp 85-99.

DiBiagio, E (1974): Contribution to discussion. Field Instrumentation in Geotechnical Engineering. Butterworths, pp 565-6.

DiBiagio, E, Myroll, F, Valstad, T and Hansteen, H (1982): Field instrumentation, observations and performance evaluations of the Svartevann dam. Trans 14th Int Congress on Large Dams, Rio de Janeiro, Q52, R49.

Gordon, J L and Duguid, D R (1970): Experiences with cracking at Duncan dam. Trans 10th Int Congress Large Dams, Montreal, vol 1, pp 469-485.

Jansen, R B (1980): Dams and public safety. U S Gov Printing Office, Denver, 332 pp.

Mallet, C and Pacquant, J (1951): Moyens et dispositifs de mesure des deformations des barrages en terre et de leurs assises Trans 4th Int Congress Large Dams, New Delhi, vol 1, pp 303-327.

Marsland, A and Quarterman, R (1974): Further development of multipoint magnetic extensometers for use in highly compressible ground. Geotechnique, vol 24, no 3, pp 429-433.

Massarsch, R (1974): Discussion on principles of measurement. Field Instrumentation in Geotechnical Engineering. Butterworths, pp 546-8.

Nagarkar, P K, Kulkarni, R P, Kulkarni, M V and Kulkarni, D G (1981): Failure of a mono-zone earth dam of expansive clay. Proc 10th Int Conf Soil Mech & Fd Engng, Stockholm, vol 3, pp 491-4.

Nose, M, Takahashi, T and Kuntl, K (1976): Results of earthquake observations and dynamic test on rockfill dams and their consideration. Trans 12th Int Congress Large Dams, Mexico, vol 4, pp 919-934.

Penman, A D M (1956): A field piezometer apparatus. Geotechnique, vol 6, no 2, pp 57-65.

Penman, A D M (1960): A study of the response time of various types of piezometer. Pore Pressure and Suction in Soils. Butterworths, pp 53-58.

Penman, A D M (1975): Earth pressures measured with hydraulic piezometers. Proc ASCE Specialty Conf on In-Situ Measurement of Soil Properties. North Carolina State Univ, vol 2, pp 361-381.

Penman, A D M, Burland, J B and Charles, J A (1971): Observed and predicted deformation in a large embankment dam during construction. Proc Instn Civ Engrs, London, vol 49, pp 1-21.

Penman, A D M and Charles, J A (1981): Assessing the risk of hydraulic fracture in dam cores. Proc 10th Int Conf Soil Mech & Fd Engng, Stockholm, vol 1, pp 457-462.

Petrovski, J, Paskalov, T and Jwukovski, D (1974): Dynamic full-scale tests of an earthfill dam. Geotechnique, vol 24, no 2, pp 193-206.

Seed, H B (1979): Considerations in the earthquake resistant design of earth and rockfill dams. 19th Rankine Lecture, Geotechnique, vol 24, no 3, pp 213-263.

Severn, R T, Jeary, A P, Ellis, B R and Dungar, R (1979): Prototype dynamic studies on a rockfill dam and on a buttress dam. Trans 13th Int Congress Large Dams, New Delhi, vol 2, pp 1075-1096.

Severn, R T, Jeary, A P and Ellis, B R (1980): Forced vibration tests and theoretical studies on dams. Proc Instn Civ Engrs, London, Pt 2, vol 69, pp 605-634.

Spangler, M G (1933): The supporting strength of rigid pipe culverts. Bulletin 112, Iowa State College of Agricult & Mech Arts, vol 32, no 37, pp 85-6.

Thomas, H S H and Ward, W H (1969) The design, construction and performance of a vibrating wire earth pressure cell. Geotechnique, vol 19, no 1, pp 39-51.

Vaughan, P R and Soares, H F (1981): Design of filters for clay cores of dams. Paper submitted to ASCE for publication.

# Grouting works in dam engineering

S.MARCHINI
*Radio S.p.A., Milano, Italy*

INTRODUCTION

A grouting work, also when properly done, can be considered satisfactory only if, after setting of the grout, it correspond to the aim required.

In other words, a consolidation grouting should improve the mechanical resistance of the soil, while a waterproofing grouting should reduce satisfactorily the permeability of the treated medium.

These considerations lead to distinguish two sorts of grouting according to the scope: Static and Hydraulic. Apparently, for static or mechanical purposes, a grout with high compressive resistance should be used while for hydraulic purposes a grout with a minimum permeability should be suitable.

The reality is more complex and sometimes it is possible to consolidate a fissured rock using a grout with a low cohesion while a grout purely impervious is useless for waterproofing purposes if the hydraulic head can displace it or punch it.

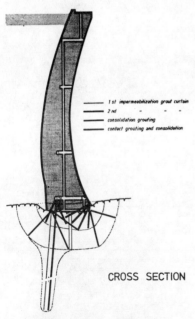

Fig. 1  Cross Section

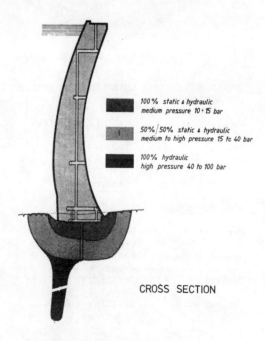

Fig. 2  Cross Section

Coming to the problem of grouting works in dam engineering we can see that both above mentioned treatments intervene and interfere (fig. 1).

We can observe that the trend is, from topsoil to the deepest point of impervious curtain a decreasing intensity and need of consolidation, and an increasing in the whole of waterproofing (fig. 2).

The geometry of this scheme of treatment depends on the RQD, the layers bedding, the dip, and type of rock, not only but also on the shape of the valley because of various considerations.

EFFECT OF SHAPE OF VALLEY

First of all let us examine the shape of valleys in various cases and its influence on the planning of grouting treatment. (fig. 3, 4, 5, 6).

IMPERVIOUS CURTAIN CONSOLIDATION SEALING CHARACTERISTICS:

Type:                  Gravity arch
River:                 Reno di Lei
Crest length:          642.00 m
Height:                141.00 m
Reservoir capacity:    197000000 $m^3$
Soil foundation:       Schist

WORK PERFORMED:

Curtain:         Drilling    24930   m
                 Grouting    2280    t
Consolidation:   Drilling    21840   m
                 Grouting    2450    t
Sealing:         Drilling    15400   m
                 Grouting    212     t
Various works:   Grouting    212     t

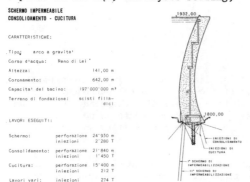

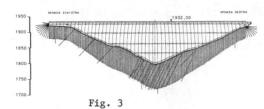

Fig. 3

IMPERVIOUS CURTAIN CONSOLIDATION SEALING CHARACTERISTICS:

Type:                  Cellular gravity
River:                 Troina
Height:                110 m
Crest length:          253 m
Reservoir capacity:    30500000 $m^3$
Soil foundation:       Sandstone

WORK PERFORMED:

Curtain:         Drilling    27163   m
                 Grouting    5861    t
Consolidation:   Drilling    43431   m
                 Grouting    5572    t
Sealing:         Drilling    8717    m
                 Grouting    818     t
Various works:   Drilling    1220    m
                 Grouting    2890    t

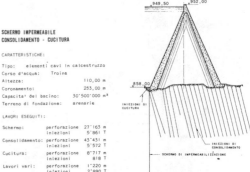

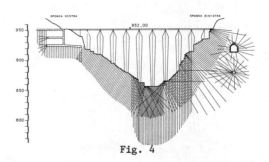

Fig. 4

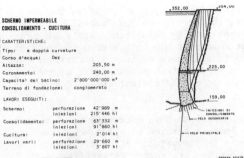

**IMPERVIOUS CURTAIN CONSOLIDATION SEALING CHARACTERISTICS:**

| | |
|---|---|
| Type: | Double Arch |
| River: | Dez |
| Height: | 203.50 m |
| Crest length: | 240.00 m |
| Reservoir capacity: | 2800000000 |
| Soil foundation: | conglomerate |

**WORK PERFORMED:**

| | | | |
|---|---|---|---|
| Curtain: | Drilling | 42989 | m |
| | Grouting | 215446 | t |
| Consolidation: | Drilling | 63332 | m |
| | Grouting | 91860 | t |
| Sealing: | Grouting | 2014 | t |
| Various works: | Drilling | 29660 | m |
| | Grouting | 3807 | t |

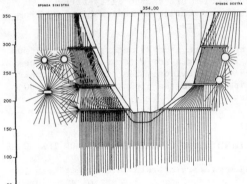

Fig. 5

**IMPERVIOUS CURTAIN CONSOLIDATION SEALING CHARACTERISTICS:**

| | |
|---|---|
| Type: | Cellular gravity |
| River: | Chiese |
| Height: | 8700 m |
| Crest length: | 561.00 m |
| Reservoir capacity: | 60000000 m$^3$ |
| Soil foundation: | diorite-tonalite |

**WORKS PERFORMED:**

| | | | |
|---|---|---|---|
| Curtain: | Drilling | 10980 | m |
| | Grouting | 1284 | t |
| Consolidation: | Drilling | 5046 | m |
| | Grouting | 231 | t |
| Sealing: | Drilling | 20452 | m |
| | Grouting | 376 | t |
| Various works: | Drilling | 2236 | m |
| | Grouting | 90 | t |

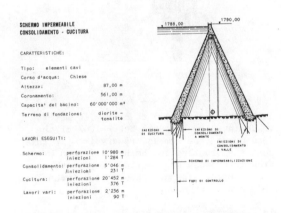

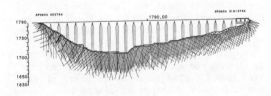

Fig. 6

Let us examine the case of fig. 3.

This is a case of a double arched dam in a wide valley. The spacing is decreasing and the depth of grouting holes is increasing from the top to the bottom.

The cross section schematized in fig. 7, shows with evidence the three zones affected by the treatment.

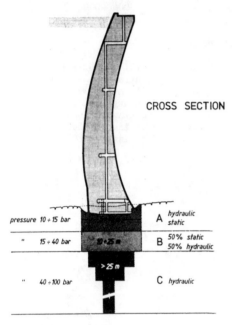

Fig. 7

A shallow zone from 5 to nearly 10 meters of depth where the soil bears 100% of the load due to the dam and reservoir: in this zone the grouting for both static and hydraulic purpose are say 100% important and the grouting pressure should be between 10 and 15 bars or more.

This treatment should not be limited to the area directly under the dam but extended upstream and downstream because of the deformation of the reservoir during first impounding.

To obtain the maximum performance, the treatment of this zone should start after ten meters of concrete is placed because, due to the natural fractures and joints open on the surface upward and mostly due to the blasting, the grout tends to come out on the surface and it is rather difficult to build up a pressure.

To avoid the leakage of grout on the surface, the process called "Dental concrete" is applied. By this method it is assumed that a patient man with a trowel should be able to seal more or less all the fissures. It is believe that a thin layer of gunite should be more efficient and more economical.

A medium depth zone – from 10 to 25 m where the load is carried by a larger volume of rock – the importance of static and hydraulic problem are shared fifty-fifty. The pressure grouting is medium-high say from 15 to 40 bars.

A third zone from 25 meters to the maximum depth required for waterproofing purposes and where obviously the problem is purely hydraulic and the pressure can reach 100 bars and more.

## DETERMINATION OF MAXIMUM DEPTH OF THE HOLES

Lugeon theory or rule is of great help for that purpose and a 60 years of experience have confirmed his validity. According to Lugeon, a rock formation can be considered reasonably impervious when a water test shows a take of 1 Lugeon unit which correspond to one liter per minute per meter of hole (1" 1/4) at a pressure of 10 bars.

APPROXIMATE CORRELATION BETWEEN LUGEON UNIT AND PERMEABILITY COEFFICIENT "K"

RADIAL FLOW THROUGH A CYLINDRICAL SURFACE :

$$K = \frac{Q \cdot 1_n (R/r)}{6.28 \cdot L \cdot H} \text{ IN CM/SEC} \quad (1)$$

WHERE :
- Q = FLOW IN $CM^3$/SEC
- R = RADIUS OF ACTION
- r = RADIUS OF HOLE
- H = HYDRAULIC HEAD
- L = LENGTH OF FILTERING SECTION

ASSUMING :
$R = 500 \cdot r$

THE LOGARITHM OF 500 BEING ABOUT 6.28, WE OBTAIN :

$$K = Q/L \cdot H \quad (2)$$

ONE LUGEON UNIT IS :
- Q = 1 LITRE/MIN = 1000 / 60 = 16.6 $CM^3$/SEC
- L = 1 M = 100 CM
- H = 10 ATM = 10'000 CM

THEREFORE :

$$K = \frac{16.6}{10^2 \cdot 10^4} = 1.66 \cdot 10^{-5} \text{ CM/SEC}$$

Fig. 8

The test should last ten minutes. This is an empirical criteria but successfully applied since 60 years ago for the **foundation** treatment of most important dam all around the world.

The value of one Lugeon, from point of view of permeability coefficient K is $1,66 \cdot 10^{-5}$ cm/sec. This value is obtained as shown in fig. 8.

The value of 1 Lugeon is valid to consider sufficiently watertight the rock foundation of a dam higher than 30 meters but for a lower dam it can be acceptable a value of 2 or 3 Lugeon. The test is carried on a 5 meters length of hole. The method applied is a down stage one and serve the purpose of fixing the depth of the impervious curtain. The depending of the test boring is stopped when the water test indicates a value of one Lugeon or more according to the height of the dam.

The grout holes, should be, as much as possible directed normal to the stratification or the dip in order to cross and interest the most important discontinuities in sedimentary and layered formation as aureoles in presence of intensively broken rock and karstified limestones (fig. 9).

IMPERVIOUS CURTAIN CONSOLIDATION SEALING CHARACTERISTICS:

Type: Cellular gravity
River: Ladhon
Height: 56.00 m
Crest length: 106.00 m
Reservoir capacity: 50000000 $m^3$
Soil foundation: limestone

WORKS PERFORMED:

| | | | |
|---|---|---|---|
| Curtain: | Drilling | 6465 | m |
| | Grouting | 755 | t |
| Sealing: | Drilling | 170 | m |
| | Grouting | 56 | t |
| Various works: | Drilling | 755 | m |
| | Grouting | 227 | t |

As it is shown in the fig. 9, following the sequence of drilling and grouting, each aureole operates also as a control of the preceding one. For example, the zone treated by the grouting holes 1,2,3,4,5,6 (left abutement) is checked by the subsequent grouting holes a.b.c.d.e.

The process was successfull and efficient. After the impound was completed, no leakages appeared downstream of the dam.

A quite special case of intervention to realize an impervious screen underneath a dam is shown in fig. 10-11.

The investigations carried out before the beginning of construction (fig. 10) had put in evidence an intensively broken rock on the left bank but further information gathered during the works had shown that the supposed broken rock was in reality a very large geological slide, slipped on the surface of in-situ rock ancient bottom of a lake.

To realize an impervious cut-off, it was necessary to build a real concrete completed by a perimetral intense grouting programme. A further consolidation programme was carried out downstream of the concrete wall in order to minimize the possibility of deformation due to the hydrostatic load.

The control of the success is normally carried out by water tests through holes directed as to cross a great number of grouting holes in order to realize a sort of map showing zones with residual permeability requiring a further grouting treatment (fig. 12).

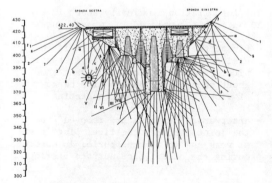

Fig. 9

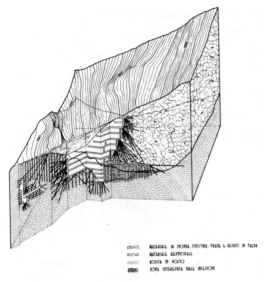

Fig. 10

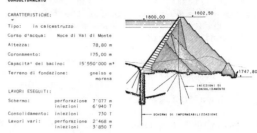

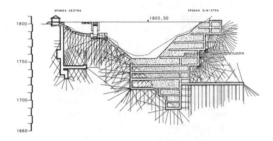

Fig. 11

A comparative test of the results is also realized by sesmic method, checking the sound velocity before and after grouting and consequently the increase of bulk density obtained by grouting.

Other control obviously is obtained by piezometric measures.

## GROUTING OF CONSTRUCTION JOINTS

Construction joints are grouted by use of the plug valve. The valve is composed by a T shaped cast iron body; the T head is flared and slightly cone shaped in order to join the valve tothe grouting pipes whereas the T stalk is decidedly cone shaped; it receives a particular rubber plug consisting of a conical hard rubber portion and of a cylindrical pure rubber part with several recesses inside which make its deformation possible.

The valve may be used separately for grouting by a small double packer, or used in conjunction with other valves placed in the same pipeline.

The peculiarity of this valve is the accessibility for cleaning the pipe from eventual foulings by metallic tools because on a complete grouting line equipped with several valves there are neither recesses nor projecting parts which makes possible washing and cleaning.

IMPERVIOUS CURTAIN CONSOLIDATION CHARACTERISTICS:

| | |
|---|---|
| Type: | Concrete gravity |
| River: | Noce di Val di Monte |
| Height: | 78.80 m |
| Crest length: | 175.00 m |
| Reservoir capacity: | 1550000 $m^3$ |
| Soil foundation: | gneiss e morena |

WORKS PERFORMED:

| | | | |
|---|---|---|---|
| Curtain: | Drilling: | 7077 | m |
| | Grouting | 6940 | t |
| Consolidation: | Grouting: | 730 | t |
| Various works: | Drilling: | 2468 | m |
| | Grouting: | 3850 | t |

The installation is clearly shown in fig. 13 and it is completed in two sequences:

- first, the cone shaped part is fitted against the form wall and it is supplied with a temporary plug to prevent some grout seeping inside the pipe because of vibration during the concreting.

- afterwards the forms are removed from the joint surface, the final plug is set at work and it remains buried in concrete during the other joint surface casting.

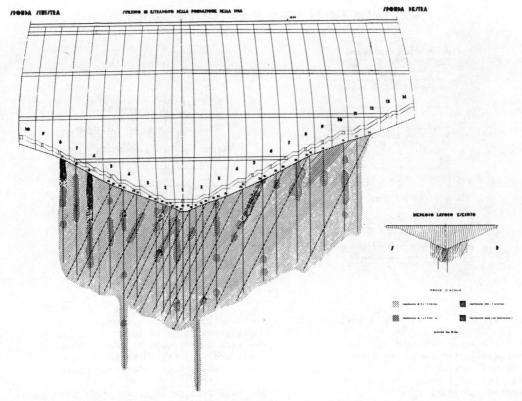

Fig. 12

During the grouting process, the grout lifts the conical plug because of pressure and this lifting is allowed by the top cylindrical part which can lose its original shape owing to several airfull recesses; when the pressure is dropped, the cylindrical part takes its outline dimensions again, pushing the conical plug into its seat, achieving the seal and allowing to grout again later on.

This valve was tested by making a joint in about one cubic meter concrete block split in two halves kept touching by some steel tension bars. The valve was grouted three times by cement mortar with a distance of about twenty days between.

GROUTING PRESSURE

In fig. 7 it is schematized which pressure grouting is suggested to adopt at various depth. Apparently these pressures could be considered too high and possibly may produce an uplift or displacement of the rock on the surface. In reality these inconveniences do not result exactly from the pressure but from a number of causes as follows:

The dip of strata: obviously the maximum possible danger of uplift exists when the bedding is horizontal or subhorizontal or parallel to the excavation surface where, if the grout penetrates in a large surface interstata, it could act as a flat jack.

In such a case the control of the grouting operation should be mostly on the take of grout and the trend of absorption; a great take and low pressure can be more dangerous than a limited absorption and high pressure. A severe control of trend, readable on the diagram of the recording manometer can give immediate signal of danger and the grouting must be stopped (fig. 14). An automatic stop of pumps can be obtained by electromagnetic or electronic devices existing in commerce.

## PLUG VALVE FOR GROUTING OF CONTRACTION JOINTS

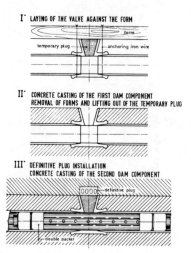

Fig. 13

Still with regard to grouting pressure and the danger of uplift, it is possible to limit the areas of grout influence and consequently the danger, by drilling a pattern of relief holes at the edges of said area and pumping the grout through a central hole (fig. 15).

By such a method, the grout can not penetrate farther of the relief holes alignment.

An effective control of uplift is possible with a meter (fig. 16) sensitive for measuring displacement of 0,5/100 of millimeters.

The device can be equipped with acoustic alarm and an electronic automatic device for stopping the injectors.

The selection of grouting pressure should take in due consideration the characteristics of grout; for a Binghamian fluid, the thickness of mix, the grain size of components, the stability, viscosity and setting time; in case of a Newtonian fluid, the density, viscosity, setting time reaction and polimerization time.

All the above mentioned characteristics have great importance from the point of view of head loss and the selection of final grout pressure.

We have examined the grouting procedure for concrete dams on rock formations but since some ten of years ago, earth and rockfill dam construction has achieved more and more importance, it is of interest to examine the grouting process in loose soils, alluvials and similars which normally are the soil foundation of said structures.

The main purpose of grouting is impermeabilisation or to reduce the permeability of the medium to acceptable values according to the height and other characteristics of the dam (fig. 19).

Due to the degree of deformability both of the dam and of the foundation we can exclude the use of a grout very rigid and with high compressive resistance and adopt a product which is stable, impermeable, and has; good plasticity after setting, the necessary rigidity to resist the hydraulic gradient, & the possibility of penetrating inside the soil structure and fill the pores.

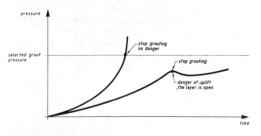

DIAGRAM OF RECORDING MANOMETER SHOWING THE TREND OF GROUTING OPERATION

Fig. 14

218

Sampling, permeability in situ test, particle size distribution analysis, porosity and other characteristics of the soil, give the necessary information for the selection of components of grout.

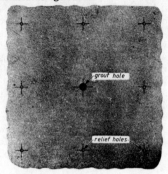

Fig. 15  Example of distribution of relief holes for the purpose of limiting the area of grout penetration

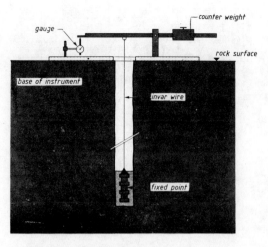

DEVICE FOR UPLIFT CONTROL

Fig. 16  Device for uplift control

The following figure (fig. 17) shows the various properties and characteristics of grout and the field of application.

The next figure (fig. 18) shows the granulometric curves relatives to the penetration limits of suspensions and colloidal solutions according to several authors.

METHODS FOR WATERPROOFING THE FOUNDATION OF A ROCKFILL OR EARTH DAM ON ALLUVIAL FORMATION

Several methods are suitable for this purpose, for instance continuous concrete wall, slurry trench, sheet piles, caissons etc. We will examine the grouting method. In-

| NATURE OF SOIL | | GRAVEL AND COARSE SAND | MEDIUM TO FINE SAND | SILTY SAND TO CLAY. |
|---|---|---|---|---|
| CHARACTERISTICS of SOIL | Grains size | $d_{10} > 0.5$ mm | $0.02$ mm $< d_{10} < 0.5$ mm | $d_{10} < 0.02$ |
| | distribution | $S < 100$ cm$^{-1}$ | $100$ cm$^{-1} < S < 1000$ cm$^{-1}$ | $S > 1000$ cm$^{-1}$ |
| | Permeability | $K > 10^{-3}$ m/sec | $10^{-3}$ m/sec $< K < 10^{-5}$ m/sec | $K < 10^{-5}$ m/sec |
| NATURE OF PRODUCT FOR GROUTING | | BINGHAMIAN FLUIDS | NEWTONIAN FLUIDS | |
| | | suspensions | colloidal solutions | pure solutions |
| SCOPE OF TREATMENT | CONSOLIDATION | water-cement (only for $K > 10^{-2}$ m/sec) aerated mix | process Joosten (only for $K > 10^{-4}$ m/sec) silica hard gel with organic reagent | phenolic resins aminoplastic resins |
| | IMPERMEABILIZATION | water-cement-clay aerated mix clay gel | silica gel with inorganic reagent emulsions lignochrome | phenoplastic resins aminoplastic resins acrilic resins |

Fig. 17  Limits of groutability of main products for grouting purposes

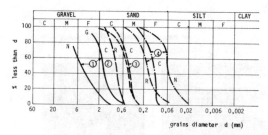

GRAIN SIZE CURVE CORRESPONDING TO THE LIMITS OF PENETRABILITY WITH SUSPENSIONS AND COLLOIDAL SOLUTIONS, ACCORDING TO VARIOUS AUTHORS (FROM CAMBEFORT, 1967)

| TYPE OF MIXES | AUTHORS |
|---|---|
| 1 cement suspensions | C = CARON |
| 2 cement-clay suspensions | G = GOLDER |
| 3 clay suspensions | R = ROCKWELL-SMITH |
| 4 sodium-silicate solutions | N = NEUMAN |

Fig. 18

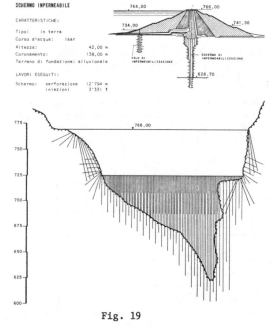

Fig. 19

IMPERVIOUS CURTAIN CHARACTERISTICS:

Type:             earth
River:            Isar
Height:           42.00 m
Crest length:     138.00 m
Soil foundation:  alluvial

WORKS PERFORMED:

Curtain:          Drilling    12794  m
                  Grouting     3331  t

stead of fixing the grouting pressure for various depth, according to the data of water tests, as in case of rock formations porosity, grain size distribution & a quantity of grout per cubic meter of soil is fixed. Said quantity or volume should be approximately 50% ÷ 60% of the percentage of voids; for instance the natural porosity being 40% it is reasonable to fill 60% of said voids reducing the porosity to 16% which is a good performance

If the selected pressure, the quantity, the flow and, let us say, the quality of grout is correct, then a good impregnation is obtained and the aim of the grouting work is attained. Let us see how to procede. Various method existing for this purpose.

We will examine one we consider the most efficient. This method is known as "Tube a manchettes" (fig. 20). Let us have a look how it works.

As shown in fig. 20, the injection holes are equipped by a tube (steel or plastic) with one-way valves (3 or 4 per metre) surrounded by a clay-cement sealing sheath - the "sleeve grout".

Each valve consists of a ring of small holes enclosed in a short rubber sleeve fitting tightly round the rube. The sleeve grout, injected through the bottom valve, fills the annular space between the hole and valved tube, sealing the same in the ground.

When the sleeve grout has reached a certain strength (defined by local conditions and prior experience), injections are made through the manchette tube by lowering into it an injection pipe fitted with a double-packer, which can be cantered over any one of the valves. When injection starts, the pressure builds up until the grout raises the rubber sleeve, breaks the sleeve grout and flows into the surrounding soil. The rubber sleeve prevents any short-circuiting of the grout back into the tube, and the sleeve grout prevents it from creeping upwards.

The great advantages of this method can be summarized as follows: the valved tube, when in place, can be used many times with a whole range of different grouts, which enable the coarser beds to be treated first and the finer ones later, with no need to know exactly their sequence beforehand; grouting can be resumed in any hole and through any valve, when necessary; and the operations of boring and injection can be carried on separately at any time, which simplifies site organization.

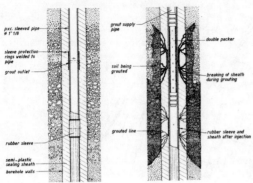

Fig. 20

These advantages are enhanced in tunnelling, which requires the greatest flexibility in design and practice.

Controls during Progress of Injection

The results of previous grout tests must be checked at a site laboratory for the most significant properties. The design specifications for these properties may require some adjustment of the composition because of the influence of the actual mixing procedure and other variable (or unexpected) factors - for example, the effect of temperature on the setting time of chemical mixes.

Control during the job involves continuous monitoring of the grout composition, flow rates and pressures at each level of injection; moreover, any uplift of the ground and any escape of grout to the surface should be recorded and prevented quickly.

Most theoretical and empirical criteria for limiting pressure are unreliable or overconservative. The best method (suitable to specific site conditions) is based on the direct determination of the hydrofracturing (claquage) threshold by increasing the flow rate until the pressure remains constant or decreases. Such tests enable allowable pressures to be derived from experimental ultimate values.

Controls on Grouted Soils

Field testing of grouted soils is not a common practice. The usual procedure for checking results is to use safeguards during the actual grouting rather than make subsequent tests in the soil. Nevertheless, some type of direct test may be needed to show whether the grout mixes have adequately permeated the soil and provided the required properties.

Field permeability tests are obviously basic for water shut-off applications. A reduction of permeability also gives an indication of strength increase if consolidating grouts have been used.

When permeability tests are carefully performed (in sufficient suitably equipped boreholes) and the results are compared with pre-grouting data, the effectiveness of the treatment can be checked. As regards the actual strength of grouted soil, undisturbed sampling or in-situ strength test are required.

The dimensions of the impervious curtain are fixed according to the abovementioned data and, similar to the case of concrete dam on rock foundation, there are three zones of different requirements.

The upper part of the foundation is treated through five or more rows of tubes a manchettes; in the medium portion, between ten and twenty meters of depth the rows are reduced to three; from twenty meters of depth and the maximum depth fixed previously and according to the in-situ tests, the rows are only two set of straggered.

OTHER METHOD OF GROUTING ALLUVIALS

Driven Point

This type of tool is a small size injection tube with a pointed end. It is driven into the ground by a hammering device and the pointed end permits the tool to penetrate into the soil without becoming plugged.

The end of the tool can be opened for grouting at the planned depth by lifting it while the point remains fixed into the soil. Injection can be made down-stages or up-stages. This method is suitable for shallow depths and loose or soft soils.

Another system uses a drill bit inside a steel casing tube. When the hole reaches

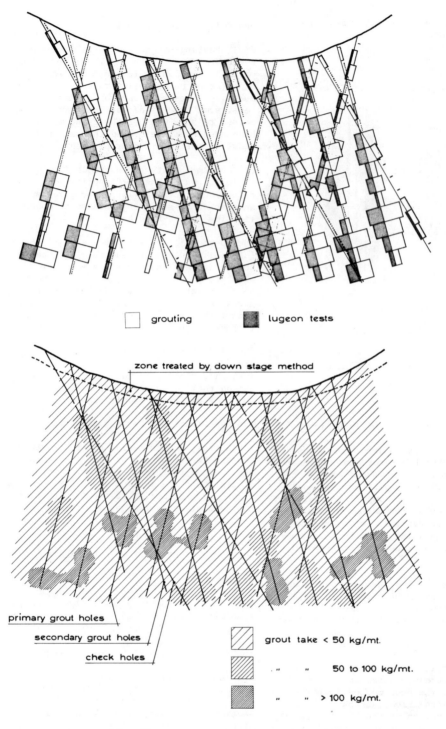

Fig. 21  Lugeon tests and grout intake data

the planned depth, the drill bit is ejected and the drill rod withdrawn. The other tube containing openings in milled slots covered with leaf springs acting as one-way valves remains in place and through a double packer similar to the tube a manchettes, grouting is accomplished.

In comparison with these methods, the tube a manchettes has many advantages, for instance: being constructed by plastics, if a portion of treated soil has to be excavated, the tubes a manchettes do not produce any difficulty as they can be easily destroyed. Also it is easy to ship & they are non corrodible by eddy currents etc.

## CHEMICAL GROUT

We have seen (fig. 17) that when the permeability K of the soil to be treated is between $10^{-3}$ and $10^{-5}$ m/sec, chemical grout (Newtonian) should be employed, for the treatment of impermeabilisation or/and consolidation.

According to the scope of work, many products are suitable, for instance, among sodium silicate based. For shallow treatment in presence of low water head a soft gel can give good result. Such gel can be obtained with sodium silicate and calcium carbonate. A more consistant gel is obtained with sodium silicate and ethil acetate.

For very quick setting, sodium silicate with calcium chloride is very efficient (Joosten process) but the two products must be injected separately (two shot) and get in contact inside the soil because the reaction starts immediately and the gel begins to harden. Specialist prefer a one shot chemical grouting based on sodium silicate and inorganic reagent.

For soils showing a K < $10^{-5}$ m/sec only pure solutions are suitable. These products are based on resins, phenolic, aminoplastics and acrilic. Caution has to be used with these products because of possible toxicity and danger of pollution.

Recent applications of mix in place methods have been perfected by Japanese and good results are obtained for consolidation of sands, fine sands, silty and clayey sands. Grout is injected at a very high pressure through a special nozzle driven into soil by rotation equipment.

## SUGGESTED GROUTING METHOD

According to the experience and taking in due consideration the technological, economic efficiency and the rapidity aspects of the method, the conclusion reached is that in the majority of cases, the best is a down stage up to 12-15 meters of depth and an upstage method from 12-15 to the full depth. Various steps in the procedure are:

a) First of all the rock surface should be covered by some meters of concrete or a layer of gunite in order to avoid grout losses in the surface and an incomplete sealing of the fissures, open mostly toward the surface.

b) Drill 5 meters, set a packer near the surface and grout according to the specifications.

c) Redrill the 5 meters and drill further 5 meters, set the packer at 5 meters depth and grout.

d) Redrill 10 meters and drill up to established full depth.

e) Set the packer 5 meters upward of the bottom of the hole.

f) Execute a Lugeon test.

g) Grout at specified pressure.

i) Lugeon water test.

k) Grout.

Carry on until grouting is completed.

Correlating the sections showing approximately the same in take we obtain a map of absorption and accordingly we intervene with secondary grouting. (Fig. 21)

After the secondary grouting operation is completed, check holes are drilled and grouted and the related data recorded indicates, by the absorption map, any improvement obtained and if necessary a third grout intervention is carried on.

The main advantage of the method is represented by the possibility of controlling the efficiency of the operation and to be a guide for economical development in connection with technical requirements.

*Symposium on Problems and Practice of Dam Engineering / Bangkok / 1-15 December 1980*

# Seepage through jointed rock
# Sealing and drainage measures for earthfill and masonry dams

K.H.IDEL
*Ruhrtalsperrenverein, Essen, Germany*

SUMMARY. Using calculatory models, which were tested on many existing dams, it is possible to estimate the effect of grout curtains, as well as the loss of seepage water and the hydrostatic pressure in rock masses, when the direction, degree of penetration and the opening widths of the system of discontinuities is surveyed adequately by means of borings and water pressure tests.
As a rule, a grout curtain is not necessary when the seepage losses of the dam are minimal and the discontinuities are not enlarged by erosion of the fillings or possibly a chemical decomposition of the rock. This is especially true for narrowly spaced discontinuities and for discontinuities under 0,1 mm width.
In large discontinuities the range of a grouting is very large so that when a large area of the rock mass is sealed a successfull grout curtain is obtained.
For varying opening widths and a danger of the erosion of the filling of the discontinuities and a chemical decomposition of the rock, the sealing of the larger discontinuities with the aid of a cement suspension, and if the flow velocity in the remaining small discontinuities of under 0,1 mm width is still too high, an additional chemical grouting, is necessary.
For static reasons the reduction of the uplift force and the hydrostatic water pressure below certain areas of a dam wall can be necessary. This can only be done by means of drainage borings, which themselves, disadvantageously increase the hydraulic gradient in the rock mass. As a compensation in erosion endangered rock in front of the drainage area, the larger discontinuities must be sealed by grouting.

The rock mass upon which a dam is founded is never homogenous, but rather is jointed by discontinuities; see Figure 1. In fissures that form a connection between the storage area and the downstream side, seepage water is lost.
An estimation of the seepage water losses as well as the flow pressure in the rock mass and the size of the hydrostatic pressure upon the dam are important prerequisites for the performance of sealing measures such as grouting of a cement suspension or of pressure relief measures such as the arrangement of

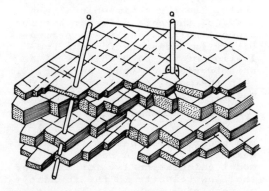

Fig. 1. Fissured rock

drainage borings that might possibly be necessary.

In addition to the engineering-geological mapping of the geometry of the discontinuity structure, water pressure tests in boreholes, also called WD-tests, have been used as a further aid in the determination of the permeability of a rock mass in the immediate vicinity of, and in the section of the borehole in which the WD-test is performed.

The width of and the number and direction of the discontinuities cannot be determined by the WD-test solely from the knowledge of the injection velocity at a certain injection pressure. Several small discontinuities can have the same uptake capacity as one correspondingly larger discontinuity. By planning the directions of the borings to include the various sets of discontinuities with varying permeability and by means of a variation of the injection pressure, the flow and stress relationships in the rock mass can be clarified using the WD-tests even before the reservoir is filled. In the last 15 years basic principles for a numerical treatment have been created using jointed body models with fissures of varying width and roughness. Using calculatory models it is possible today, to estimate the effect of a sealing measure or of the drainage of a rock mass. We are indebted to Prof.Dr.Wittke and his students for important results in this area of research. In his dissertation "Determination of the Water Permeability of Joint Rock", RWTH Aachen, 1977, Dr.Rißler reported on the evaluation of WD-tests. The three-dimensional analysis of the sets of water bearing discontinuities and their permeability is necessary for the evaluation of the permeability of an anisotropic rock mass; see Fig. 2.

In the model calculation the permeabilities

$$k_{11}, k_{22} \text{ and } k_{33}$$

are grouped together to form a permeability tensor. A result of the model calculation is, for example, the flow velocity in the foundation rock of a dam wall after a grout curtain has been installed; see

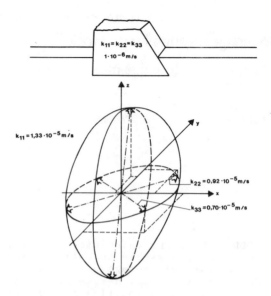

Fig.2. Joint directions underneath Listerdam

Figure 3.
In the upper picture the rock is assumed to have uniform $k_G$ value, while in the lower picture a schistosity in the rock mass is assumed. Here, the grout curtain is bound to a more compact layer. The flow velocity in the lower picture is increased in the upper layer of the foundation rock.

Figure 4 shows a comparison of the seepage water losses and rock mass without a grout curtain using the same model, with the two variations of the grout curtain shown in Figure 3. On the basis of the results, one could come to the conclusion that only the slight reduction of the total seepage water loss from $Q = 0,61$ l/s . m to $0,51$ l/s . m and the reduction of the uplift force from $A = 7$ MN/m to $A = 6.9$ MN/m does not justify the effort of rock sealing measures.

A drainage directly behind the grout curtain, see Figure 5, has a considerably more favorable effect upon the reduction of the uplift force from $A = 7$ MN/m to $A = 3,79$ MN/m than the grout curtain itself does. This is, however, accompanied by an increase in the amounts of seepage water from $Q = 0,61$ to $Q = 1,05$ l/s . m.

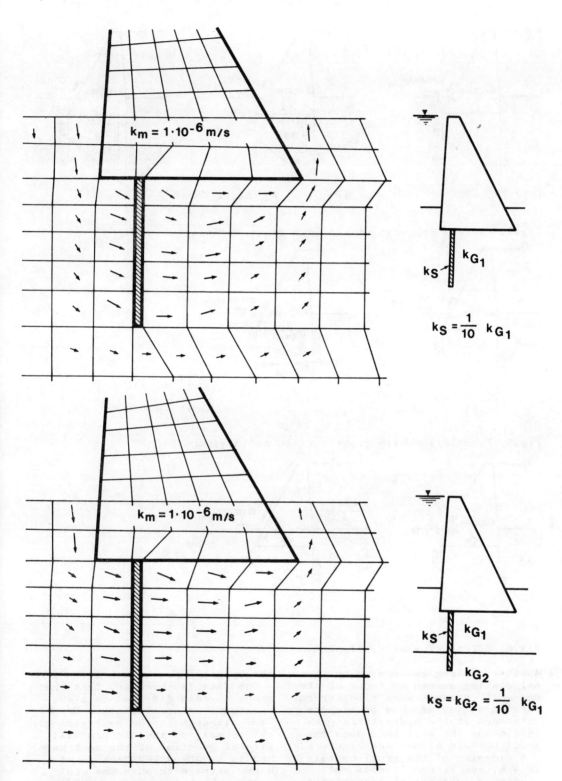

Fig. 3. Percolation velocity underneath Listerdam

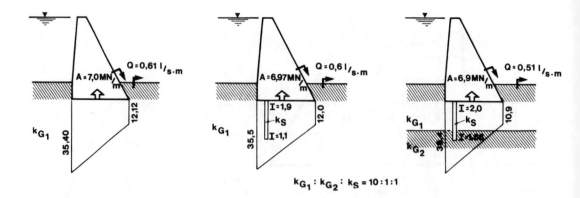

Fig.4. Percolation without and with grout curtain

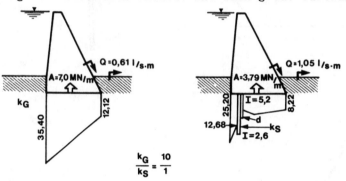

Fig.5. Percolation with grout curtain and drainage

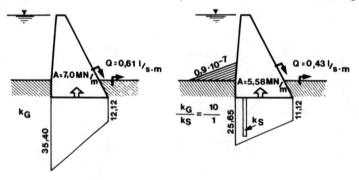

Fig.6. Effect of sealing blanket

Another sealing measure, a horizontal sealing screen in front of the dam wall, demonstrates a better reduction of the amounts of seepage water and of the hydrostatic pressure below the wall than does the grout curtain alone; see Figure 6.

A widening of the grout curtain to 4 m, see Figure 7, does not achieve a substantial reduction in the amount of seepage water and the uplift force.

The fact is, however, that the grout curtain did lead to a successful sealing at the new Henne Dam, although in the same geological formation, the old dam wall without grouting of the rock mass had to be abandoned because of heavy seepage through the abutments, which even endangered the stability of the wall. Thus, there

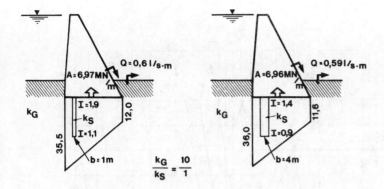

Fig.7. Percolation with small and wide grout curtain

are more factors that must be taken into consideration of the effectiveness of grout curtains, than just the amount of seepage water and the hydrostatic water pressure. Often, the prerequisites made for the evaluation of the Figures 2 - 7 are not fulfilled. The discontonuities are completely or partially filled with clay and silt, or the rock is not chemically inert, so that the fillings of the discontinuities can be eroded by the seepage flow or the discontinuities can be enlarged by chemical decomposition of the rock. In this manner very large flow velocities had developed below the old Henne Dam, which led to a rapid destruction of the rock structure. As will be shown later using a calculated example, the successfull sealing of the rock mass of the new Henne Dam is due to the sealing of the large discontinuities. The flow velocity in the many small discontinuities was only slightly influenced by the sealing measures since the discontinuities smaller than 0,1 to 0,2 mm are not capable of taking up any of the cement suspension. By plugging the larger water channels, it was also possible to prevent the erosion of the fillings of the discontinuities.

To make a calculated estimation of the effect of a grout curtain we must, in addition to the average permeability tensor of the anisotropic rock mass, also know the proportion of the discontinuities, the width of which, allows the entrance of the cement grout. After these larger discontinuities have been sealed, the approximation only depends upon the flow velocity in the smaller discontinuities of 0,1 mm and less that remain open. From systematic model tests, supplemented by hydraulic calculations, it was possible to determine the actual flow velocity in discontinuities of differing width with rough and smooth surfaces; see Figure 8.

In a rock mass block with an edge of 1 m the same coefficient of permeability

$$k_f \text{ of } 1,7 \cdot 10^{-5} \text{ m/s}$$

results for smooth discontinuities according to Figure 8 for a penetration of 1 discontinuity that is 0,3 mm wide, or of 10 discontonuities that are 0,14 mm wide, or even of 30 discontinuities that are 0,1 mm wide. The actual flow velocities are: in the large discontinuity

$$V_K = 5,4 \cdot 10^{-2} \text{ m/s;}$$

in the 10 discontinuities of 0,14 mm width

$$V_K = 1,4 \cdot 10^{-2} \text{ m/s;}$$

and in the 30 small discontinuities only

$$0,6 \cdot 10^{-2} \text{ m/s,}$$

that is only 1/10 the flow velocity in the large discontinuity. After the sealing of the large discontinuities, the true flow velocity in the small discontinuities still open is recuded even more. Experience has shown

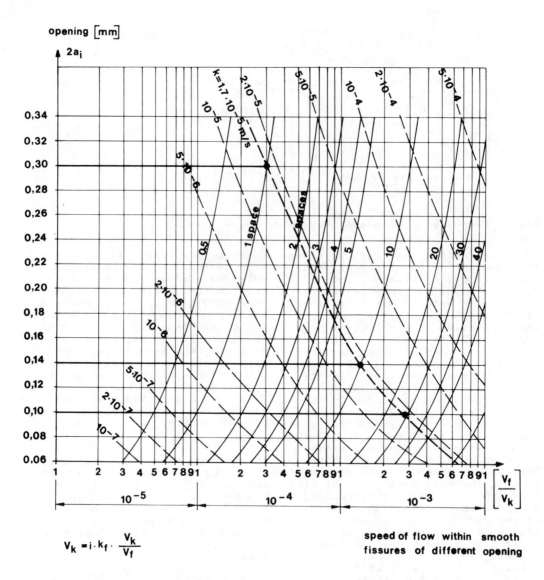

Fig.8. Percolation through fissured rock

that sealing measures reduce the permeability to 1/10 of the original value.

This cannot be proven in the calculatory model without estimating the number of the smaller discontinuities, which are no longer subject to flow as a result of the sealing of the large discontinuities; see Figure 9.

From Figure 9 we can see that the effectiveness of the grout curtain is considerably greater than that proven in the calculation. Controls of bored cores and recalculation of the amount grouted, compared with the measured volume of the discontinuities show of the Möhne Dam, that the grouting of the grout curtain in the area of

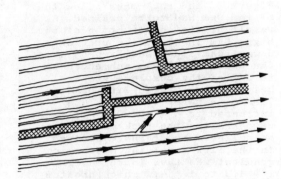

Fig.9. Percolation through fissured rock wide joints sealed

the large discontinuities has sealed almost the entire rock mass below the wall.

The question of at which flow velocity in the discontinuities the filling of the discontinuities will be eroded has not yet been systematically investigated. Observations in the dam wall of the Lister Dam have made an estimation of the hydraulic gradient at which the clayey silt was moved by the seepage water in the system of discontinuities possible; see Figure 10.

Under the calculatory assumptions as are valid for Figure 2, the hydrostatic water pressure under the dam wall and the seepage water loss are given in the picture on the right side of Figure 10. Measurements, that the Ruhrtalsperrenverein has performed since 1912 have shown from the very beginning, a smaller hydrostatic pressure, refer to the measurements of 1912. This initial measurement will then be used as a basis for a new calculatory model, which corresponds better with the reality of construction than does the picture on the right side. In actuality a sealing screen of clayey silt was deposited in front of the wall. Under the assumption of

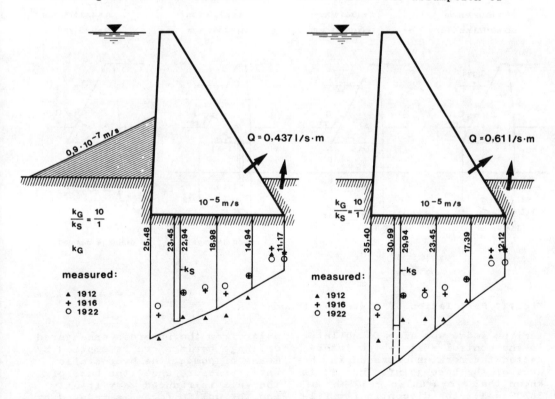

Fig.10. Percolation of Listerdam foundation

$$k = 0{,}9 \cdot 10^{-7} \text{ m/s}$$

for this triangular sealing wedge a good correlation results between the measurement for 1912 and the calculation. From 1912 to 1916 and then again from 1916 to 1922 the hydrostatic water pressure decreased considerably. Since according to our investigations a natural sealing of the basin floor can be excluded - in contrast, the permeability of the sealing wedge has certainly increased due to small particles being washed away - the decrease in permeability of the rock mass must be due to the washing of clayey silt into the system of discontinuities. The proof of this migration could be found 50 years after the structure was completed during the excavation of a construction pit at the foot of the wall. The typical clayey silt of the sealing wedge was found in the discontinuities of 0,5 - 1,5 mm. The special clayey silt of the

1,5 mm are eroded.

For reasons that are shown in Figure 5, the rock mass below the Lister Dam had to be drained. As a result of the construction of the Bigge Dam in 1965, the previous downstream side, see Figure 11, of the Lister Wall was submerged up to 2/3 of the complete storage height. Thus, the uplift of A = 7 MN/m - refer to Figure 4 - is increased to A = 10,09 MN/m. As a compensation the rock mass had to be relieved by drainage borings. In the flow model the hydraulic gradient was thus increased from i = 1,2 to i = 4,1 which creates the danger of greater erosion of the fillings of the discontinuities. This danger is reduced again by the grouting of a cement suspension in the larger discontinuities. At the present time the mathematically determined and the actually measured hydrostatic water pressure with the drains open is still within the allowable range. The drainage adit for removing

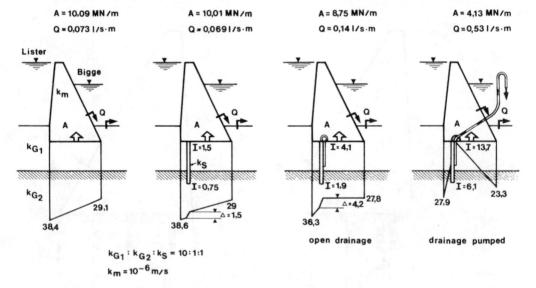

Fig.11. Percolation of Listerdam foundation

sealing wedge was also found later in the control adit, which intersected the rock underground in the area of the base joint. Thus it is known that a hydraulic gradient of 36/30 = 1,2 the discontinuity fillings in discontinuities of 0,5 -

water from the numerous endangered drainage borings can be emptied by means of pumps; the hydrostatic water pressure under the base of the wall is reduced very greatly and the uplift force is reduced by more than half.

Section C:
Environmental considerations, geologic control
and case histories of dam construction

*Symposium on Problems and Practice of Dam Engineering / Bangkok / 1-15 December 1980*

# Environmental parameters in planning and operation of multi-purpose reservoir projects

H.F.LUDWIG
*Seatec International Consulting Engineers, Bangkok, Thailand*

## INTRODUCTION

The situation on water resource development in Thailand may be said to be crucial primarily because of the population crunch. It is expected the population will approximately double before it stabilizes, and because the country has very little unutilized agricultural lands for absorbing new people, and quite limited fossil fuel resources, the population crunch is bringing with it both food and energy crises. Studies over the past decade, including those by the Mekong Committee, indicate that Thailand must proceed promptly to harness all of its available water resources for maximizing energy and food production, and moreover must even push to harness its share in international streams, especially the Mekong river. Both shortages, in energy and food are critical, and as serious as the energy problem is, in the "long-run", meaning two or three decades only, the food crunch may be even more serious.

For example, the Mekong studies (Mekong, 1976,1978) indicate that even if all the indigenous streams of the Northeast and Central basins of Thailand are harnessed, some additional 28 billion $m^3$ of water must be furnished from somewhere just to support enough irrigation to grow enough rice to meet year 2000-2020 food needs. The only feasible source of this water is the Mekong river, and the only feasible dam site which can furnish this water, on an environmentally acceptable basis, is the proposed Pa Mong project.

## IMPORTANCE OF ENVIRONMENTAL ANALYSIS

Why is Environmental Analysis important in this scheme of things here in/Southeast Asia? It is because experience here over the past decade has shown that much more care must be expended in planning for use of the limited water resources so that truly "comprehensive multi-purpose use" is obtained. Also, Environmental Analysis is perhaps the only present approach to project evaluation which can be said to be truly independent and comprehensive, hence readily able to analyze a project in the light of the changing situation rather than in traditional terms. These days the international financing assistance agencies, in analyzing a project, still require a sound economic prospectus but also place greet emphasis on social equities, especially on helping the low-income people including the rural poor.

To achieve such projects will require a real broadening of the basic concept of the appropriate scope of a dam/reservoir project in a developing country. It means, in addition to using the dam/reservoir complex for the usual purposes of furnishing power, irrigation and municipal water supply, and flood control, the project planner must broaden his vision to include every other possible use of beneficial value to the people. These include, for example, fishery production, agro-industrial and mineral development, plus much more attention to socio-economic impacts in the region derectly affected by the project, and much more attention to rural water supply and electrification. This is a problem because, traditionally, the Governmental institutions empowered to plan and implement dam/reservoir projects are often single-purpose oriented and thus hardly ready to undertake such analyses. Nor are some of the civil engineers who usually have primary responsibility for project planning.

## THE FISHERIES PARAMETER

One good example of this need for change in

the concept of dam/reservoir project planning is in fisheries. Most civil engineers have had little if any training in the subject, but because dam/reservoir projects usually exert profound influence on the stream hydrologic regime, they drastically alter the stream's role in fishery production. The effects are both bad and good, and in many projects in the past they have been predominantly bad simply because of lack of attention to the problem. With good planning the end result can not only be good but can very appreciably contribute to enhancing the project's benefit/cost ratio and in enhancing the benefits to the poor people.

The damaging effects of dam/reservoir projects on the existing natural fisheries may be quite profound and, unfortunately, quite damaging in a "silent way" (unappreciated by Government statistics) to rural farm populations because of the influence of the reservoir storage in levelling out flood flows. While this brings about very desirable flood protection it also depreciates the natural yield of the "innundation fishery", i.e., the natural production of fish in the fields flooded by the annual river overflows. This annual heavy fish production has been characteristic of many rivers in rural tropical monsoon countries, in the plains areas where most of the farmers live (Mekong, 1976). The Mekong studies show that the innundation fisheries have traditionally been a major source of fish protein for these rural families, that dam/resevoir projects will appreciably reduce this productivity, but dam implementing agencies rarely even thing about it. The poor farmer simply loses part of his food supply, and he has had no redress.

In addition to reduction in the innundation fishery, dam/reservoir projects also may exert drastic impacts on downstream estuarine fisheries, and on offshore marine fisheries which may be dependent on nutrient discharges in the streams or on use of estuarine areas as nursery zones for the early life stages of the species later caught commercially offshore. The proposed Palasari dam project in Bali, for example, would probably exert adverse effects on valuable offshore fisheries for both these reasons (Ludwig, 1981). Another adverse fishery impact from dam/reservoir projects in the loss of most of the natural fishery in the newly-created zone of turbulence which may reach for a considerable distance downstream.

However, to offset these losses, and in many cases much more than offset them, is the fishery created in the new reservoir, which can be relatively large and even lucrative. Two good examples of this in the region are the Nam Ngum project in Laos, built in 1971, and the Nam Pong project in Northeast Thailand, built in 1966 (Mekong 1976). The natural reservoir fisheries at these reservoirs have proven to be of the same magnitude of value as the hydropower which was the primary economic benefit considered in the financial planning of these projects. Moreover, because the fish products are consumed locally, the regional diet, notoriously dependent upon and short of fish protein, has obviously significantly improved, and the need for importing ocean fish significantly reduced. It is remarkable that the financial planning for these projects assigned quite limited value to the reservoir fisheries, and that the "operations planning" paid no significant attention to any fishery aspects of the projects, even though both projects involved international agency financing.

Little attention has been given also to the actual operation of the reservoir fishery since the reservoirs were built, despite the fishery values, with unfortunate and unnecessary results. Due to lack of appropriate reservoir fishery management at Nam Pong, for example, the reservoir zone has been invaded by "aliens" who have moved in on the lake fishery, the number of fishermen is far too many, middlemen make most of the money, destructive dynamite fishing is practiced, and the new fishing villages at the lake are virtual slums as well as socio-economic and socially unstable messes.

The plea here - the environmental plea - is to recognize that fish, along with power, irrigation, and flood control, should represent another key parameter in dam/reservoir planning, not only to enhance the benefit/cost ratio but to do much for the rural farmers. In addition to nutrition improvement by good reservoir fishery management, set up as an integral part of the project plan, it should be possible to provide employment for the downstream fishermen families who would otherwise suffer from the project, and also for resettled families, and to establish and maintain good quality fishing villages at the lake thus contributing to local stability. Also, by use of special fishery technologies, the fishery productivity, high as is it is by nature alone, can be significantly increased, sometimes even doubled (Pantulu, 1977-79). Also, the year-around water supply made available by the irrigation system sets the stage for development of aquaculture in the irrigation zone. At places where the topography enables economical pond construction, the gains from aquaculture per unit of land can be several times those from paddy. In addition many reservoirs offer potential for attractive waterfowl production through appropriate management.

A current illustration of the civil engineers role in this type of situation is in the Citanduy basin in Java in Indonesia (Ludwig, 1980b). At the lower end of this basin is a large estuarine mangrove swamp ("Segara Anakan"), which serves as a nursery supporting a very valuable offshore marine prawn fishery. While the swamp is large, due to deforestation in the watershed in recent decades siltation has already filled in a considerable portion of it, and the civil engineering group doing project planning for water resource development in the basin has given serious consideration to use of engineering measures to accelerate/complete the filling in order to create more land for paddy. This seemed logical to them, but most illogical to the estuarine biologists of Java who think the engineers should also explore the alternative of using engineering measures to limit the filling and to preserve the swamp resources for both commercial and aesthetic reasons.

FORESTRY AND WILDLIFE

The example of fisheries is but one of many of the environmental issues involved in major dam/reservoir projects in this region, and each of them could be discussed in the same level of detail. There isn't time to discuss them all here today, but it should be worthwhile to discuss briefly some more examples of the key issues which thus far, like fisheries, have tended to be neglected by project planners. These include the forestry/wildlife issue, the potential for agro-industries, and community socio-economic impact including resettlement.

With respect to forestry and wildlife the key issue is not the impact of the reservoir on drowning out the relatively small portion of the watershed which is innundated, but the fact that the project creates ready access to portions of the upstream watershed previously difficult to reach, mostly by roads but also sometimes by boat. Also the project speeds up population concentration in the project area. The population encroachment of course tends to accelerate loss of forest habitat through encroachment for farming, illegal logging, etc., and to intensify hunting.

In the long run it may be argued in many cases that this population encroachment is bound to occur, sooner or later, with or without the project, and that the forest habitat is more-or-less doomed. This hardly means, however, that the planners of dam/reservoir projects should do nothing about it. Rather studies made in the past dacade of the entire subject of protection of forest habitat-wildlife have shown (Mekong, 1976,1978) that it is at just such moments, when a major project is being planned, that it is most feasible to introduce and include some provision for wildlife conservation. For example some part of the more remote upper watershed may be set aside as a habitat/wildlife reserve, with the funds for doing this considered an integral part of the project. Another possibility is to identify and quantify the key wildlife species in the watersheds to evaluate their chances for survival under both with and without project conditions, and to recommend appropriate correction and enhancement measures. Usually the funds needed for such measures are small in terms of the overall project budget, but they are very large in terms of the budgets of most national wildlife protection agencies, hence are unobtainable by them. If the civil engineer and others concerned with planning of dams/reservoirs are unconcerned with conservation of wildlife, who is to be concerned?

LOCAL INDUSTRIES

Agro industries and other local industries represent another potential for doing a lot of good for the country and especially for the rural poor. Study after study in the past decade have emphasized the need for decentrallizing, for keeping every thing and everybody from moving to the capitol, for promoting local industries in the rural areas, and with generally quite limited success. Again, it should be recognized that a new dam/reservoir project presents a real opportunity, with the newly available power and infrastructure, for promoting local industries including various agro industries and in some cases mineral development. However, also to be recognized is the fact that special planning including possible changes in traditional institutional relationships between the capitol and regional areas, are needed to realize these potentials.

The intention here is not that planners of major dam/reservoirs should be made responsible for local industrial development. The intention is that the project analysis, perhaps prompted by the Environmental Analysis, should identify such potentials and set the stage for appropriate follow-up measures. Such thinking should be an integral part of the project.

An interesting example of such a potential is the concept of a new major "Pa Mong City" that could readily be established in the vinicity of the proposed Pa Mong dam. Properly planned, taking advantage of plentiful local power and raw materials (for example, vast rock salt and potash deposits), this could become a major metropolis and significantly reduce the "converge-on-Bangkok" syndrome (Mekong, 1978).

RESETTLEMENT

Resettlement of families living in the zone of innundation has been the most painful of all environmental issues in the implementation of dam/reservoir projects in developing countries, and the record of not attending to this responsibility has been as bad here in Southeast Asia as elsewhere (Mekong, 1978). This record of sorry performance stems from the fact that proper management of resettlement is expensive, hence is easier to ignore, it or to manage it superficially or partially, knowing that the rural people involved are hardly in a position to gain appropriate attention and compensation. Appropriate attention means provision of funds for rehabilitation as well as for property compensation, and many project planners have avoided the rehabilitation issue by leaving this up to the Government's Public Welfare Agency—but with full knowledge that these agencies will not be able to obtain the needed monies from regular Government coffers. The only honest solution is to include provision for all appropriate costs as an integral part of the project budget.

This "novel" principle has been vigorously expounded by the Mekong Committee, and here in Thailand the Electrical Generating Authroity (EGAT) has shown real leadership in implementing this concept in its projects over the past decade—for example the resettlement program at Pattani in Southern Thailand. Generally, however, the record is sad. A good example is the Nam Ngum project in Laos where, after almost a decade, the resettlement problem still has not been solved despite the fact that the project has proven to be a "gold mine" for the Government. The Mekong Committee is much to be commended for its continuing efforts to clean up this problem, to demonstrate it can be done, as a desirable preliminary step before proceeding with the Pa Mong project (which might be termed a Nam Ngum type project on a very large scale).

SOCIO-ECONOMIC IMPACTS

In the same sense, it should be recognized that virtually all major dam/reservoir projects in developing countries, such as in this region, exert profound impacts on the entire socio-economic spectrum in the region affected by the project. The included not only new irrigation zone villages and new village around the reservoir but all existing villages. The idea here is, by astute planning, to take advantage of the new project to maximize the benefits to local communities as well as, say, producing power for export. This might include not only enhancement of village water supplies but many other improvements in village infrastructure and in employment opportunities. Again, the job for the dam/reservoir project planner is to recognize, identify, and describe the potentials and set the stage for Government follow-up. As it is now, such inputs are not part of project planning and it is left to the regional economics to adjust to the new project impact on its own. This generally results in an inadvertent loss in a good part of the project benefits and in undesirable inequities in distribution of project benefits.

ENVIRONMENTAL ANALYSIS OF DAM/RESERVOIR PROJECTS IN SOUTHEAST ASIA

Among the countries of Southeast Asia, Thailand has taken the lead in generating attention to the environmental consequences of dam/reservoir projects and in getting Environmental Impact Assessment studies and reports done for some of the major projects built and being planned in recent years. The National Environment Board (NEB) has made good progress in implementing the EIA concept for all types of project development, and EGAT also should be commended for its efforts which began even before NEB was created. The NEB has produced a "Manual of EIA Guidelines" (NEB, 1979) which has been valuable not only here but throughout Southeast Asia. Similar progress is being made elsewhere in the region, for example in Indonesia, the Philippines, and Malaysia.

The approach used for EIA studies in Thailand is that developed by the U.S. Corps of Engineers/Battelle (Battelle, 1974) for analysis of the Columbia River project in the USA. This classifies environmental values into four main groups, (i) natural physical resources, (ii) natural ecological resources, (iii) economic development, and (iv) quality-of-life values including socio-economic, public health, recreation, aesthetics, and historical/cultural treasures. A detailed listing of supplemental guidelines for EIA studies, specifically for dam/reservoir projects, has been prepared by NEB (NEB, 1979).

While all of us are familiar with such

classifications, and the fact that an Environmental Analysis is intended to look into all such parameters, it is not generally recognized that this is really only the first step, and much remains to be done in utilizing the findings of the EIA for productive engineering. This means modifying the project plan to accomodate the findings of the environmental study (those that have the merit). This means that the EIA must be considered as an integral part of the overall project planning operation. In other words, environmental integrity should be considered as an appropriate parameter to be taken into account in all project planning, along with structural integrity, hydraulic integrity, economic integrity, etc. It is predicted that within another decade or so the EIA analysis will come to be regarded as simply another routine component of project planning, not as it has been, as a "separate" analysis sometimes not much related to real project planning. When this happens the engineer will again have stature as a competent project planner, with all sectors of the public including the conservationists.

Another point to made here relates to the appropriate cost of EIA studies in the region, i.e., to recognizing that EIA studies here must be carried out at budget levels far below accepted levels in countries like the USA. The NEB experience (Ludwig, 1978) has shown that this is feasible, and NEB has developed Terms of Reference for EIA studies which can be realistically absorbed within project budgets. This is done by limiting the analysis to project aspects of significant economic or socio-economic importance, and by deleting costly attention to relatively unimportant minor issues.

MONITORING

A final plea is to note the great importance of continuing monitoring of project impacts following completing of construction. Again, breaking with tradition while monitoring has not been considered an appropriate part of project planning and budgeting, the experience of the past decade indicates overall project benefits could be significantly enhanced if this were routinely done. Not as an isolated effort (such as assessment made one or two years after project completion) but on a continuing basis as needed for permitting continuing evaluation of the important impacts. The monitoring program would not be limited to conventional "environmental issues" but would included all issues of real importance such as economic gains and their distribution.

An illustration of an environmental parameter to be included in the monitoring program for a dam/reservoir project is that of morbidity incidence for water-oriented communicable diseases that might result from the project (but which might take a number of years to develop). Again, it might be said by the Project Planner that this should be left to the Government's Health Ministry, but this is hardly realistic because most Health Ministry budgets now and in the future usually cannot cover these expenses.

It is of interest to call attention here to the "post-mortem" analysis made in 1978-1979 of the actual impacts of the Nam Pong multipurpose dam/reservoir project built in Northeast Thailand in 1966 (Ludwig, 1979a). This study, conducted by the National Energy Administration together with the Mekong Committee, showed vast differences between the project planning prospectus and what actually happened with respect to many of the project features, e.g., irrigation and fisheries. This type of analysis, a post mortem conducted a decade or so after completion of construction, is rarely done but it should be done routinely for all major dam/reservoir projects. The results should be of very great value in reorienting project planning to be far more realistic and thus in achieving the desired maximum benefits. It seems obvious that it is the major dam implementing agencies which should take the lead in planning and implementing these studies. They will be the primary beneficiaries.

CONCLUSIONS

The discussion above has dealt with only a few of the many environmental parameters involved in water resource development projects, i.e., with some of those found from experience over the past decade to be of special concern in trying to get the attention of civil engineers and other dam/reservoir project planners to be more aware of an interested in the environmental issues. The object is not simply to protect environment, but to promote project planning to maximize benefits including economic and socio-economic gains as well as environmental protection. Hopefully this Symposium will help stimulate those responsible for project planning to recognize that "New Terms of Reference" are in order for such planning, to recognize that the water resources in tropical monsoon zones are often not plentiful but quite limited and in ever-increasing short supply, and that "comprehensive multipurpose planning" is the only way to ensure maximizing of benefits urgently needed for the country's welfare, including the rural poor, and that all of this must be realized must more rapidly in a new order of time frame. As recently noted by President Robert Mcnamara of the World Bank on the occassion of his retirement we no longer can take our time for such planning and project implementation. All of us, nationalists and international assisters, must shift into a higher gear, we must do it now, and attention to environmental parameters must be an integral part of the process.

REFERENCES

Battelle/Pacific (1974): "Environmental Assessment Manual, Columbia River and Tributaries", for U.S. Corps of Engineers, May 1974.

Mekong (1976): "Environmental Effects of Pa Mong", Working Paper No. 1 Supplement to Pa Mong Optimization and Downstream Effects Study, Mekong Committee Publication MKG/36, August 1976.

Pantulu (1977-79): Publications of Mekong Committee (Bangkok) on proposed projects for improved management of reservoir fisheries in Northeast Thailand and in Laos (prepared by V.R. Pantulu of Mekong Secretariat), 1977-79.

Mekong (1978): "Pa Mong Optimization and Downstream Effects Study", Mekong Committee publication MKG/45 Rev. 1, June 1978.

Ludwig (1978): "Appropriate Budget Allowances for EIA Studies and Reports for Projects in Thailand", H.F. Ludwig, National Environment Board, June 1978.

NEB (1979): "Manual of NEB Guidelines for Preparation of Environmental Impact Evaluations", National Environment Board of Thailand (NEB), Jan. 1979.

Ludwig (1979a): "Post Mortem Analysis of Environmental Effects of Typical Multipurpose Dam/Reservoir Project in Thailand", H.F. Ludwig, Thai National Environment Board, April 1979.

Ludwig (1979b): "Preliminary Environmental Assessment of Proposed Irrigation Package Project in Chao Phya Basin of Thailand", H.F. Ludwig, Asian Development Bank, 14 Dec. 1979.

Ludwig (1980a): "Preliminary Evaluation of Loei Multipurposed Project:, H.F. Ludwig, Asian Development Bank, 4 May 1980.

Ludwig (1980b): "Environmental Aspects of Lower Citanduy Project", H.F. Ludwig, Asian Development Bank, 24 August 1980.

Ludwig (1981): "Environmental Analysis of Bali Irrigation Project", H.F. Ludwig, Electroconsult/ADC, for Government of Indonesia/ADB, 1981.

*Symposium on Problems and Practice of Dam Engineering / Bangkok / 1-15 December 1980*

# Construction works and geology

HIKOJI TAKAHASHI
*Kajima Institute of Construction Technology, Tokyo, Japan*

## 1 Tunnels

### 1.1 Geological Consideration for Tunneling

In constructing tunnels, the geological conditions of the site must first be examined to make sure whether the ground is suitable for tunnel construction. In other words, such a tunnel driving method must be chosen as may be deemed the best suited to the specific geological conditions of the site.

In most of tunneling works in Japan, a standard tunnel driving method, which is deemed to be suitable to the average geological conditions of Japanese Islands (Fig. 1), is in extensive use. According to this typically Japanese style method, excavation is first carried out for the bottom heading, next the upper half of the tunnel is excavated and then concrete is placed on the arch (the inverted lining works). After that the lower half is excavated (the enlargement), and finally concrete is placed on both side walls to complete the entire lining.

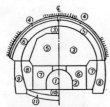

*Fig.1 Japanese Style Tunneling Method*

Standard ( Right Side )
① Bottom Heading Excavation
② Heading Support
③ Upper-half Excavation
④ Arch Support
⑤ Arch Concrete
⑥ Sidewall Bench Excavation
⑦ Sidewall Excavation
⑧ Side Wall Concrete
Remark = Left Side shows for Soft Rocks

The first problem of the geological conditions lies in whether or not the cutting face, the most advance part of tunnel driving, is in the state of self supporting. It will be the most desirable condition if the cutting face can sustain itself vertically to the floor of tunnel so that the full face can be excavated without the necessity of providing any temporary support.

If the cutting face is not in the state of self supporting, the tunnel section will be divided into parts and excavations are advanced part by part. Since, in such cases, some appropriate measures must be taken for protection of the crown and face of excavation, excavation procedures will have to get inevitably more complex, and due attention will have to be paid to safety execution of tunneling. In case of the cutting face in poor geological conditions, the time to prevent the excavated sections from yielding until the completion of concrete lining works must be shortened as far as possible. Also, unlined sections must be reduced to minimum. In this case, additional excavation and concreting will be required. Thus, the environment for project operation and the construction procedures will become complex. As the number of the working items increase, so does the length of the round time increase.

The second problem is how to break rock ground. Holes will be drilled on cutting face for charging with dynamites, and then blasting will be performed. This is the rock tunneling method. In breaking and crushing rock ground, the rate of drilling operation is the most important. Such drilling speed is directly affected by the physical properties of rocks including their hardness and toughness. Such drilling rates are given in unit of cm/min.

The third problem concerns the earth pressure and seepage of the ground water. The earth pressure varies not only with the softness of the ground and the size of the cutting face, but also with the time the lining operation starts after excavation and with the distance between the site of lining works and the cutting face, as well. Unless these elements are properly dealt with, deformation of tunnel and rupture of temporary supports occasionally occur. The seepage is a phenomenon of ground water seeping out to the tunnel. The seepage water seriously worsens the geological conditions of the site and will cause the earth pressure to increase dangerously.

1.2 Geological Problems in Tunneling

Geological requirements and treatments for tunneling with respect to the shape and dimension of the section of heading, the order of excavation for divided sections, drilling method, and lining will be as summerized below. (Ref. Fig. 2.)

1.3 Tendency of Tunnels Towards Greater Length

Recent tunnel construction in Japan tends toward greater length, as shown in Table - 1. Increase of construction experimences, improvement of the construction techniques, and progress of mechanized construction operations have made it possible to plan and execute the construction of longer tunnels without any hesitation. Long length tunnels have various advantages, such as good linearity, ease of route selection, the construction of cross-mountain rock tunnels, and the possibility of bypassing a residential district. Thus, long tunnels have made, and will continue to make, a great contribution toward the development of transport and economy. On the other hand, the construction of long length tunnels has given occasion to some unexpected problems.

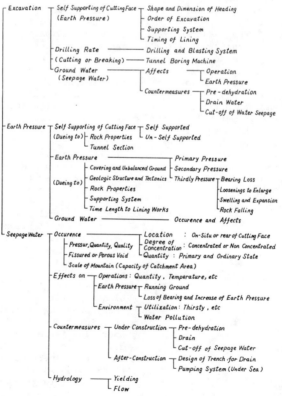

Fig. 2  Geological Problems on Tunneling

Table.1  Long Length Tunnels in Japan

| NO | Site | Name | Length(m) | Geology | Construction Period |
|---|---|---|---|---|---|
| S.1 | Under Sea (Railway) | Seikan | 53.815 (23.300) | Tertiary Pyroclastic and Volcanic Rock | 1971 ~ under construction |
| S.2 | (Railway) | Kanmon | 3.614 (1.300) | Granite, Porphyrite and Mesozoics | 1943 |
| S.3 | (Highway) | Kanmon | 3.640 (780) | Porphyrite and Diorite | 1958 |
| S.4 | (Railway) | Shin-Kanmon | 18.713 (880) | Granite and Mesozoics | 1973 |
| R 1 | Mountain Railway | Daishimizu | 22.235 | Granite Paleozoics and Hornfels (Popping Rocks) | 1979 |
| R 2 | | Rokkō | 16.250 | Granite | 1967 |
| R 3 | | Nakayama | 14.790 | Tertiary and Pleistocen Pyroclastics | 1972 ~ under construction |
| R 4 | | Haruna | 14.350 | Pleistocene Pyroclastics | 1972 ~ 1980 |
| R 5 | | Hokuriku | 13.870 | Paleozoics and Granite | 1961 |
| H 1 | Highway | Kanetsu | 10.900 | Granodiorite and Hornfels | 1977 under construction |
| H 2 | | Enasan | 8.489 | Granite and Grano-porphyry | 1975 |
| H 3 | | Shin-Kobe | 6.910 | Granite | 1976 |
| H 4 | | Shin-Sasago | 4.417 | Granite | 1977 |
| H 5 | | Tsuruga | 3.175 | Paleozoics | 1978 |

In the cross-mountain route, which often contains long length tunnels, it is not advisable to drive through deep into mountains. As the tunnel goes deeper, the seepage water will increase, resulting in wastage of large quantities of water otherwise consumable by residents in the neighborfood. Moreover, it will cause the earth pressure to grow greater under the poor geological conditions. This happens also with a tunnel under a basin.

In cross-mountain long length tunnels, it is desirable that the tunnels be so routed that they will run along and close to mountainside slope above the groundwater level, and the depth of tunnel should be shallow enough to prevent unbalanced earth pressure or landslide. (Fig. 3.)

Even if the length of the tunnel should be greater by routing so, additional travelling time for vehicles would be negligible.

1.4 The NATM (New Austrian Tunneling Method)

Use of the NATM is recently gaining greater popularity in Japan. The ground containing a cavity has a physical nature to generate the arch action inside the ground surrounding the cavity by its own internal forces. The arch thus formed inside the ground is called "the ground arch". (Fig. 4.)

Fig.3  Tunneling Above Groundwater

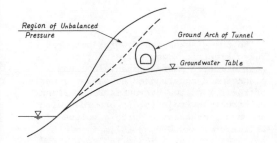

Fig.4  Forming Ground Arch Around Tunnel

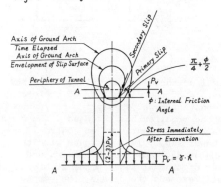

NATM aims that forming the ground arch surrounding the cavity as one body by construction of a flexible thin lining, using shot-crete and rock bolts, on the cavity surface immediately after excavation. This method complies with the principle of rock mechanics and makes it possible to construct rational, safe and economic tunnels. Therefore, this method is also attracting the attention of dam engineers for its wide applicability regardless of rock conditions.

1.5 Large Artificial Underground Cavities

The pumped storage power generating system was introduced to Japan some thirty years ago to achieve more efficient operation than the hydraulic, thermal or nuclear power plants, and many underground power stations have since been built. (Table-2)

the upper reservoir by operating water wheels in reverse turning. Thus, pumped storage power plants usually installed in man-made underground cavities are provided with the upper and the lower reservoirs. (Fig. 5.)

Fig.5 Schimatic Profible of Pumped Storage Power Plant

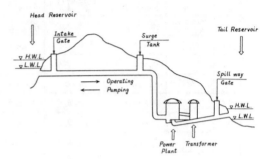

Table. 2  Underground Power Plants in Japan

| NO | Construction Period | Power Station | Location | Dimension (m) width × hight × length | Geology |
|---|---|---|---|---|---|
| 1 | 1963-11 | Kawamata | Tochigi | 17.1 × 34.1 × 33.2 | Rhyorite |
| 2 | 1973-5 | Numappara | Tochigi | 22 × 45.5 × 131 | Diorite |
| 3 | 1979- | Shin-Takase | Nagano | 32.5 × 59.5 × 139.5 | Granite |
| 4 | (1981) | Oku-Yahagi | Aichi | 25 × 55 × 90 | Granite |
| 5 | (1982) | Tambara | Gumma | 26.6 × 49.5 × 116.3 | Conglomerate |
| 6 | (1982) | Honkawa | Kōchi | 22.3 × 45.4 × 96 | Black-schist |
| 7 | (1984) | Imaichi | Tochigi | 33.5 × 52 × 160 | Rhyorite |
| 8 | (1982) | Shimogo | Fukushima | 22 × 45.5 × 171 | Diorite |

Thermal or nuclear power plants have the advantage of the capability of performing the constant power supply in accordance with the required basic demand. The pumped storage power plant aims at controlling the power supply in accordance with the demand at day-time peak loads. Since the output of thermal power plant at mid-night usually exceeds the demand, the surplus power can, therefore, be used to pump up the water from the lower reservoir into

The Shin-Takasegawa Power Plant that Tokyo Electric Power Co., Inc., the biggest electric power company in Japan, planned and constructed as one of a series of the Takasegawa Hydraulic Power Plant Construction Projects, is provided with an underground cavity with a volume of a little over 240,000 m³ (27 m in width x 59.5 m in height x 165 m in length).

The ground that can hold such a large cavity is required to consist of perfect firm bed rocks. However, such bed rocks cannot easily be found anywhere in the world. In particular, it is hard to find such ideal site within Japanese islands that are formed of complicated geological structures. Therefore, actually the site is usually selected at the areas with less geological defects. In excavating large cavities, primary consideration should be given to setting up a long time stability of excavated surface of wall during and after construction.

According to the report of the construction of the Shin-Takasegawa Power Plant, initial stress distribution inside the rock ground was measured, and as a result, the initial design specification for the cavity was modified and the modification was adopted before the construction of the cavity.

(1) The initially specified longitudinal direction of the cavity was altered to comply with the direction of the maximum principal stress of the ground.

(2) Cavity's sectional area was reduced to the maximum extent[5].

## 2 Dams

### 2.1 Geological Requirements for Dam Site

In construction of dams, bed rocks are used as foundations, which are required to be capable of bearing dam body and reserved water safely. Since excavation for the removal of overburden has an aim to expose the bed rocks required as the foundation of the dam, utmost care should be exercised to assure the stability of slopes of the bed rocks at each work stage of the excavation. (Fig. 6.) It is not very rare that exposed bed rocks have steep slopes exceeding 70° in gradient. After the bed rocks are completely exposed, concreting operation or embankment of dam body will be commenced and the bed rock stability will increase gradually.

There are many types of dams, such as arch dams and gravity dams if structurally classified, and concrete dams and fill dams if classified by the materials used. Selection of the type of dam depends primarily upon the strength of bed rocks, although in case of fill type dams the availability of suitable materials is also important.

When bed rocks serve as foundation of a dam, their function must be divided into two parts, one is the lower elevation part which receives mainly vertical loads, and the other is the inclined part reaching up to the crest of dam which receives mainly loads in the vertical and axial direction of dam.

Since, in the arch dams, the pressure of reserved water is transmitted mainly to bed rocks of both banks, loads at the abutment in the axial direction of arch dam is much greater than that applied in case of gravity dams. On the gravity dams, bearing and sliding resistances at the dam basement are more important than the loads at the abutment in the axial direction of dams. No matter what type of dam body is adopted, principal requirements for bed rocks are bearing resistance and deformation characteristics. Furthermore, the watertightness is also required to prevent leakage of the reserved water. In this connection, uncemented sand strata or cracky rocks are not suitable for the foundation ground unless some

Fig-6 Excavation Lines and Function of Bed Rocks.

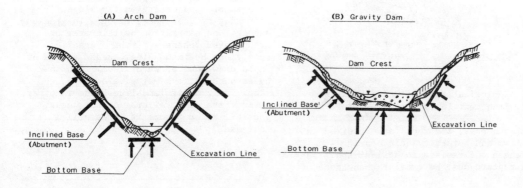

effective countermeasures are taken for their improvement. Especially when water permeation is likely to occur in the abutment area where dam body is fixed to the foundation ground, it may often cause the increase of seepage water, the piping phenomenon, or the occurrence of up-lift, which may even give occasion to the collapse of dam body. For the arch dam or the concrete gravity dam, relatively firm and intact rocks are used as the foundation ground. However, in the earth fill dam, the foundation is often put on relatively new strata which may contain uncemented or weakly cemented sand strata. In this case, problems about water leakage are likely to occur.

If the bed rocks are weak in bearing resistance and are permeable, such suitable treatments are indispensable as replacement of weak part with concrete plug or carrying out carefully the grouting of cut-off water leakage. Moreover, it is not unusual that slope failure or landslide occurs in a water reservoir, when the natural slopes of banks surrounding the reservoir become saturated or submerged by rising up of water level, changing drastically the hydraulic conditions at the site. (Fig. 7.) Damages can be restored if its scale is small. However, there are sometimes severe damages to the dam body or to the function of dam. When such damages are feared, suitable measures should be taken by providing some slope protection or making the slope grade more moderate.

*Fig. 7  Variable Hydraulic Condition in Banks by Impounding*

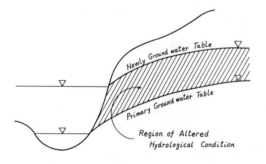

### 2.2 Geological Problems on Foundation Treatment in Dam Construction

For arch dams, bed rocks are required to be firm and intact. Their bearing strength and resistance against the sliding rupture must be greater than those of earth fill dams. The foundation of earth fill dams does not have to consist of rocks, if only proper impermeable strata are available. Thus, degrees of requirements for the foundation ground vary with types of dams. The problem of dam foundation treatment depends all the more upon degrees of improvement that can be done on the foundation ground in the selected site.

The problem of excavation for dam foundation is the same as that of slope construction. Therefore, that will be explained under the chapter 3, "Slopes". Since lining works by concrete or other materials belong to the construction works of dam body, only the bed rock treatment has been discussed here. In general, the improvement and strengthening of the foundation ground is classified in two categories -- (1) one for permeability and (2) the other for mechanical properties of the bed rocks. (Fig. 8. and Fig. 9.)

## 3  Slopes

### 3.1  Geological Problems for Slopes

Construction works of slopes are not confined only to forming of slopes, but they also include the construction of approach to tunnels and entrance of tunnels, as well as excavation of the banks for dam abutment and various open cutting works. Therefore, the construction work of slope represents a fundamental element common to all aspects of construction operations.

An already formed slope, no matter whether it is natural or artificial, is considered stable in itself. It will keep stable, when its height and gradient are kept in good harmony with various elements of the earth or rock ground. However, actions that may cause some changes to the present conditions of slope must be regarded as one of the factors affecting the slope stability. Such actions include cutting of the lower part of slopes, changes of load acting on the slope, loosening of rock mass, changes in gradient, seepage and saturation by rainwater or groundwater, changes of the hydraulic conditions by impounding, sudden uprise or rapid draw-down of water level. In some cases such actions may give occasion to slope failures or landslides, and in other cases where such failures do not occur, possibility of future failures will increase.

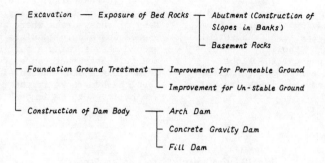

Fig. 8 Geological Problems in Dam Construction

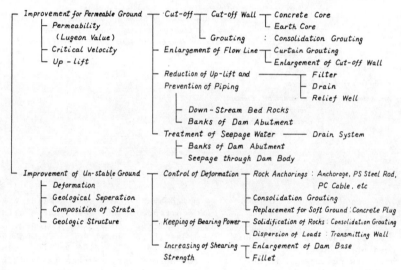

Fig. 9 Problems on Foundation Treatment in Dam Construction

When a dam is completed, the groundwater level inside the slopes will rise correspondingly to new reservoir water table. Before the dam construction, the groundwater level was at a lower level as controlled by the river water table. The physical properties of earth or rock ground will change by water penetration or saturation. Also, their soil constants may change. There are questions regarding the test method and conditions for obtaining such soil constants. Our experiences obtained from construction of many dams tells that the soil constants could be modified by the different surroundings which newly developed.

## 3.2 Geological Problems on Slopes

As to two kinds of slopes, natural and artificial, both having similar problems, the most important treatment undoubtedly lies in keeping such stability of slopes as is adequate to prevent the slope failure or landslide from occurring.

Geological problems on slopes are as shown in Fig. 10.

encountered during the excavation of tunnel in the neighborhood of its entrance consisting of loose geologic conditions. Many of those areas where such geological phenomenon occurs consists of a moved ground mass forming a part or whole of a mountain body that can be considered to move along a deep seated sliding surface. Even the bed rocks in the dam site which are believed to be immovable are often

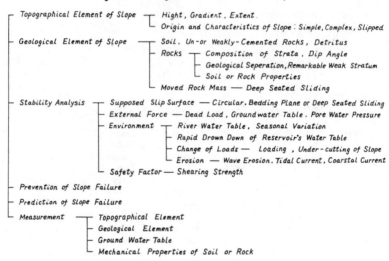

Fig.10  Geological Problems on Slopes

## 4  Loosening of Ground Mass due to Deep Seated Sliding

There are not a few cases where a ground mass is loosened by the deep seated sliding. Since this phenomenon is not peculiar to Japan but is common to wide localties, the author wants to call your attention to this problem which has direct bearing on all kinds of construction works including slopes, dams and tunnels. It is quite imperative that we should take absolutely optimum measures in dealing with this problem.

There is a case where an open crack is found in the rock ground during the cut work on a slope or a terrace, or a case where, especially on mountain sides, the groundwater level is unusually low or hard to be located. There is also a case where an unexpectedly high earth pressure is moved by the deep seated sliding. Occurrence of the deep seated sliding is not all rare in ordinary slopes. In the areas where landslides occur habitually, a deep seated sliding is often mistaken for an ordinary surface landsliding. It is essential to treat the deep seated landslide in a way different from the ordinary landslide.

In many cases, the deep seated sliding occurs on old large scale ground masses. The sliding of ground mass may be dormant at pressent, but can be made active easily if only the present conditions of ground mass are somehow changed.

The topographical feature of moved ground mass that seems to be induced by deep seated sliding can easily be identified by aero-photography, and there are many places having such geological conditions in Japanese islands.

There may be a great number of routes of highways or railways located along valleys between mountains having the deep sliding, as well as tunnels passing through the deep sliding area. The Oriwatari Tunnel and the Ashitani Tunnel are typically indicative of such conditions. (See Table 3).

The basic factors for, and mechanism of, the deep seated sliding which create loosening of ground mass can be considered as follows:

(1) Basic Factors and Mechanism

A stratum consisting of loose rock, such as Tertiary deposited clayey rock or sandstone, moves within a layer and then gets loosened, upon receiving such a folding action relating to tectonic movement. After that, the cracks which had occurred on rock mass were concealed by the surface ground later deposited. Then a large slope was formed, becoming unstable and movable by erosion or other actions. Its process is as shown in Fig. 11.

order from the oldest -20 m, -90 m, -100 m, and -90 m in comparison with the present age. The sea level in the last 4th glacial age is lower by 90 m than the present one. In fact, the waste-filled valleys having deep cuts are aggraded by alluvium.

(3) Present Conditions of Moved Rock Mass

Visual aspects of moved rock mass are macroscopically not much disturbed. Loosening of rock structure is formed by many cracks and displacement of blocks of various size. Cracks are of various regular or irregular shapes, and are open or closed, and some are partly crushed into lumps of clay which fill voids in some cases. Moreover, some cracks partly hold cleft water in their open spaces, or there are some open cracks from which the cleft water has leaked out.

The characteristics of the ground mass moved by deep seated sliding are as stated above. When a tunnel is driven or cut is

Fig. 11 Mechanism of Creating the Loosening of Moved Rock Mass

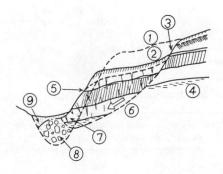

① Original Ground Surface (Before Sliding)
② Present Ground Surface (After Sliding)
③ Scarp
④ Unmoved Side (Original Stratification)
⑤ Moved Side (Structure of Stratification in this Area is Macroscopically not disturbed)
⑥ Slip Surface with Clay
⑦ Moved Toe Part of Rock Mass
⑧ Moved, Ruptured and Deposited Tip Area (Ruptured Area)
⑨ New Deposit

(2) Period of Deep Seated Sliding: During and after pleistocene (Approx. 2 million years ago)

The time during which topographical feature was formed is the one preceding to the present, and then, such a topography was later so deformed that upper parts of the ground were chipped off, and lower parts were aggraded, and finally the present topographical feature was formed. As long as the present topographical conditions and geological environments remain unchanged, the apparent stability of the ground is maintained.

In addition, Pleistocene has four glasical ages. The sea levels in those ages are roughly estimated respectively in

carried out in such a moved ground mass, the ground is geologically activated and an abrupt earth pressure can be induced. That is why a special attention should be paid to this problem beforehand.

Moreover, there were possibilities that such geological phenomena might have occurred in Paleozoic or Mesozoic strata. Although it is of little doubt that loosened structure of old strata must have become fairly compact under the load of the later laid deposits, it should be noted that there is a chance of re-loosening which may grow on the open cut free surface.

## 5 Fairly Low Seismic Velocity Zone
(A comment on the seismic exploration)

Seismic exploration is most commonly used in Japan as a means of geological survey in the planning of a dam or tunnel. This section is intended to make the author's comment on the relationship between the elastic wave velocities and the geological conditions of bed rocks.

Regarding elastic wave velocity, the author has the following view. Assuming the possible geological structure, the elastic wave velocities (the primary wave or longitudinal wave) should be classified into three zones, namely;
(1) Normal velocity zone,
(2) Fairly low velocity zone,
(3) Low velocity zone.

As explained before, the author believes that the fairly low velocity zone has direct bearing on loosening of ground mass caused by deep seated sliding.

Table 3 shows the fairly low velocities along the tunnels. Taking the Ashitani Tunnel as an example, velocities which are supposed to belong to normal velocity zone are in a range from 4.3 to 4.5 km/sec and those for the fairly low velocity zone are in a range from 3.7 to 4.0 km/sec.

Therefore, while excavating in the ground condition of the fairly low velocity zone, an unexpectedly abrupt earth pressure may well be encountered. Even if the bottom heading should have been driven successfully in such ground conditions, the possibilities of falling of ceiling rocks during the excavation of the upper half of the tunnel would increase to a dangerous extent.

Thus, on the seismic exploration, the author recommends that the fairly low velocity zone be treated with great care. In construction of the dams in geologically disturbed areas, proper attention should be paid to the treatment of the bed rock conditions in a fairly low velocity zone. In the absence of such attention or care, the circumstances will inevitably require employment of work procedures drastically different, more complex, hence more expensive, to cope with the new adverse situation.

REFERENCES

(1) HIKOJI Takashi: "GEOLOGY FOR CIVIL ENGINEERS"; Kajima-Shuppan-Kai, 1974-4.
(2) Hikoji Takahashi: "ALL ABOUT TUNNELING FROM THE GEOLOGICAL STANDPOINT", Seko-Gijutsu. 10-1, 1977-1.

Table. 3  Fairly Low Velocity Zone on the Seismic Exploration

| | Tunnels | Geology | Normal Zone (Vp km/sec) | Fairly Low Velocity Zone (Vp km/sec) |
|---|---|---|---|---|
| 1 | Oriwatari Uetsu-line (Ry) | Tertiary mudstone (Miocene) | 2.3 ~ 2.7 | 1.2 ~ 1.9 |
| 2 | Ashitani Uetsu-line (Ry) | Diorite | 4.3 ~ 4.5 | 3.7 ~ 4.0 (2.0) |
| 3 | Shin-Takabayama Iiyama-line (Ry) | Tertiary mudstone (Miocene) | 2.2 ~ 3 piece sample | 1.8 ~ 2.0 whole area |
| 4 | Yoneyama Shinetsu-line (Ry) | Tertiary mudstone (Miocene) | 2.2 ~ 2.4 | 1.5 ~ 2.0 (0.8 ~ 1.2) |
| 5 | Ōsugi Dosan-line (Ry) | Black and Green Schist (Paleozoic) | 4.5 ~ 5.0 | 2.5 ~ 3.8 |

(3) Hisashi Yoshimura: "AN OUTLOOK ON THE NATM", Lecture at the Construction Engineering Conference for the NATM sponsored by the Japan Tunneling Association, 1977-7.

(4) Reiji Amano: "THE STUDY ON THE NATM", Other, Special Issue for all about the NATM, Seko-Gijutsu, 1977-11.

(5) Susumu Ikeda, Tohru Inagawa: "EXCAVATION OF AN UNDERGROUND CAVITY AND ITS BEHAVIOR", Tunnel and Underground, 9-6, 1978-6.

(6) Toshio Taniguchi: "LANDSLIDE IN THE NEIGHBORHOOD OF A DAM RESERVOIR", Jour. of Japan Soc. of Civil Egs., 49-8, 1964-8.

(7) Seizo Miyazaki, Hikoji Takahashi: "THE CIVIL ENGINEERING GEOLOGY", Kyoritsu-Shuppan, 1975-4.

(8) Masasuke Watari: "DAM WATER POURING AND LANDSLIDE", Large Dams, 80, 1977-6 (pp 41 to 53).

(9) Toshio Fujii: "THE STUDY ON DAM COLLAPSE", Large Dams, 80, 1977-6 (pp 1 to 28).

(10) "ROCK MECHANICS FOR CIVIL ENGINEERS", Japan Soc. of Civil Egs., 1966-11 (also, revised edition, 1975-7).

(11) Takahashi, Ikeda, Shirai, Iizuka: "TUNNEL DEFORMATION AND ITS MAINTENANCE", Dobokukogakusha, 1977-4.

(12) Hikoji Takahashi: "GEOLOGICAL SURVEY OF TUNNEL AND ITS CHECK SYSTEM", Report of Railway Tech. Res. Inst., JNR, 71-133, 1971.

LIST OF FIGURES AND TABLES

Fig. 1   Japanese Style Tunneling Method.
Fig. 2   Gelogical Problems in Tunneling.
Table-1  Long Length Tunnels in Japan.
Fig. 3   Tunneling Above Groundwater.
Table-2  Underground Power Plant in Japan.
Fig. 4   Forming Ground Arch Around Tunnel.
Fig. 5   Schimatic Profile of Pumped Storage Power Plant.
Fig. 6   Excavation Line of Dam.
Fig. 7   Variable Hydraulic Condition of Banks by Impounding.
Fig. 8   Geological Problems in Dam Construction.
Fig. 9   Problems on Foundation Treatment in Dam Construction.
Fig. 10  Geological Problems on Slopes.
Fig. 11  Mechanism of Creating the Loosening of Moved Rock Mass.
Table-3  Fairly Low Velocity Zone on the Seismic Exploration.

# Systematic weak seams in dam foundations

RANJI CASINADER
*Sir William Halcrow & Partners, Colombo, Sri Lanka*

## INTRODUCTION

The presence of systematic weak seams in certain rock formations which are otherwise very adequate foundations for dams, have necessitated special measures in the design and construction of dams on such sites. Although this phenomenon is not a new discovery, there has probably not been sufficiently wide dissemination of knowledge on this subject, with the result that the occurrence of systematic weak seams in otherwise competent dam foundations is not always anticipated and investigated. The aim of this paper is to draw attention to the effect that this phenomenon has on dam design rather than to give an authoritative exposition of it. Although a number of - published papers have dealt with the occurrence of such seams at a particular site, there are few papers that deal with the subject more generally. Notable contributions to a general discussion of this matter have been made by Deere (1973).

The importance of discovering systematic weak seams early in the foundation investigations for a dam cannot be stressed too highly. In cases where the discovery has not been made until construction is underway, methods of dealing with the problem have usually been expensive. In some cases, earlier discovery may have led to a different design of dam being selected. In cases where the phenomenon has been discovered and allowed for in the initial designs, solutions have often not been unduly costly. Anticipation and discovery of systematic weak seams is likely to be more successful if the nature of their occurrence is fully understood.

The systematic weak seams referred to in the title are those which occur parallel to bedding in sedimentary rocks or foliation in metamorphic rocks. These represent a significant weakness in the rock mass because they are continuous over considerable distances. Other systematic joints in metamorphic and sedimentary rocks are usually normal to bedding and not always continuous over any appreciable distance. They do not therefore normally contribute a significant structural weakness in the foundations. Singular shear zones extending over appreciable distances do, of course, occur in many foundations and have to be specifically investigated and treated.

This dissertation is confined to systematic weak seams parallel to bedding or foliation.

The treatment of the subject will be in three parts. Firstly, the geological nature of the seams will be described. Secondly, the engineering properties and significance to the dam design will be discussed. And thirdly, some case histories will be given of sites where such seams were encountered.

## GEOLOGICAL NATURE OF WEAK SEAMS

A knowledge of the mode of formation, mineralogy and physical characteristics of bedding plane and foliation shears is a pre-requisite to assessing the engineering significance of these weak seams.

The fact that the main types of rock in which these seams occur, sedimentary and metamorphic, have a 'layered' structure is directly responsible for the formation of these planes of weakness. In both bedded and foliated rocks relatively strong layers are interlayered with weaker layers and it is with these weaker layers that the shears are associated.

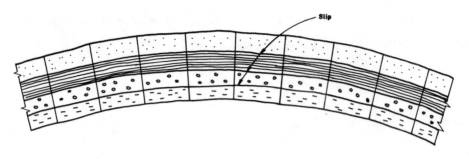

FOLDING

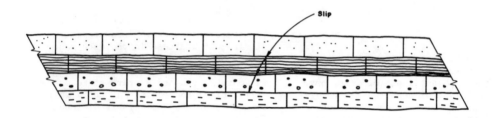

THRUSTING

Fig. 1 Diagrammatic illustration of the formation of bedding plane and foliation shears

Two possible methods of formation are predicated:

- by folding or thrusting of the rock mass

- by relatively shallow rock movements caused by valley erosion and undercutting dip-slopes.

In the first case, the inevitable sliding between adjacent layers of rock during folding of the rock mass takes place in the weaker layers forming sheared zones within them. Fig. 1 illustrates diagrammatically how such tectonic movements could give rise to shearing. The sheared zones vary between seams a few millimeters thick bounded by hard relatively intact rock to zones containing gouge and adjacent crushed rock tens of millimetres thick within a generally heavily jointed and sheared band of rock a few metres thick. Because of the nature of their formation, they are continuous over tens of metres and occur at considerable depth below ground surface.

The presence of slickensided surfaces within the weak seam is a firm indication of the mode of formation of these seams, but the slickensided surfaces are unusually difficult to observe, especially in the presence of seeping water which often occompanies the seams. It is likely also that in the upper more weathered layers of the rock mass, the weaker layers of rock within which these shears occur, have weathered to a greater extent than the adjacent harder rock layers, thus masking the presence of the shear zones within it.

At greater depths, beyond the limit of weathering of the weak rock layers, brecciated rock accompanying the sheared seam forms a preferential path for seepage water and weathers differentially to the adjacent intact rock. The presence of the sheared seams is again masked. The direction of slickensiding in seams formed by folding or other tectonic movements are not related so much to the current attitude of the rock layers, but more to the direction of previous tectonic movements. For instance, slickensiding formed by tectonic movements could be oriented parallel to strike rather than down-dip.

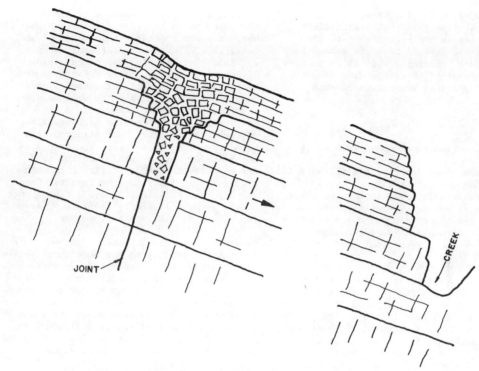

Fig. 2 Illustration of probable origin of gravity drop fold (after D.H. Stapledon)

In contrast, sheared seams caused by valley erosion and undercutting of slopes, have slickensiding orientated down-slope, i.e. in the direction of the recent movement causing them. These seams are generally at shallow depth and do not extend below the base of the valley. Their formation appears to be associated with stress relief during valley formation which causes a reduction of shear strength on potential sliding planes. Fig. 2 illustrates the formation of such shallow shears giving rise to features termed "gravity drop folds" by D.H. Stapledon (Casinader and Stapledon, 1979 : 603). The "drop folds" are formed by slumping of the upper layers of rock into widened joints created by the sliding of the lower layers. Because of the relatively shallow depth at which weak seams of this type occur, they are more easily dealt with than the tectonically formed seams referred to earlier.

In the descriptions given above of the mode of formation of these weak seams, no differentiation has been made between those formed in metamorphic and sedimentary rocks, as it is believed that the mode of formation of these seams is identical in both types of rock.

However, the chemical and physical characteristics of the seams do depend on the lithology of the rock layers in which they occur.

The range of seam types is probably smaller in sedimentary formations. In most sedimentary formations the end product of which the seams are formed is usually basically a clay with greater or lesser percentages of silt and sand size particles which affect the physical properties of the seam. The seams are typically non-homogeneous, in that thin layers of almost pure clay, coinciding with the actual plane of shearing, as evidenced by slickensiding, could exist within seams generally formed of silty or sandy clay. The significant physical feature is the low shear strength of the thin slickensided clay layers.

In metamorphic formations, a wider range of seam types are found, depending on the mineralogy of the weak layers in which shearing has occurred. Typically, the seams contain mica, chlorite, talc, kaolinite, etc. It would appear that insufficient research has as yet been done into the mineralogy of sheared seams in metamorphic rocks. Such research would assist in a better understanding of the mode of occurrence and physical properties of such seams.

## ENGINEERING SIGNIFICANCE AND PROPERTIES

The feature of engineering significance in foliation shears are their ubiquitousness, their considerable areal extent and their low shear strength. The attitude of the seams in the foundations at any particular site also has a great bearing on the design of the dam. These features will be discussed in turn.

The fact that the sheared seams are ubiquitous means that precise identification and location of all seams in a dam foundation becomes a difficult task. The pattern of seams is further confused if there is minor faulting at the site. In practical terms, therefore, it is difficult to design the dam foundations assuming particular location of seams; it is usually necessary to assume that as they are ubiquitous, a seam could exist at the most critical location in any section of dam under consideration.

The areal extent of the seams depends to some extent on whether they have been formed by tectonic movements, or by recent movements. In the latter case, the areal extent can be relatively limited and the design of the foundation can take into account specific limits of the weak planes. With tectonically formed seams, which extend for tens of metres or further if they are not displaced by faulting, the engineering design must usually assume that they are effectively continuous.

The most important engineering property of these seams is their shear strength. Where the seams are accompanied or formed within bands of softer or weathered rock, the compressibility of these bands of rock may also have to be considered in assessing the foundations. This is likely to be more important in the design of rigid (concrete) dams than in the case of flexible (fill) dams. However, it is the shear strength of the seams that is their most significant property affecting the design of all types of dams.

Determination of the shear strength of the seams is in itself a difficult task. Where the seams occur within bands of soft rock, careful excavation of small samples for laboratory testing may be feasible. Alternatively, large scale in-situ shear tests could be carried out. Each procedure has its drawbacks which will be discussed later. Where seams only a few millimetres thick occur in hard unweathered rock, testing for shear strength is a much more formidable task. The Author is not aware of any successful method that has been developed for testing such seams either in the laboratory or in-situ. Careful dry core drilling, to avoid disturbance of the seam, followed by laboratory testing in a shear box may be feasible.

Undisturbed sampling of seams in soft rock for laboratory testing has been successfully carried out. Nevertheless, various problems arise. Field sampling can either be in the form of a large undisturbed block sample from which test specimens can be carefully cut in the laboratory for shear box or triaxial testing, or individual undisturbed test specimens cut from the seam in the field. In either case, since it should be the intention to test the strength of the slicken-sided planes within the seam, in the direction of the slickensiding, extreme care has to be taken in orientating the sample in the testing apparatus. For this reason in laboratory testing, the shear box is to be preferred to the triaxial apparatus. However, even in the shear box, the slightest misalignment of the slickensided plane to be tested will give an apparent higher peak strength than actually exists. This is illustrated in Fig. 3 where the lower line indicates the true shear strength/deformation plot, whereas the upper line could result if the specimen is incorrectly mounted in the shear box. In such a case, the degree of deformation required to reduce the shear strength to be residual may assist in indicating whether the peak is apparent or real. The advantage of laboratory testing is that close control can be exercised on the test procedure, especially with respect to saturation of the specimen and rate of strain.

In-situ testing avoids the possibility of sample disturbance associated with laboratory testing, and enables somewhat larger sample to be tested. Nevertheless, careful excavation of the sampling site is necessary. The testing technique used is to use hydraulic jacks to apply normal and shearing loads to the sample to obtain a shear strength vs normal stress relation-

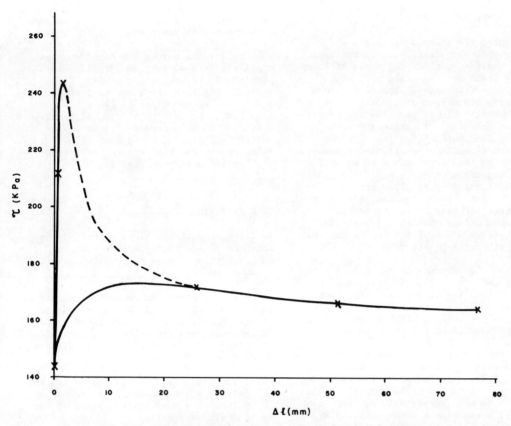

Fig. 3 Shear strength vs deformation of sheared seams

ship. Although, the larger size of the in-situ test specimen may give a more representative result, neither the size of the laboratory sample nor in-situ test sample can be large enough to enable the effective shear strength due to the natural asperities along the sheared seam to be evaluated. On the other hand, the in-situ shear test is difficult to carry out under close control with respect to saturation and rate of strain. The complexity of the testing apparatus and cost are other factors mitigating against the use of in-situ testing. It could also be argued that a number of results from small scale laboratory tests give greater confidence than two or three results from larger scale in-situ tests.

In all types of test shearing should be carried out to large strains (using reversal techniques if necessary) to obtain 'peak' and 'residual' values. Interpretation of the 'peak' value should be carried out with caution as misalignment of the specimen with respect to the pre-existing plane of sliding will give a false peak as noted above. Tests often give no peak value and it is likely that careful testing of these sheared seams will always show that the material exists at its residual strength. Testing of re-moulded samples, when undisturbed samples cannot be obtained, will certainly give false peak values of shear strength, and may also give inaccurate values of residual strength as the properties of thin slickensided layers within the sheared seams are lost by the re-moulding.

Typically, the shear strength (residual) of clayey bedding plane seams in sedimentary rock expressed in effective stress parameters is $C' = 0$ and $\emptyset' = 10$ to $15°$. In metamorphic rocks, it is likely that the strength of the seams will depend on their mineralogy and values of $\emptyset'$ may range from $10°$ to $25°$ with $C' = 0$. Foundation design will nevertheless have to be based on the strength of the weakest seams likely to exist on the site.

In nature, the sheared seams are not planar, but have gentle undulations. As noted earlier, neither the scale of laboratory tests nor the scale of in-situ tests is sufficiently large to obtain a realistic evaluation of the effect of these undulations on the shearing resistance of the seam. To allow for this effect, some authorities increase the measured $\emptyset'$ value by an arbitrary amount. The author considers that such an arbitrary adjustment renders carefully obtained test results meaningless and recommends that the more prudent course is to make allowance for this and any other conservative assumptions made in the analysis, by accepting a somewhat lower factor than would otherwise be accepted.

Finally, the attitude of the beds in the foundation has a very significant effect on the stability of the dam. In general, the seams having a steep dip and a dip direction coinciding with the axis of the dam have least effect on stability.

Where the rock and the seams have a low dip upstream, problems arise in ensuring the stability of the dam in the downstream direction as the seams 'daylight'. This problem is more severe with concrete dams than with embankment dams. With this orientation of beds upstream sliding of an embankment dam must also be checked, especially if the foundation profile is such that the seams can daylight upstream as well. Fig. 4 illustrates possible failure mechanisms with downstream dipping shears.

Where the seams have a low dip downstream, stability in a downstream direction is a problem if the foundation profile falls more steeply than the seams so that the seams tend to 'daylight'. Special attention has to be paid to sliding of embankment dams in an upstream direction with this orientation of the seams. Fig. 5 illustrates possible failure mechanisms with upstream dipping shears.

CASE HISTORIES

To illustrate the problems caused by the presence of systematic weak seams in a dam foundation, three case histories will be described.

The first of these, in chronological order, is the case of the Muda Dam, constructed in N.W. Malaysia in the late 1960s as part of the Muda Irrigation Project, for which the consulting engineers were Sir William Halcrow and Partners.

The second case is Sugarloaf Dam which has recently been designed and constructed by the Melbourne & Metropolitan Board of Works in Victoria, Australia.

The third case history that will be described is that of the Kotmale Dam in Sri Lanka, which will shortly be constructed as part

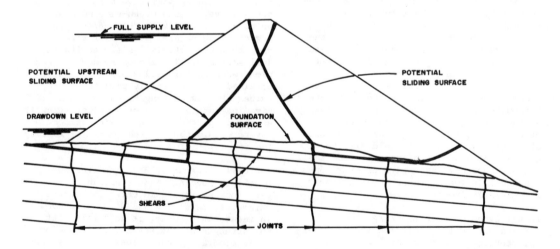

Fig. 4 Stability of dam on foundation with downstream dipping seams

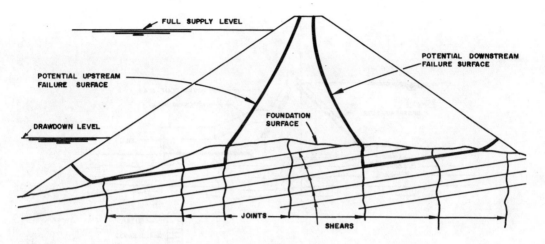

Fig. 5 Stability of dam on foundation with upstream dipping seams

of the Accelerated Mabaweli Development Programme. The consulting engineers are Sir William Halcrow and Partners and Kennedy & Donkin in collaboration with the Central Engineering Consultancy Bureau.

MUDA DAM

Muda is a concrete buttress dam 33 metres high whose foundations are massive quartzites with partings and thin beds of mudstone at numerous levels, many of which were shaley and weathered.

Although it was anticipated that the foundation rocks at the site might not provide adequate sliding resistance, it was not until excavation of the foundations that the shaley layers were observed.

The geological structure at the site was such that the foundation strata and in particular the shaley beds were subhorizontal. (See Fig. 6). This condition was critical for the sliding stability of the buttress dam.

In-situ shear tests were carried out on a 2 ft (0.6 m) square block cut in the foundations. (James, 1970). Three-stage tests were carried out, and shearing was carried out to large strains by the multiple reversal technique. A controlled rate of strain was used in the tests, and the rate of strain was based on an estimate of the consolidation characteristics of the material. Three in-situ blocks were tested.

The results of these tests are shown in Fig. 7. These tests showed that the residual shear strength of the shaley layers was between $\emptyset = 18°$ and $20°$ with $C' = 1$ to $6$ p.s.i. ($7$ to $42$ KN/m$^2$). In two cases out of the three, the material existed at residual strength as the 'peak' shear strength value was effectively no higher than the residual shear strength. In the third test block, a peak shear strength of $\emptyset' = 26°$ with a very low $C'$ value was observed. Laboratory tests on undisturbed samples gave residual shear strengths of $\emptyset' = 17.5°$ to $19°$ with no cohesion intercept. There was, therefore, good agreement between laboratory and field results as regards the residual angle of shearing resistance. It is likely that the cohesion intercept measured in the in-situ tests was due to testing technique. In any case, the cohesion value was not significant in relation to the forces transmitted from the dam.

On the basis of the tests, the dam was analysed for sliding stability assuming the shear strength of the shaley mudstone layers as $C' = 0$, $\emptyset' = 18$. To ensure stability high tensile steel cable anchorages were employed. (Taylor and James, 1967).

SUGARLOAF DAM

Sugarloaf is a concrete faced rock-fill dam 85 metres high. Foundations are siltstones, with some interbedded sandstone of

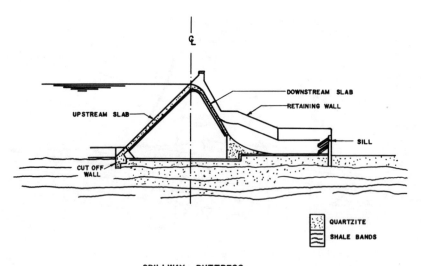

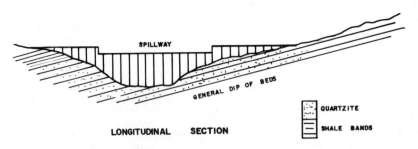

Fig. 6 Sections showing simplified geological conditions

Silurian and Devonian ages. The dam site is located on one limb of a broad syncline so that the general dip of the strata is upstream. On the right bank, the valley is relatively steep but the left wing of the dam is constructed on a ridge. The detailed geological structure and site topography are such that dip-slopes or near-dip-slopes occur on both right and left banks of the dam site.

One of the major defects in the foundation rock mass discovered in the pre-construction site investigations was a series of crushed seams ranging from 2 mm to 20 mm thickness, parallel to bedding and typically at 1 to 5 m spacing. In the upper weathered rock zone, the crushed seams have b-en extremely weathered to form a low plasticity clay. Slickensided surfaces were found within and alongside some of these seams. The crushed seams are believed to have been formed due to inter-bed movements during folding. (Casinader and Stapledon, 1979).

As the stability of the upstream shoulder of the dam depended greatly on the strength of these seams, laboratory direct shear tests were carried out on 60 mm x 60 mm specimens cut directly from the seams in the field. All samples obtained were from the weathered zone because of the practical difficulties in sampling the crushed seams in fresh rock.

The test specimens were mounted in the shear box apparatus so that shearing took

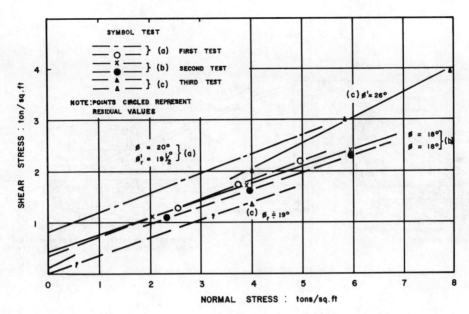

Fig. 7 Muda dam foundation seams. Shear test results

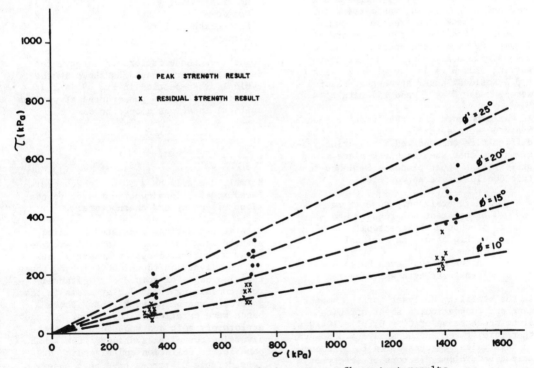

Fig. 8 Sugarloaf dam foundation seams. Shear test results

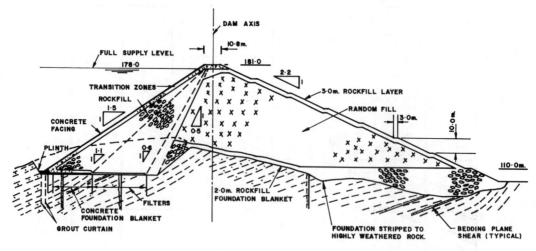

Fig. 9 Sugarloaf dam, typical cross-section

place as nearly as possible along slickensided surfaces. Reversal techniques were used and stage testing was carried out to determine 'peak' and 'residual' shear strength. Results of the tests are shown in Fig. 8. The seven specimens tested gave a wide scatter of 'peak' strength varying from $\phi' = 15°$ to $\phi' = 27°$ with $C' = 0$. Residual shear strengths measured were between $\phi' = 9°$ and $15°$ with $C' = 0$.

It was concluded that the 'peak' strengths measured were apparent only, due to the difficulties encountered in mounting specimens so that shearing took place along pre-existing shear planes. Evidence for this conclusion were two fold:

:: in general. only some 25 mm of movement was required to produce a shear strength within 5% of the residual value
:: in one of the seven tests no 'peak' strength was measured

In the stability analyses for the embankment and foundation, a shear strength along these seams of $C' = 0$, $\phi' = 10°$ was, therefore, taken in the analyses. However, as this value was considered to be a conservative minimum, factors of safety of the following values were considered acceptable for the up stream shoulder, whose stability was the more critical:

| | |
|---|---|
| End of construction | 1.1 |
| Reservoir partly full | 1.3 |
| Drawdown | 1.2 |
| Earthquake with reservoir partly full | 1.1 |

To achieve these factors of safety, the foundations were shaped as shown in the typical cross-section in Fig. 9, and stabilising fills were constructed at various critical locations along the dam.

KOTMALE DAM

Kotmale dam will be a 90 m high concrete-faced rockfill dam constructed on foundations of gneiss and charnockite.

Site investigations consisted of adits, trenches and boreholes. Initial excavations were carried out at an early date for the diversion tunnel portals and these revealed bands of completely weathered rock containing sheared clay seams in the generally weathered rock zone. These bands were parallel to foliation. In one exploratory adit and subsequently in the diversion tunnel excavations, weak seams, parallel to foliation and extending over considerable distances have been observed within the fresh rock zone. Initial excavations for the dam have confirmed these findings. A study of these seams is as

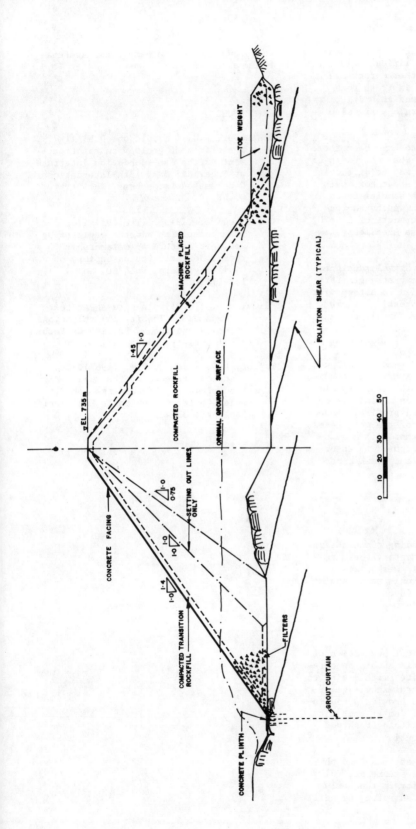

Fig. 10 Kotmale dam - Typical cross-section (Preliminary)

yet incomplete, but it would appear that seams of different compositions exist. Whilst the seams in weathered rock are clayey in nature, seams in fresh rock appear to be micaceous and chloritic. Seams of different mineralogy are also likely.

The seams are believed to have been formed by tectonic movements. Shear strength of the seams in fresh rock may be difficult to test, but laboratory shear box tests have been carried out on samples taken from the clayey seams in weathered rock. These give a residual strength of $C' = 0$, $\phi' = 10°$ to $12°$ with some specimens showing no 'peak' strength.

The dam is being conservatively designed using these residual shear strength parameters, but it is intended to attempt further investigations and testing before construction begins to ascertain the exact nature and physical properties of the seams in fresh rock. The foliation at the site typically has a dip of $12°$ to $14°$ downstream and the weak seams are aligned parallel to foliation. A critical condition for the embankment is stability of the downstream shoulder as a combination of low dip and the ground profile, caused potential 'daylighting' of the weak seams. Foundation shaping together with a downstream toe weight are proposed to ensure stability as shown in the typical cross-section in Fig. 10.

CONCLUSIONS

Foliation shears and bedding plane shears in metamorphic and sedimentary rocks respectively form systematic weak seams in dam foundations. Their existence should be anticipated and site investigations carefully designed to discover them or to prove conclusively that they do not exist at any particular site.

The mineralogy and physical properties of the various seam materials should be carefully studied at each site, with a view to contributing to the general knowledge of these seams, and also to provide a realistic design basis for the dam at that site. Means of testing seams within fresh rock requires particular attention.

Providing the location, attitude and physical properties of these seams are fully understood and allowed for in the early stages of design, dams can safely be built on foundations containing these systematic weak seams.

REFERENCES

Casinader, R.J. & Stapledon, D.H. (1979), The effect of geology on the treatment of the dam-foundation interface at Sugarloaf dam, 13th International Congress on Large Dams, New Delhi, Q. 48, R. 32.

Deere, Don U. (1973), The foliation shear zone - an adverse engineering geologic feature of metamorphic rocks, Journal, Boston Society of Civil Engineers, Vol. 60, No. 4, : 163-176.

James, P.M. (1970), In-situ shear tests at Muda dam. In-situ investigations in soils and rocks, British Geotechnical Society, London.

Taylor, R.G. & James, P.M. (1967), Geotechnical aspects of the Muda irrigation project. Proceedings 1st S.E. Asian Conference on Soils Engineering, Bangkok : 33-42.

# Use of Roller Compacted Concrete in dam construction

JACK C.JONES & GARY R.MASS
*Harza Engineering Co., Chicago, Ill., USA*

## INTRODUCTION

Cement treated materials have been used in construction around the world for many years. We have utilized cement treated aggregate base (CTB) in road and highway construction. Use of soil-cement for road construction began in 1933 and has since been used for slope protection in dams and waterways. However, it is only recently that a concrete type fill material has been considered for use in dams as the primary construction material. This method of dam construction developed concurrently in Great Britain, Canada, Japan, and the United States. In North America, this material is referred to by some as "Rollcrete;" in Japan as "Roller Compacted Dam" (RCD) concrete and in Great Britain as "Dry Lean Concrete" (DLC). "Roller Compacted Concrete" (RCC) is the nomenclature used by the mass concrete technical committee of the American Concrete Institute, ACI Committee 207, to represent a concrete of no-slump consistency which is transported, placed, and compacted using the same construction equipment used for earth and rockfill dams.

In May, 1962 John Lowe of TAMS, New York, in an unpublished discussion of a paper entitled "Utilization of Soil-Cement as Slope Protection for Earth Dams" by Holte and Walker, described the use of a material termed rollcrete as the impervious core for a 210 foot (64 m) high cofferdam for the Shihmen Multipurpose Dam Project in Taiwan. The cofferdam had just been completed at that time. In the conclusion to his discussion Mr. Lowe states, "In the future we may be able to use these materials for an earth dam overflow section."

Lean concrete has been used in the construction of hydroelectric projects for many years, primarily as a foundation material when high strengths were not required.

In 1970, Professor Raphael of the University of California, Berkeley, proposed the construction of a fill dam with steep side slopes using bank run sand and gravel mixed with cement and placed and compacted with earth moving equipment as a way to reduce the cost of fill dams.

In a follow-up conference in 1972, R. W. Cannon of the Tennessee Valley Authority (TVA) presented a paper, "Concrete Dam Construction Using Earth Compaction Methods." Later that year at an ACI symposium he also presented a paper, "Compaction of Mass Concrete with Vibratory Roller." In these papers Cannon presented the test results on samples taken from roller compacted concrete.

The U.S. Army Corps of Engineers at their Waterways Experiment Station (WES), Vicksburg, Mississippi, and in the Portland District have also studied the feasibility of no-slump concrete for mass concrete construction. A fairly comprehensive study was made at Lost Creek Dam near Trail, Oregon to investigate bond of lift joints and permeability qualities of roller compacted concrete.

Test results from a field placed section constructed for these studies indicated that roller compacted concrete has properties equivalent to those possessed by conventional mass concrete. The material is not just a coarse aggregate soil-cement.

Perhaps the first massive use of roller compacted concrete (RCC) was in early 1975 when 350,000 m³ were placed in 42 working days at Tarbela Dam in Pakistan following failure of Tunnel 2 in the outlet works. With a maximum daily placement of 18,000 m³/day and average of 8,400 m³/day, RCC was the only placement method suitable for the repairs which could be successfully accomplished in the restricted time period before the start of the rainy season and refilling of the reservoir.

RCC was again used at Tarbela in 1978 to stabilize the highly eroded plunge pool in the service spillway. This work involved placement of roughly 518,000 m³ of RCC in approximately 4 months, mid-February to mid-June. Average daily placement was bout 5,180 m³ in two-10 hour shifts per day.

Milton Brook Dam located near Plymouth, England is currently awaiting financing by the South West Water Authority and may be the world's first roller compacted concrete dam.

During the summer of 1980 the American Concrete Institute, Technical Committee 207 on mass concrete, published the most complete report to date on roller compacted concrete. Since the time of preparation of this report, two major projects in the U.S.A. have been planned for construction using RCC. The first will be Willow Creek Dam, a flood control dam located near Hepner, Oregon, scheduled for construction in 1981 by the U.S. Corps of Engineers. The second will be Upper Stillwater Dam in Utah, a 90 m high by approximately 870 m long major structure under the U.S. Bureau of Reclamation. Construction of the dam is scheduled for 1984 and will consist of placement of 1 million m³ of RCC.

The intent of this presentation is to highlight the important points of the ACI 207 report, of which courtesy copies are available at the registration desk, and to show a few slides of the construction work at Tarbela during the service spillway plunge pool lining which demonstrate utilization of RCC. Later this morning Mr. G. LaVilla of Impregilo (Tarbela Joint Venture/Indus River Contractors) will make a presentation on the practical aspects of roller compacted concrete from the Contractor's viewpoint.

CONCEPTS OF ROLLER COMPACTED CONCRETE FOR DAMS

The following are exerpts from ACI 207.5R-80, "Roller Compacted Concrete":

Mixture Proportioning and Materials

"For effective consolidation, RCC must be dry enough to support the weight of the vibratory equipment but wet enough to permit adequate distribution of the paste binder throughout the mass during the mixing and vibration process. Concrete suitable for compaction with vibratory rollers differs significantly in appearance, in the unconsolidated state, from that of normal concrete having a measurable slump. There is little evidence of any paste in the mix until it is consolidated. Consolidation time can be used as a measure of the consistency of the concrete and the efficiency of compaction equipment. Investigations by ACI 207 committee members indicate acceptable vibration equipment should be capable of fully consolidating the stiffest mixes within 60 seconds for a broad range of mix proportions.

Laboratory studies of mix consolidation are shown in Fig. 2.2.1 and 2.2.2. In those studies a vibration table was used to consolidate the concrete in a standard air content meter container.

Increases in water significantly reduce the vibration requirement for full consolidation but also increase the possibility of inadequate stiffness to support the weight of the vibratory roller.

Table 2.2.1 provides optimum water contents for manufactured aggregate mixes without pozzolans. The differences in water requirements for mass mixes and bedding mixes illustrate the basic difference in compactive effort required for full consolidation for the two types of mixes. Mass mixes are recommended where continuous placement can be made without cold joints. Bedding mixes are required for bedding and bonding fresh concrete to hardened concrete at lift or cold joints. The data given in Table 2.2.1 is based on actual mixes utilizing 3/4 - through 4-1/2 - in. (19 - 114 mm) maximum size aggregate (MSA). Although the 4-1/2 - and 6 in. (114 and 152 mm) MSA mixtures are difficult to handle without segregation, they are included because both reduction in heat development and economy may dictate the extraordinary precautions necessary for their use.

Although the quality requirements of the aggregates used in concrete are not directly influenced by the strength requirements of the concrete, aggregate variability does significantly affect cement and water requirements of the mix which in turn affect strength and yield. In massive concrete structures the practice of using unrealistically high strength requirements unnecessarily increases the cost of the structure, and is a principal contributor to the cracking problems associated with heat of hydration of the cementitious materials.

Large vibratory rollers are capable of compacting quarry-run rock in layers as thick as 2 feet (0.6 m) in rockfill dams. Limitations on aggregate size is therefore not controlled by the lack of suitable compaction equipment. For manufactured aggregates there does not appear to be sufficient material cost savings in aggregate sizes larger than 3 in. (76 mm) to offset the added batching costs and costs of handling the increased segregation problems associated with the larger aggregates. When the placement layers are more than three times the maximum aggregate size, the size has little influence on compactibility with the type of vibratory equipment presently utilized in rolled rock fills.

Although the use of uniformly graded aggregates such as specified by ASTM C33 is desirable for best results, gradation is not as important for achieving the desired compaction in RCC as in slumpable concrete because of the differences in equipment used to densify the two types of concrete. Aggregate gradation does, however, affect the relative compactability of the concrete and may influence the minimum number of vibrating passes required for full consolidation of a given layer thickness of RCC. It also affects the water and cementitious material requirements needed to fill the voids in the aggregate and coat the aggregate particles to produce a solid concrete volume. The aggregates used in most of the research work to date were graded in accordance with accepted standards, and even the bank or pit run materials at Tarbela Dam and the U.S. Corps of Engineers Zintel Canyon studies were relatively free of any size gaps in the gradation curve. Using the normal distinction between fine and coarse aggregates, the grading of fine aggregates has the greatest influence on paste requirements. When no pozzolan is used, a 5 percent increase in percentage passing the No. 100 sieve may be beneficial in reducing paste requirements.

The properties of concrete made with potentially economic sources of aggregate should be determined, and their suitability judged for RCC application, in preference to rejection on the basis of nonconformance with standard specifications.

When gradations are controlled by screening and dividing the aggregates into separated size fraction, the void content may be controlled within limits.

The absolute volume of coarse aggregate per unit volume of concrete will normally fall within the limits of Table 2.3.2.

The void content of fine aggregate as determined in dry rodded weight measurements normally ranges from 34 to 42 percent.

Pozzolans used in RCC as compared to the usual structural application relates to the utilization of larger volumes of pozzolans and reduced emphasis on the effect of the pozzolan on workability. One principal function of a pozzolan or added fines is to occupy space which would otherwise by occupied by cement and water. The fact that even a small amount of free lime liberated from cement is sufficient to react with large volumes of pozzolans has been demonstrated by agencies such as TVA which has for many years used fly ash to replace fine aggregate as well as cement. Their tests show continuing pozzolanic activity at extended age in years.

At ages beyond 28 days the pound for pound difference in strength contributions for the various cement types decreases, with the slower strength developing cements ultimately producing the highest strengths. For most Type I or II cements, Fig. 2.4.1 can be used to proportion equal strength concrete for varying proportions of cement and fly ash.

Both air-entraining and water-reducing and retarding admixtures have been routinely used at normal dosage rates in most RCC applications and test placements made to date.

RCC Properties

The significant properties of conventionally placed concrete are also important in RCC. These are compressive strength, modulus of elasticity, tensile strain capacity, Poisson's ratio, triaxial shear strength, volume change (thermal, drying, and autogenous), thermal coefficient of expansion, specific heat, thermal conductivity, diffusivity, permeability, and durability. Any differences in the properties of RCC and conventional concrete are probably primarily due to differences in mix proportions. There is generally 40 percent less water and 30 percent less paste in RCC than in conventional concrete.

Typical mixes and strength data obtained from cores drilled out of the test fills or structures are given in Table 3.2.1. Since there are no early age strength requirements for massive placements of RCC, strengths should be based on strengths for later ages, such as 6 months to 1 year in age, unless the structure is to be put in service sooner. Results of experiments indicate close correlation between the strengths of cores and cylinders when the molding procedure achieves full consolidation. This can readily be accomplished by extended vibration of an overfilled cylinder. When the test specimen is fabricated using a vibrating table or modified Vebe apparatus, consolidation is complete when paste flows out around the edges of a surcharge retained on top of the concrete deposited in the cylinder mold. Various small mechanical and pneumatic tampers ("pole tamper," etc.) have been used to consolidate larger test specimens, such as beams, slabs, etc.

A properly proportioned and consolidated RCC mixture would be expected to have elastic modulus values similar to its conventional counterpart batched with the same aggregate.

Polivka et al report in "Studies of Creep in Mass Concrete" that "the magnitude of creep was observed to be directly proportional to the volume of cement paste contained in a unit volume of concrete." On this basis, it would appear that minimum paste mixes compacted to 98 percent of the air-free densities would have approximately 20 percent lower creep for a given loading situation than values obtained for slumpable concrete. Generally aggregates of high modulus of elasticity will produce low creep concrete.

If the induced strain due to volume change is tensile and it exceeds the strain capacity of the concrete, cracking will occur. Strains in concrete can be developed by volume reductions induced by drying and autogenous shrinkage or by cooling of the concrete. Crushing or addition of crushed material will improve strain capacity. Increase in cement content improves strain capacity by increasing tensile strength; however, this improvement is generally offset by heat dissipation problems caused by the increase in heat generated by higher cement content. The strain capacity of most RCC would be expected to be low because it usually will be batched with lower cement contents and/or higher pozzolan replacement of cement. This lower strain capacity should be offset by lower temperature-induced strain resulting from the leaner mix and thinner lift placement. The thermal properties of concrete are influenced principally by the concrete aggregates and moisture content. These properties for RCC should differ from those of conventional concrete only to the extent of these mix differences (See Table 3.4.1).

When there is sufficient paste to minimize the air void system, and compaction equipment is capable of fully consolidating the mass, the RCC should be relatively impervious. Cracking and cold joints represent the greatest potential for water passage through all types of concrete. This points up the necessity for the RCC mix covering the cold joints to have an excess volume of paste to bond and seal the joint, thereby preventing leakage. Joint treatment has always been a major concern in conventional concrete placement, and testing of RCC placed to date substantiates a need for special consideration of construction joint treatment in this type of construction.

Soil-cement slope protection on earth dams and massive soil cement embankments still in service constructed by processes similar to those utilized for RCC are exhibiting good service performance under sometimes severe exposure conditions. On this basis no durability problems should be expected with similar lean-mix RCC. Furthermore, wear resistance is benefited by the use of high concrete strengths, smaller maximum size aggregates, and smooth surface textures. Erosion studies by the Corps of Engineers for Zintel

Canyon Dam proposed near Richland, Washington, showed 1-1/2-in. (28 mm) MSA RCC mixes to be resistant to erosion at water velocities as high as 70 ft. (20 m) per sec. No studies or experiments have been performed on resistance to chemical attack; however, the principal factors governing resistance to deterioration from chemical attack should be similar to those for conventional concrete.

## Construction

A 6ft. (1.5 m) wide vibratory roller of appropriate design (drum size, compactive force, amplitude, frequency, etc.) making four passes and travelling at a speed of 2 mph (3¼ Km/hr) can compact more than 350 cu. yd. (260 cu m) per hour of no-slump concrete in 10-in. (25 cm) layers. A conventional batch plant would require approximately four 4-cu yd (3 cu m) mixers or one large 12-cu yd (9 cu m) mixer to match the compaction capabilities of one large roller on a continuous placement operation. These conditions appear to suggest the desirability of continuous mixing to offset the time lost in batching operations.

At Tarbela Dam, for the outlet works repair, the concrete was continuously mixed by using a series of rock ladders strategically placed to blend the ingredients. The plant had a maximum production output of approximately 1000 cu yd (750 cu m) per hour.

It is just as important for the plant operator to see the batch in the mixer for no-slump concrete as it is for slumpable concrete. An operator can train his eye to judge the consistency of this concrete as readily as he can judge slump. For proper mix control, his problem is completely reversed from that of conventional concrete. The placement crew will complain about wet batches instead of dry batches. Variations in free moisture on the aggregates can be particularly troublesome in initial batches because of the low water requirements of the mix. Most operators make the mistake of overestimating free moisture and provide too little water in the initial mixes. This is particularly undesirable because most initial mixes will be used for covering construction joints and should be on the wet side for adequate bond. It is much better to start with a wet batch and reduce batch water in subsequent batches to achieve the desired consistency. Excess water rises in subsequent layer placements, thereby eliminating any strength concerns for the wet batch.

The principal of utilizing RCC as a mechanized placement system for pavements would be similar to placing it in layers or lifts during the construction of a dam. Principal advantages of continuous mixers are the elimination of batching time from the production time cycle and the significantly lower cost of plant in relation to the output capacity. Conventional nontilting drum type truck mixers are not recommended for RCC either for mixing or transportation because of serious segregation problems anticipated during the discharge cycle.

The volume of material to be placed and access to the placement area will generally be the controlling factors in the selection of equipment to be used for transporting RCC from the mixing location to the placing area. Frequently, the same equipment will perform both the functions of transporting and placing. Essentially, there are two methods for transportation of RCC: batch and continuous. To some extent, selection may be influenced by the type of mixing equipment utilized. However, with the use of proper controls and accessories, such as holding hoppers, continuous mixer can be utilized with batch transportation and batch mixers with continuous flow transportation equipment.

Field experience indicates that RCC can be compacted at rates of 1000 cu yd (750 cu m) or more per hr. with several presently available self-propelled vibratory rollers operating. The concepts developed for "main line" highway paving and volume earth moving appear to be more applicable for RCC than equipment previously used for mass concrete. Full development of the economic potential of RCC is also dependent on the maintenance of adequate stockpiles of raw materials to sustain the high rates of continuous mixing and placing.

Experience indicates that 1-1/2-in. (38 mm) MSA concrete can be transported and placed in nonagitating haul units designed for aggregate hauling and earth-moving without objectionable segregation. The 3-in. (76 mm) MSA concrete and larger aggregate concrete has a distinct tendency to segregate when it is dumped from this type of equipment onto hard surfaces such as the conditions which prevail when starting a new lift or placing on a previously compacted surface. To date, the problems of segregation occurring during the transportation and placing of 6-in. (152 mm) MSA concrete have been so severe that its use with presently designed equipment does not appear to be practical.

Batch transportation equipment is almost always some form of container mounted on a rubber tired haul unit or trailer. The most common units are conventional end or bottom dump trucks but they can range in size from power buggies handling several cubic feet to enormous off highway haulers that will handle up to 38 cu yd (29 cu m) per load cycle. The selection of equipment size should be based upon required placing capacity, cycle time (elapsed time for one round trip for one unit), and total volume to be placed. Additional considerations are cost of transportation unit, accessibility of placement area, spreading requirements, and cost of preparation and maintenance of haul roads and access points. Cement is extremely adhesive and will tend to cake or build up on parts exposed to it. For this reason, most units are designed to be unloaded by end or bottom discharge dumping with little or no control of the flow of the materials out of the unit. Because of the consistency of RCC, equipment which relies on mechanical blades or paddles to move through the concrete probably will prove to be underpowered and susceptible to extreme wear and mechanical failure. Continuous transportation of RCC seems most applicable to use of belt conveyors.

To date field tests have indicated that optimum placement layers range from 8 to 12 in. (20 to 30 cm). This contracts to normal layers in conventional mass concrete of 18 to 24 in. (46 to 60 cm). Considering the much higher placing rates attainable with RCC in layer depths one-third to one-half that of conventional concrete, it is apparent that spreading of the concrete before compaction is an important operation in the production process and must be efficient to keep pace with compaction capabilities. Several pieces of spreading equipment may be necessary to keep pace with one large vibratory roller. Equipment which deposits batches in piles of concrete must be supplemented with equipment which will knock down and spread the piles in thin layers 8 to 12 in. (20 to 30 cm) thick at the same rate at which concrete can be transported and compacted. Dozer type spreaders and road graders appear to be the most practical. Rubber tired equipment is preferred because concrete materials do not have the tendency to build up on tires as it does on tracks, rollers, and sprockets of crawler type equipment. Spreading is facilitated by consistent and symmetrical placing. Depositing the concrete in a window configuration for spreading is desirable. Rubber tired hauling equipment can operate on the freshly compacted surface with no adverse affects on the concrete. However, provision must be made to keep the haul equipment from tracking in mud and dirt which would have an adverse effect on the quality of the concrete and could cause problems with joint integrity and consequent leakage in structures subject to hydrostatic pressure.

When placing concrete with belt conveyors in a continuous operation, the best rule is to have as little equipment as possible actually in the placing area and to minimize as much as possible the actual movement of equipment while concrete is being placed. With some modification slipform equipment may be used for spreading RCC. Also equipment designed for spreading asphaltic concrete or mechanical devices, such as the "Jersey Spreader" which handles only aggregates, may be adaptable for spreading RCC in thin layers for efficient compaction.

A wide range of vibratory roller equipment has been used for test and construction applications of RCC. Generally any vibratory roller which has been used successfully to compact rock fills will compact RCC. Self-propelled rollers, with power-driven vibrating drums, have proven to be more suitable for RCC than rollers which only vibrate and require other vehicles or means for propulsion. The power drive on the vibrating drums enables the roller to reverse directions without disturbing the compacted concrete. Maneuverability, compactive force, drum size, frequency, amplitude, and operating speed are all considered to be principal parameters in the selection of a roller. Rollers larger than 4 or 5 tons usually cannot operate closer than 6 to 9 in. (15 to 23 cm) to formwork or obstacles, so that rollers smaller than 1 or 2 tons are usually needed to consolidate the concrete in these areas. The minimum number of passes for a given vibrating roller to achieve full consolidation (maximum obtainable density) depends primarily on the concrete mix and layer thickness. Layer thickness will be governed more by spreading efficiency than by compaction requirements, since the large rollers are fully capable of compacting layers of quarry-run rock up to 24 in. (60 cm) and more. Most placements to date of RCC have used 8 to 12-in. (20 to 30 cm) layers. Test should be performed in "test fills" prior to or during the early stages of construction to determine the minimum number of passes for full consolidation, using the correct mix and the planned layer thickness.

Recognition of fully consolidated concrete is also somewhat dependent on the concrete mix. If the mix has been proportioned for paste volumes in excess of minimum, then fully consolidated concrete will exhibit plasticity and a discernible pressure wave can be detected in front of the roller, particularly when two or more plastic layers have been placed. The consolidated mass of a mix with adequate paste volume will respond to working of the surface like still molding clay or jelly. This action enables scattered rocks on lift joints to be worked into the surface under the compactive effort of the vibrating rollers, making a solid mass of the layered construction. Some observers viewing test placements expressed concern for the "pore-pressure" inherent in a concrete of excess paste volume. If pore pressure is not relieved in soils work, it will contribute directly to a reduction in shear strength. In concrete work this pore pressure will dissipate with hydration of the cement and has no affect on strength. If the surface of concrete with an excess paste volume is "worked," water will rise to the surface. The surface moisture immediately following consolidation evaporates rapidly, before hardening takes place, so it should not be confused with bleed water or laitance.

The only sure way to determine full compaction in the field is by measurement of the compacted density and comparison with the same mix fully compacted in the laboratory by extended vibration. The nuclear measuring devices now used in soil density measurements may be applicable for plastic density measurements of compacted layers.

Anytime a placement layer is not covered by the time is reaches initial set, it will no longer be workable and becomes a cold joint, or unplanned construction joint. This time element is very sensitive to temperature, the amount of portland cement in the mix, and the type and set retarding characteristics of the admixtures, if any. On the basis of time element there appears to be a definite advantage in utilizing set-retarding admixtures, thereby extending the setting and working time, providing flexibility against unscheduled delays, and allowing enlargement of the working surface of the lift. Treatment of horizontal lift on construction joints differs from conventionally placed mass concrete in that there is no surface water gain during set of the surface. If the construction joint has been kept clean and moist throughout its exposure, no joint treatment is required. Experiments to date have shown that satisfactory bond at horizontal construction joints can be assured by utilization of a bedding mix of 1-1/2 in. (38 mm), or smaller, maximum size aggregate provided the mix has a paste volume at least 20 percent greater than the minimum volume of paste which produces maximum density. The consistency of the bedding mix should ideally be "too wet" to be used in full layer thickness and should be spread in a layer approximately 3 in. (8 cm) thick and covered with a sufficient layer of the regular mass mix to make a full thickness layer, prior to rolling. The most essential requirements of the bedding mix appear to be excess paste, excess wetness, and excess strength over that of the mass mix.

The required curing and protection of RCC is generally consistent with the treatment necessary for conventional concrete, whereby the concrete should be maintained in a moist condition and at a temperature favorable for cement hydration. No curing of the compacted surface may be necessary under construction conditions where succeeding layers are placed rather rapidly as long as surface drying does not occur. It should be noted that most of the problems generally associated with hot weather concrete work, such as increased water demand, slump loss, and plastic shrinkage cracking, are not severe in RCC construction primarily because of the low water requirement and methods of transportation, placement, and compaction used.

Side slopes can be controlled to any desired shape by the application of conventional formwork; however, special consideration must be given to the design and anchorage of conventional formwork. The height of overhanging sloping forms will restrict the working areas of the vibratory rollers. Small rollers can be operated within 1 in. (2.5 cm) of vertical formwork; however, large rollers cannot get closer than 6 to 9 in. (15 to 23 cm). Good compaction can be achieved at vertical surfaces with 1-1/2 in. (38 mm) and smaller maximum aggregate size mixtures with careful attention to details. After compaction there is always a small projection of uncompacted concrete above the surface which the roller cannot reach. If this is kept raked away from the form, layer lines will not stand out on removal of forms and a surface comparable with conventionally placed concrete is possible. The handling and raising of conventional formwork may become the principal slowup in the operation unless an efficient system can be developed. However, the cost of conventional formwork should be less with RCC because of the elimination of transverse joints.

The most economical means of controlling side slopes appears to be the utilization of power curbing machines with conventional concrete to slipform curbs some 24 to 30 in. (60 to 75 cm) high against which the RCC placement can be made the following day. By controlling the curb shape (see Fig. 4.7.2), it is possible to maintain an average production rate of 1.5 to 2 ft (45 to 60 cm) per day for the full length of placement and still maintain the distributed pressure of the compacting rollers within curb concrete at least 3 to 4 days old. These machines are commonly used in highway and street construction and special forms can be fabricated to fit almost any shape curb desired. They are capable of slipping curbs at rates of 15 ft. (5 m) per min or more to very precise alignments. Overall costs of the curbing system would depend on the size of structure. Costs have been estimated as low as 3 to 5 percent of conventional form costs for a waterworks structure in Great Britain.

If no attempt is made to compact edges of an RCC placement, the sides will assume a natural angle of repose estimated to be somewhere between 40 and 45 deg. Any means of containing the loose concrete at the edges long enough for even partial compaction could result in side slopes as steep as 60 deg. A dam built with upstream and downstream slopes of this steepness would meet stability requirements without installation of the usual foundation drainage systems used in conventional gravity dams with formed surfaces. Although the RCC section would contain approximately 20 to 30 percent more volume as compared with the conventional gravity section, elimination of the drainage system could make RCC economically feasible.

Contraction Joint Spacing and Waterstops

One deterrent to utilization of the rolled concrete method of construction has been the reluctance of designers to deviate from the traditional block construction concept. The principal function of contraction joint spacing is to limit the zones of significant foundation restraint to the lower regions of the dam. Control of hydration temperatures is still required to prevent cracking in these zones of restraint; however, contraction joints do serve to relieve cracking in the upper regions of the dam. The principal concern for cracking in gravity dams without contraction joints is appearance and leakage control. Surface cracking is generally caused by internal restraint rather than foundation restraint and is therefore limited in depth. Foundation restraint can contribute to surface cracking in a dam built without contraction joints; however, the propagation of surface cracks relieves the internal restraint condition, thus requiring a continuing decrease in volume for further propagation. The cooling down of the interior mass of a large dam may therefore require years to reach the critical volume change for cracking to progress throughout the full height. In most instances the critical volume change in the lower portions of the structure can be prevented by cold weather placement and utilization of low heat generating concrete.

The Alpa Gera and Quaira Della Miniera Dams were constructed in Italy without formed contraction joints. The construction differed from roller compacted concrete construction only in the method used for compacting concrete. Concrete was transported and deposited in dump trucks, spread with bulldozers, and compacted by means of tractor mounted internal vibrators. The concrete was of low slump consistency. Contraction joints were cut approximately 12 hours after consolidation by means of vibrating blades mounted on a tractor. The same type of equipment was used in a similar manner to cut contraction joints in the RCC concrete for Okawa Dam in Japan. To prevent the propagation of uncontrolled cracks from the upper surface, such cut joints need not be spaced closer than the average height of the dam above the foundation.

Studies of the heat generation and temperature rise of massive RCC placements indicate that the uniform sequential placement of lifts may have a beneficial effect on crack reduction due to the uniform temperature distribution throughout the mass and the reduced restraint between lifts. Average placement rate has a more significant effect on maximum temperature rise than height of lift, and ambient temperature affects peak temperature more than placing temperature. Fig. 5.4 shows the effect of placing rate and lift height on temperature rise. For the same average placement rate, temperature rise decreased with decreased lift height. Several hours of exposure of the 10 to 12 in. (25 to 30 cm) plastic layers to ambient temperature reduces the effectiveness of the difference in placing temperature and ambient temperatures. This works to an advantage in cool weather, when the air temperature is less than the placing temperature of the concrete, but to a disadvantage in hot weather, when the temperatures are reversed and the layers are subject to radiant heat absorption from the sun.

Waterstops can be installed either externally or internally to control leakage in contraction joints cut by vibrating blades. An external application might simply be to attach a waterproof membrane to the upstream face at the joint. Flexible membranes are commercially available which are self-adhering and easy to install; however, some method of protection would be required for external systems. Conventional waterstops can be installed internally prior to placement of succeeding RCC lifts by using regular concrete within a tapered form and joint material to align the waterstop with the joint. The system can be utilized with conventional formwork or with the curbing system. Waterstop should be located as close to the upstream face as practical to minimize interference with the RCC placement and reduce the amount of required protection. Waterstops may also be installed in holes drilled out of the completed structure or may be situated in precast upstream panels which serve as upstream forms during placement of RCC."

In concluding my excerpts from ACI 207.5R-80, "the utilization of vibrating rollers as a method of compaction does not change the basic concepts of design for dams, locks, or other massive structures. However, it does affect construction procedures. Therefore, construction planning, layout of appurtenant structures, and treatment of joints should consider the full utilization of the rapid construction possibilities with RCC construction. Any massive structure of sufficient length and width to accommodate the rollers and spreading equipment will be economically benefited."

TARBELA REPAIRS

The repair work at Tarbela Dam in Asia is the most impressive application of roller compacted concrete placed to date. This work serves as an excellent example of the construction operations involved, particularly the most recent work in the service spillway as shown in Fig. 6.1. The RCC plant shown in Fig. 6.2 was a substantial improvement over that used during the 1975 tunnel repairs. Aggregate grading was controlled, to some extent, by separating the pit run material into size ranges and recombining in uniform proportions by vibrating feeders. The quantity of cement entering the mixture was controlled by a variable speed, cement auger. Three 130 cu yd/hr (100 $m^3$/hr) continuous pugmill type mixers (revolving drum), similar to that shown in Fig. 6.3, were installed for efficient mixing. Batch transportation, consisting of 70-ton end-dump trucks, was used to deliver material from the plant to the placement. Material flow from the continuous mixing operations into the truck was controlled by a temporary holding hopper as shown in Fig. 6.4.

In placing the mixture, very little segregation was noted even though the mixture contained aggregate up to 6 inches (152 mm) in maximum size. This can be seen in Fig. 6.5 which shows the truck dumping. The discharge pile was spread to a depth of roughly 13 inches (330 mm) by dozer as shown in Fig. 6.6 Each layer was compacted by a tractor drawn, smooth drum, vibratory roller with 4 passes. Compaction equipment is shown in Fig. 6.7. Fig. 6.8 shows a laborer spraying water on the compacted surface to prevent surface drying before covering with the next layer.

Conventional 4000 psi (27.6 MPa) concrete and an improvised stairstep forming system were used on the face immediately behind roller compacted concrete placement. This is shown in Fig. 6.9.

Fig. 6.10 shows the 1978 service spillway repairs nearing completion.

In conclusion, additional roller compacted concrete work has been completed on the service spillway and is under construction for the auxiliary spillway at Tarbela. Without doubt, RCC is the material for, and method of, rapid and economical dam construction where usable materials are readily available and layout of the structure will accommodate a RCC operation. Properties of the material and techniques for construction have already been proven. Therefore, only the designer's judgment based on site conditions, or failure to consider RCC, will limit more worldwide utilization of RCC in dam construction.

REFERENCES

ACI Committee 207, July-August 1980, "Roller Compacted Concrete," ACI Journal, Vol. 77, No. 4, pp. 257-285.

Cannon, Robert W. Oct. 1974, "Compaction of Mass Concrete with Vibratory Roller," ACI Journal, Proceedings Vol. 71, No. 10, pp. 506-513.

Cannon, Robert W. 1972, "Concrete Dam Construction Using Earth Compaction Methods, "Economical Construction of Concrete Dams, American Society of Civil Engineers, New York, pp. 143-152.

Hall, Donald J. and Houghton, Donald L., June 1974, "Roller Compacted Concrete Studies at Lost Creek Dam," U.S. Army Engineer District, Portland, Oregon.

Johnson, H. A., and Chao, P. C., Nov. 1979, "Rollcrete Usage at Tarbela Dam," Concrete International: Design and Construction, Vol. 1, No. 11, pp. 20-33.

Moffat, A.I.B. and Price, A.C., July 1978, "The Rolled Dry Lean Concrete Gravity Dam," Water Power and Dam Construction, pp. 35-42.

Raphael, J. M., 1971, "Optimum Gravity Dam," Rapid Construction of Concrete Dams, American Society of Civil Engineers, New York, pp. 221-247.

"Rolled Lean Concrete Debut on Devon Dam," New Civil Engineer, September 11, 1980, page 8.

Shimizu, S. and Takemuda, K., April 1978, "Field Test for Construction of the Upstream Cofferdam of Okawa Dam by R.C.D. (Roller Compacted Dam) Concrete Method," Concrete Technology, Vol. 16, No. 4, pp. 8-16.

Tynes, W. O., Oct. 1973, "Feasibility Study of No-Slump Concrete for Mass Concrete Construction," Miscellaneous Paper C73-10, U.S. Army Engineer Waterways Experiment Station, Vicksburg.

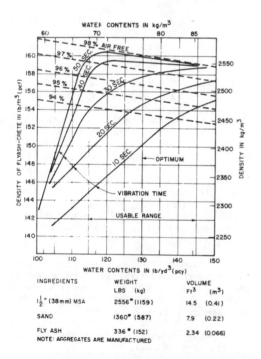

Fig. 2.2.1 — Effect of vibration time and water contents on compacted density of fly ash-crete

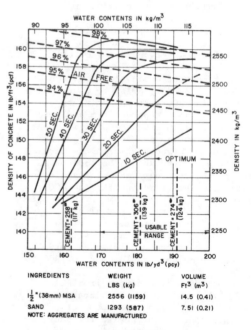

Fig. 2.2.2 — Effect of vibration time and water contents on compacted density of portland cement concrete

### Table 2.2.1 — Approximate mixing water requirements for RCC mixes without pozzolans — Manufactured aggregates

| Mix type | Maximum size aggregate, in. (mm) | | | | | |
|---|---|---|---|---|---|---|
| | ⅜ (9.5) | ¾ (19) | 1½ (38) | 3 (76) | 4½ (114) | 6 (152) |
| | Unit water content, lb/cu yd (kg/cu m) | | | | | |
| Interior mass | 195 (116) | 180 (107) | 165 (98) | 145 (86) | 135 (80) | 130 (77) |
| Bedding mix | 215 (128) | 200 (119) | 185 (110) | — | — | — |

*Note:* Although aggregates utilized in tests represented in this table and Fig. 2.2.1 and 2.2.2 were manufactured and some change in unit water contents can be expected when natural aggregates are used, the same basic relationships of compactive effort can be expected with both aggregate types.

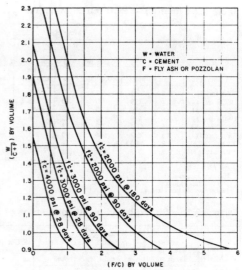

Fig. 2.4.1 — Proportioning curves for equal strength concrete

### Table 2.3.2 — Absolute volume of coarse aggregates per unit volume of concrete $C_v$

| Maximum size aggregate, in. | 6 | 4½ | 3 | 1½ | ¾ | ⅜ |
|---|---|---|---|---|---|---|
| Maximum size aggregate, mm | 152 | 114 | 76 | 38 | 19 | 9.5 |
| Absolute volume, percent of unit concrete volume | 63-64 | 61-63 | 57-61 | 52-56 | 46-52 | 42-48 |

### Table 3.2.1 — RCC mix information

| | Mix data | | | | | | | Core strength data | | |
|---|---|---|---|---|---|---|---|---|---|---|
| | | | | | | | | Unconfined | | |
| | | Weights per cu yd concrete, lb | | | | | | Compressive strength, psi | Shear strength* | |
| Source | MSA, in. | Cement | Pozzolan | Water | Fine aggregate | Coarse aggregate | Age, days | | Mass, psi | Joint, psi |
| TVA test #1 | 3 | 94 | 130 | 130 | 1020 | 2780 | 138 | 3300 | 640 | 230 |
| USCE test #1 | 4½ | 235 | 0 | 135 | 1042 | 2991 | 72 | 3720 | 695 | 130 |
| USCE test #1 | 3 | 235 | 0 | 145 | 1151 | 2850 | 66 | 3280 | 810 | ‡ |
| USCE test #2 | 3 | 235 | 0 | 140 | 1140 | 2700 | 120 | 3280 | 895 | 395† |
| USCE test #2 | 3 | 70 | 132 | 140 | 1140 | 2700 | 120 | 2300 | 555 | 205 |
| TVA test #2 | 1½ | 127 | 277 | 150 | 1255 | 2403 | 90 | 3810 | ‡ | 388 |
| TVA — BLN§ | 1½ | 75 | 300 | 142 | 1226 | 2424 | 90 | 2580 | 265 | 290† |
| TVA — BLN§ | 1½ | 195 | 235 | 174 | 1107 | 2424 | 90 | 5980 | ‡ | ‡ |

*By CRD C 89.
†Bedding mix used on joint.
‡No test data reported.
§Bellefonte Nuclear Plant.

### Table 3.2.1 (metric) — RCC mix information

| Source | MSA, cm | Mix data - Weights per cu m of concrete, kg | | | | | Age, days | Core strength data - Unconfined | | | | | |
|---|---|---|---|---|---|---|---|---|---|---|---|---|---|
| | | Cement | Pozzolan | Water | Fine aggregate | Coarse aggregate | | Compressive strength | | Shear strength* Mass | | Joint | |
| | | | | | | | | kgf/cm² | MPa | kgf/cm² | MPa | kgf/cm² | MPa |
| TVA test #1 | 7.6 | 55.8 | 77 | 77 | 605 | 1649 | 138 | 232 | 23 | 45 | 4 | 16 | 2 |
| USCE test #1 | 11.4 | 139 | 0 | 80 | 618 | 1774 | 72 | 262 | 26 | 49 | 5 | 9 | 1 |
| USCE test #1 | 7.6 | 139 | 0 | 86 | 683 | 1691 | 66 | 231 | 23 | 57 | 6 | ‡ | ‡ |
| USCE test #2 | 7.6 | 139 | 0 | 83 | 676 | 1602 | 120 | 231 | 23 | 63 | 6 | 28† | 3† |
| USCE test #2 | 7.6 | 41.5 | 78 | 83 | 676 | 1602 | 120 | 162 | 16 | 39 | 4 | 14.5 | 1 |
| TVA test #2 | 3.8 | 75.3 | 164 | 89 | 745 | 1426 | 90 | 268 | 26 | ‡ | ‡ | 27 | 3 |
| TVA — BLN§ | 3.8 | 44.5 | 178 | 84 | 727 | 1438 | 90 | 181 | 18 | 18.6 | 2 | 20.4† | 2† |
| TVA — BLN§ | 3.8 | 116 | 139 | 103 | 657 | 1438 | 90 | 420 | 41 | ‡ | ‡ | ‡ | ‡ |

*By CRD C 89.
†Bedding mix used on joint.
‡No test data reported.
§Bellefonte Nuclear Plant.

### Table 3.3.1 — Strength and elastic properties of Zintel Canyon RCC concrete studies

| Mix | Cement content, lb/cu yd | Water-cement ratio, by weight | Age | Compressive strength, psi | Splitting tensile strength, psi | Shear strength, psi | Modulus of elasticity, psi × 10⁶ | Poisson ratio |
|---|---|---|---|---|---|---|---|---|
| Interior | 100 | 1.95 | 3 days | 220 | — | — | 0.31 | — |
| | | | 7 days | 340 | — | — | 0.70 | — |
| | | | 28 days | 610 | 90 | 120 | 1.31 | 0.20 |
| | | | 90 days | 1090 | 165 | 200 | 2.15 | 0.21 |
| | | | 1 year | 1580 | — | — | 2.57 | 0.17 |
| Exterior | 200 | 0.98 | 3 days | 880 | — | — | 1.35 | — |
| | | | 7 days | 1170 | — | — | 1.31 | — |
| | | | 28 days | 1920 | 200 | 290 | 2.20 | 0.20 |
| | | | 90 days | 2280 | 255 | 395 | 2.47 | 0.17 |
| | | | 1 year | 3180 | — | — | 3.28 | 0.21 |

Note: 1. Aggregate: Natural pit or bank run sand and gravel, 3-in. MSA.
2. Cement meets both Type I and II requirements.

### Table 3.3.1 (metric) — Strength and elastic properties of Zintel Canyon RCC concrete studies

| Mix | Cement content, kg/m³ | Water-cement ratio, by weight | Age | Compressive strength | | Splitting tensile strength | | Shear strength | | Modulus of elasticity | | Poisson ratio |
|---|---|---|---|---|---|---|---|---|---|---|---|---|
| | | | | kgf/cm² | MPa | kgf/cm² | MPa | kgf/cm² × 10⁵ | MPa | kgf/m² | MPa | |
| Interior | 59.3 | 1.95 | 3 days | 15.5 | 2 | — | — | — | — | .22 | 2,135 | — |
| | | | 7 days | 23.9 | 2 | — | — | — | — | .49 | 4,825 | — |
| | | | 28 days | 42.9 | 4 | 6.3 | 1 | 8.4 | 1 | .92 | 9,025 | 0.20 |
| | | | 90 days | 76.6 | 8 | 11.6 | 1 | 14.1 | 1 | 1.51 | 14,125 | 0.21 |
| | | | 1 year | 111 | 11 | — | — | — | — | 1.81 | 17,710 | 0.17 |
| Exterior | 118.7 | 0.98 | 3 days | 61.9 | 6 | — | — | — | — | .95 | 9,200 | — |
| | | | 7 days | 82.3 | 8 | — | — | — | — | .92 | 9,025 | — |
| | | | 28 days | 135 | 13 | 14.1 | 1 | 20.4 | 2 | 1.55 | 15,160 | 0.20 |
| | | | 90 days | 160 | 16 | 17.9 | 2 | 27.8 | 3 | 1.74 | 17,020 | 0.17 |
| | | | 1 year | 224 | 22 | — | — | — | — | 2.36 | 22,600 | 0.21 |

Note: 1. Aggregate: Natural pit run sand and gravel, 76-mm MSA.
2. Cement meets both Type I and II requirements.

### Table 3.4.1 — Strain, creep,* and flexural properties of Zintel Canyon RCC

| Mix | Cement content, lb/cu yd (kg/m³) | Water-cement ratio, by weight | Loading, age | Creep coefficients | | Autogenous volume change, millionths | Rapid load test | | | |
|---|---|---|---|---|---|---|---|---|---|---|
| | | | | 1/E | F(K) | | Tensile strain capacity, millionths | Modulus of rupture, | | |
| | | | | | | | | psi | (kgf/cm²) | (MPa) |
| Interior | 100 (59.3) | 1.95 | 7 days | 1.43 | 0.086 | −7 | 39 | 55 | (3.9) | (0.5) |
| | | | 28 days | 0.76 | 0.082 | 0 | — | — | — | — |
| | | | 90 days | — | — | −3 | 61 | 150 | (10.5) | (1) |
| | | | 1 year | — | — | −8 | — | — | — | — |
| Exterior | 200 (118.7) | 0.98 | 7 days | 0.76 | 0.054 | −5 | 64 | 165 | (11.6) | (1) |
| | | | 28 days | 0.45 | 0.033 | −4 | — | — | — | — |
| | | | 90 days | — | — | −4 | 89 | 275 | (19.3) | (2) |
| | | | 1 year | — | — | −17 | — | — | — | — |

*ASTM C 512

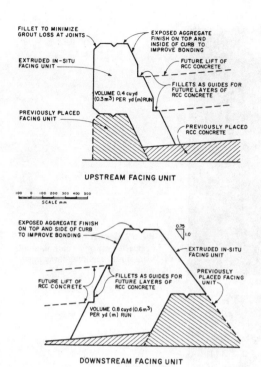

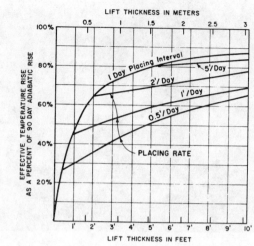

Fig. 5.4 — *Natural cooling of mass concrete maximum temperature rise*

Fig. 4.7.2 — *Details of upstream and downstream facing units*

TARBELA DAM

Fig. 6.1. Service spillway plunge pool erosion protection. Roller compacted concrete during initial placement.

Fig. 6.2. General view of the roller compacted concrete continuous mix plant for the service spillway repair work.

Fig. 6.3. Continuous pugmill type revolving drum mixer. Material flow is right to left.

Fig. 6.4. Temporary holding hopper to retain continuous mix flow in order to fill the batch transporting 70 ton end-dump trucks.

TARBELA DAM

Fig. 6.5. End-dump truck discharching roller compacted concrete mixture for spreading and compacting. Conventional concrete facing placement in the lower right.

Fig. 6.6. Dozer spreading the discharge pile for compaction.

Fig. 6.7. Smooth drum vibratory roller compacting the layer of RCC.

TARBELA DAM

Fig. 6.8. Laborer moistening the surface of a compacted layer to prevent drying before the surface is covered by the next layer.

Fig. 6.9. Conventional concrete and step formwork for the plunge pool facing.

Fig. 6.10. 1978 service spillway placement of roller compacted concrete nearing completion.

# Geotechnical aspects of the Larona Hydroelectric Project in Sulawesi, Indonesia

RICHARD L.KULESZA
Bechtel Civil & Minerals, Inc., San Fransisco, Calif., USA

SYNOPSIS. The construction of a hydropower dam and canal in mountainous terrain in Sulawesi, Indonesia presented difficulties caused by wet climate and weak, highly compressible residual soils. The adopted solution consisted of a rockfill dam with upstream concrete facing founded on rock stripped of overburden, and concrete canal supported on rockfill embankments constructed at a controlled rate and surcharged to minimize long-term settlements. The selected canal route minimized deep cuts to avoid slope stability problems. The paper describes difficulties of geotechnical explorations in the dense, hilly jungle and the properties of residual soils having an unusually high void ratio. Investigation of the slide in fissured claystone that occurred during powerhouse site excavation is also described.

INTRODUCTION

The Larona Hydroelectric Project is located in central Sulawesi (Indonesia) and provides 1.3 billion KW-hours of energy to a nickel processing plant near Soroako on Lake Matano. Larona River drains an area of 2400 km$^2$ which includes major lakes Matano and Towuti.

Lake Towuti, having a surface area of 585 km$^2$, forms the reservoir for the hydro-electric project, which consists of the following facilities (Figure 1).

- Butubesi Dam, located on Larona River 4 km from its origin at Lake Towuti. The forebay created by the dam has a volume of approximately 10 million m$^3$. It is a 32 m high zoned rockfill structure with concrete facing.

- Reinforced concrete canal, 14.4 m wide and 6.9 km long, roughly paralleling the Larona River and carrying the average basin yield of 137 m$^3$/sec. The canal is constructed over a succession of side hill cuts and rockfill embankments traversing hilly terrain underlain by deep and compressible soils.

- Penstock intake structure, sited on 70 m of residual clayey silts. An overflow spillway directs excess flow back

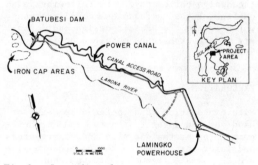

Fig.1. Location plan.

to Larona River.

- Three steel penstocks with diameter varying from 3.7 m at the intake structure to 2.6 m at the powerhouse, having a length of 1300 m.

- Lamingko powerhouse, located in a deep open excavation on the left bank of Larona River, consisting of three 55 MW units under a generating head of 142 m.

- 20 km long, 150 KV transmission line delivering the power to the nickel plant near Soroako.

The design and construction were carried out by Bechtel under a cost-plus contract to P. T. International Nickel, Indonesia. Geologic exploration commenced in the virtually inaccessible jungle in May 1975, while the first electric power was produced in January 1978. Apart from an extensive exploration and the construction of all access roads and support facilities, the work accomplished in a little over 2½ years included excavation and controlled disposal of 3.2 million m³ of soft silty overburden; the quarrying, hauling and compaction of 2.6 million m³ of rockfill and gravel; placement of 115,000 m³ of reinforced concrete; installation of a 20 km long transmission line through hilly jungle terrain; and construction of the powerhouse, penstocks and penstock intake structure.

The project area is located in a hilly terrain with elevations varying from 170 to 320 m and nearby peaks rising to over 1000 m above the sea level. Climatic conditions are characterized by a mean annual rainfall of approximately 300 cm, and a mean annual temperature of 25°C.

Apart from the climate, the subsurface conditions characterized by thick deposits of soft residual soils having an average void ratio of 3.0 had an important impact on the design philosophy. The following is an outline of the geotechnical considerations which led to the adopted design and a summary of the experience obtained during construction.

EXPLORATION

Site exploration was conducted by means of hand dug and dozer test pits, and coreholes commencing at the dam site and quarry areas and later extending to the powerhouse site and the canal alignment. It was soon established that the residual overburden soils were weak and highly compressible, presenting serious problems both as embankment foundation and fill materials. Because of the undesirable properties of these soils and the wet climate, it was decided at an early stage to construct the Batubesi Dam of quarried rockfill on firm bedrock foundation stripped of all overburden and weak rock. The exploration for the dam concentrated therefore on estimating the extent of fault zones and investigating the permeability of rock. During the design exploration, 22 coreholes and 44 test pits were put down at the dam site, while further 26 coreholes were added during construction to better define the geologic conditions.

Early decision to construct the dam and canal fills out of rockfill necessitated a concentrated effort to locate suitable quarry sites. It was known that overburden was generally very deep in flatter areas and thinner on steeper slopes, but no geologic mapping and few outcrops were present. The only logical place to start looking for rock was the high, steeper ridges.

In some cases, test pits were dug by helicopter-transported crews. Drilling at quarry sites was limited because of the necessity of transporting drilling water, and because of a need for the drill rigs elsewhere on the project. Much of the information required to determine overburden conditions at quarry sites was obtained in dozer trenches, but this was extremely difficult work. No trees had been cleared due to the steepness of the slopes, up to 45° in places. The dozers had to winch their way up slope, then cut trenches transverse to the slope. Until the job was completed the dozers stayed on the ridge and fuel was carried up by hand. The sites chosen for rockfill quarries proved to be adequate and fairly well located in regards to haul distance. Stripping of thick overburden and its disposal prior to starting rockfill production was the biggest problem encountered at the quarries.

For the power canal, it was initially intended to limit the exploration to test pits and seismic surveys, as preliminary data indicated that overburden thickness may not exceed 12 m. Hand-dug test pits tended to confirm this, terminating in what appeared to be the top of rock at a depth of 3 to 11 m. However, when equipment access was developed, dozer trenches demonstrated that the overburden was much thicker than was thought. The subsurface profile included a zone of relatively hard weathered rock blocks, which caused refusal in hand-dug test pits, but which were underlain by a considerable thickness of saprolite, often decreasing in density with depth. In fact, some of the highest void ratios (up to 7.7) were determined on samples taken near the bottom of saprolite, which frequently extended to over 30 m depth. An interesting observation was unusually small size of the heaps of excavated material in relation to the volume of the pits, caused by the exceptionally high void ratio of the in-place soils.

A test pit crew usually consisted of four to five men, who used pick and shovel. Material was removed from the pit by rattan basket on rope attached to a wooden tripod erected over the pit. Walls of the test pits held up quite well in the clay-silt saprolite overburden. The time required to dig a typically 5 m deep test pit in the saprolite was 2 to 3 days. The test pits provided useful information on the nature of the upper part of overburden, and permitted an early start of laboratory testing on Shelby tube samples taken by jacking into the sides of the pits.

Dozer trenches penetrated through the rocky upper layer, but also encountered difficulties, particularly in achieving access through the rough jungle, and rainy weather frequently hampered dozer work. While the trenches gave a good indication of some of the soil characteristics, overburden was often too deep to permit bedrock to be found. The deepest dozer trench was approximately 12 m.

A third method of initial exploration of the canal alignment was by means of seismic survey, which it was hoped to correlate with test pit information. A crew of about five was needed to cut trails, string lines and carry instruments and equipment. Helicopter was used to transport the crew to helipads, followed by walking through dense jungle to each site. A Nimbus enhancement seismograph with oscilloscope-waveform readout and a small Nimbus pocket seismograph with digital readout were used. The operation consisted of measuring shock wave velocities along a 100 foot line at 10 foot intervals using a single geophone. Shock waves were created by using a 14 pound hammer and steel plate with the exception of one line on the dam left abutment where 1/3 to 1/4 stick of dynamite and cap was used to create a seismic shock. Results of the seismic survey were disappointing. The low density of the overburden and the thick root mat prevented determination of the overburden depths or bedrock contours.

The above difficulties plus the discovery that low-density overburden was much deeper than was anticipated necessitated recourse to exploration of the entire 7 km length of the canal alignment by drilling, which was neither considered necessary nor practical during the initial planning of the exploration. The number of boreholes along the canal finally reached 100, with a total of 2007 m length. Drilling of overburden was carried out either with use of tricone roller bits 2.5 to 4 inch diameter or by using a core barrel. The exploration included disturbed and undisturbed soil sampling. Disturbed sampling consisted of the standard penetration tests, while undisturbed samples were obtained using 3 inch I.D. Shelby tubes approximately 3 feet in length. Usually when bedrock was encountered, AX double, NX double or triple tube diamond core barrels were used. The triple tube barrel was found to give generally very good core recovery. It consisted of a standard inner swivel-head barrel containing a split-tube. This system worked well to protect the core. The core barrel was 57 mm, slightly bigger than a standard NX barrel. A single barrel advance by dry drilling was very successful in several instances, such as in fault zone in the dam foundation, where use of water prevented adequate core recovery. Drilling was carried out by two Tone UD-5 skid-mounted drill rigs of Japanese manufacture. The drilling contractor was an Indonesian company from West Java. The overall performance of this drill was quite good, and the major problem encountered was in moving the drill from one drill site to another. Before access roads were developed, the rig had to be disassembled and moved by helicopter to the work areas.

Apart from moving from site to site, the provision of drilling water in the rugged, densely wooded terrain was a major problem, and much repair work was required on pumps and long water lines. Although the drilling was under a subcontract, frequent assistance by Bechtel mechanics and machine shop was given in order to expedite the drilling program. This logistic support was instrumental in achieving an adequate rate of progress of the exploration.

Laboratory testing was performed in a laboratory set up in Malili under a separate subcontract. Malili is a small town at the northern end of Bay of Bone, located about 30 km from the jobsite, which served as supply port for the project. The laboratory was equipped to carry out most of the required tests, including consolidation and effective stress-triaxial tests, but some residual strength and aggregate tests were done in Australia. The availability of a well equipped laboratory located near the project site helped to minimize sample disturbance and time loss in transportation, and assured a smooth flow of soils data for the design of the earthworks.

The exploration was supervised by Bechtel geologists and soils engineers, with results transmitted weekly by telex and pouch mail to the design offices in San Francisco and Toronto. Adjustments to program were made as required on the basis of encountered subsurface conditions and design considerations.

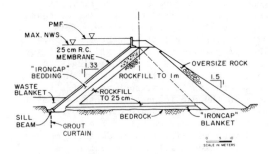

Fig.2. Batubesi Dam - cross section.

## BATUBESI DAM

A cross-section of the dam is shown on Figure 2. The dam embankment is a zoned rockfill structure with an upstream concrete face, topped with a concrete parapet wall. The embankment crest width is 6.0 m, and the slopes are 1.33:1 on the upstream side and 1.5:1 on the downsteam side. On the left abutment where the embankment is founded on sedimentary rock, a berm was provided in order to reduce the overall slope angle. The outlet works consist of the canal intake structure, a low level gated outlet constructed in hard rock foundation, and an emergency spillway slab constructed over a part of the downstream slope behind a "fuse plug" included in the crest of the dam.

Residual soil originally covered the dam foundation area to a depth varying from zero at the river channel to over 25 m on abutments. This residual soil cover was removed from the entire embankment foundation and the dam was founded on bedrock. On the right bank of Larona River, the bedrock consists of igneous peridotite and peridotite conglomerate, both of which provide a strong foundation. A fault zone occurring at the riverbed separates these rocks from silty conglomerate and silty sandstone and siltstone, which form the foundation on the left bank.

The sedimentary rocks are generally well indurated, but extensive softer zones were encountered. Dam foundation preparation included removal of the weaker materials. The surface of the rock was very irregular, containing deep fissures and pockets eroded by current and chemical action. These features, up to 15 m deep, proved very difficult to clean out during foundation preparation. Mass and formed concrete were used to fill the larger fissures and cavaties, and "iron cap" gravel was hand-compacted in lesser depressions. Along the upstream toe of the dam embankment, cutoff trench was excavated to firm groutable bedrock. A concrete sill beam constructed along the cutoff trench connects to the toe of the concrete face of the dam. Grout holes were drilled through casing included in the sill beam into the foundation rock to form a barrier against seepage beneath the dam embankment. The barrier consists generally of two lines of shallow consolidation grout holes in the central portion of the cutoff, extending generally to depths of 25 m. On the left abutment the grout curtain terminates at the west edge of the canal intake structure. Beyond the canal intake along the east slope of the ridge a blanket of low permeability materials originating from the overburden stripping was constructed to impede potential seepage. The concrete facing, 25 cm thick, ties into a parapet wall at the crest and into a concrete sill beam anchored into the rock foundation at the upstream toe. Detail of the upstream toe is shown on Figure 3,

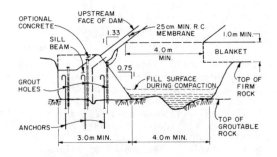

Fig.3. Batubesi Dam - upstream toe detail.

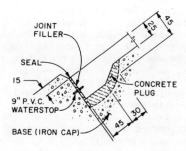

Fig.4. Batubesi Dam - joint of face slab and sill beam.

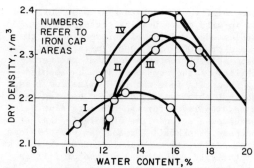

Fig.6. Compaction curve for "iron cap" gravel.

and detail of the connection between the face slab and the sill beam is shown on Figure 4. Special precautions taken to ensure tight compaction of fill behind the sill beam included providing at least 4.0 m width of work area in the trench between the sill beam and excavated bedrock to permit compaction by heavy roller, placement of fill in thin lifts and sloping each layer of fill against the concrete and rock face, as shown on Figure 3.

The embankment consists of the following zones. Zone 1 is a minimum 4 m wide bedding material supporting the upstream concrete membrane and the downstream emergency spillway slab and consisting of "iron cap" lateric gravel. Selected coarser Zone 1 was also used as blanket material on the dam foundation. Zone 2, which is minus 3 inches crusher product, was used in certain areas as a substitute for Zone 1 or as a blinding course where required between

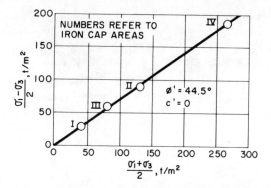

Fig.7 Shear strength of "iron cap" gravel.

Zones 1 and 4. Zone 3 is quarry rock up to 1 m in size. Zone 4 is select quarry rock up to 25 cm in size and was used as a transition between Zones 1 and 3. Zone 5 is oversize rock, placed in the downstream part of the dam. "Iron cap" material is an induration product of the laterite, and was found at some high ground areas in the vicinity of the right abutment of Batubesi Dam. It was found to be a desirable material for use as compacted fill. Typical grain size distribution curves are shown on Figure 5, while compaction and shear strength test results are included on Figures 6 and 7. The specification required the "iron cap" in the zone behind the concrete facing to be compacted to at least 98% of maximum dry density obtained in a compaction test using an energy rating of 20,000 ft-lb per cu. ft. This was achieved using Ingersoll-Rand SP-60 vibratory roller on the horizontal fill surfaces, and SP-54 and Bros 6.6 mt vibratory rollers adapted to travel on the 1.3:1 face of the dam. Low compressibility of the "iron cap" zone supporting

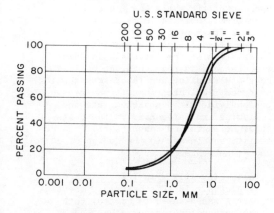

Fig.5. Particle size distirbution of "iron cap" gravel.

the reinforced concrete slab was confirmed by in-situ plate bearing tests, which indicated a modulus of deformation in the range of 6,000 to 13,000 t/m$^2$. In-situ permeability tests showed that the permeability of the compacted iron cap behind the concrete face was $10^{-5}$ cm/sec or less, so that this zone, apart from serving as a bedding layer for the concrete slab, also provides a second line of defense against seepage. Rockfill for the dam was obtained from peridotite quarries located within 3 km of the site. The total volume of fill in the dam was 650,000 m$^3$.

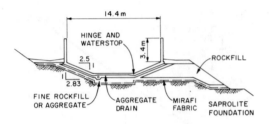

Fig.8. Power canal - cross section.

POWER CANAL

Canal description

A 14.4 m wide by 5.0 m deep reinforced concrete canal with a trapezoidal section and a total length of 6.9 km long, connects the intake at the dam to a concrete forebay near the powerhouse. A cross section of the canal is shown on Figure 8. The canal was constructed on 23 m wide cut-and-fill bench roughly paralleling Larona River, requiring 1,207,000 m$^3$ of excavation and 1,890,000 m$^3$ of rockfill. There are over twenty rockfill embankments up to 30 m high at ravine crossings, with volumes of up to 231,000 m$^3$. Corrugated metal culverts placed under the embankments vary from only one with a 1.22 m diameter to three each with a 2.44 m diameter. In several locations, these culverts were used to measure settlement of the embankments. The canal has a flow capacity of 153 cubic meters per second and an invert slope of 0.00035. The penstock forebay is 48 m wide by 210 m long and provides storage for load changes on the turbines, and also incorporates a spillway which discharges any excess canal flow into the Larona River via a 1000 m long U-shaped reinforced concrete conduit.

The concrete slab was laid on fine crushed rock placed over Mirafi fabric, which stabilized the soft silt foundation and acted as filter. In foundation preparation it was found that 2 m or more of unintended excavation occurred because of the difficulty of stripping the root mat. To avoid the unnecessary work, it was decided to leave the root mat in place in relatively flat portions of the terrain traversed by the canal, provided that there would be at least 2 m of rock fill between the foundation and the canal invert slab.

Subsurface profile

The subsurface exploration disclosed that the soil conditions were very similar along the entire length of the power canal from Batubesi Dam to the penstocks. The bedrock, which was essentially peridotite and conglomeritic siltstone, was overlain by residual soils up to about 40 m thick. Yellow-brown to red lateritic clayey silts formed the upper part of the soil profile, grading with depth into silty and clayey saprolites having a similar color. As the weathered bedrock was approached, the frequency of rock fragments increased and the saprolites graded into a material consisting of sand, gravel and boulder sizes in a silty and clayey matrix. The transition to sound rock was occasionally abrupt but more frequently gradual. Characteristically, the soil overburden was thicker at ridges than in ravines so that the portions of the power canal requiring high cuts generally traversed a greater thickness of soil than the portions requiring high fills.

The variation with depth of the lateritic and saprolitic soils was found to be significant only in terms of the content of rock fragments, while the properties of the clay-silt matrix remained constant. Observation of the materials in dozer trenches and in road cuts, as well as borehole sampling and laboratory testing evidence, indicated that it was reasonable to subdivide the overburden into the following two categories on the basis of shear strength and consolidation characteristics relevant to the design of cuts and embankments.

(a). Clay-silt with low content of rock fragments. In this material, the fine matrix was assumed to control both the unconsolidated-undrained and drained shear strength. This stratum was critical for stability of cuts and fill foundations analyzed in terms of total stresses using the undrained shear strength. For the purposes of stability analyses, the soil was assumed to belong in this category when the proportion of rock fragments consistently did not exceed about 40 percent.

(b). **Clay-silt with high content of rock fragments.** In this material, a large proportion of the shear stresses resulting from the excavation or embankment loading was expected to be taken by the rock fragments. The silt-clay matrix between the rock fragments was assumed to consolidate rapidly due to internal drainage within the generally loose, coarse-grained and pervious skeleton material. The drained shear strength was therefore assumed to control slope stability in this material, and its upper boundary was taken as the lower limit of slip surfaces to which undrained shear strength applied. For simplicity and because of difficulty of sampling and testing of the material containing many rock fragments, the effective stress parameters of this stratum were assumed to be the same as those of the material with a low content of rock fragments, since an effective angle of friction of $33°$ determined for the clay-silt was thought to be reasonable also for a gravelly material with a relatively loose structure.

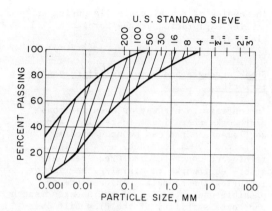

Fig.9. Particle size of distribution of saprolite

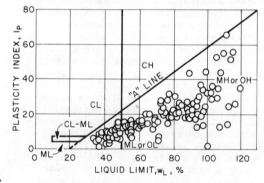

Fig.10. Atterberg limits of saprolite.

## Soil Classification

Particle size distribution and plasticity properties are shown on Figures 9 and 10. Where no rock fragments were present, the residual soils typically included 5 to 45 percent clay fraction (minus 2 microns). Plasticity properties plotted considerably below the "A-line" corresponding to ML and MH soil classifications, and were similar for all the material directly derived from peridotite bedrock, occurring over most of the canal. Higher plasticity materials derived from siltstone-claystone occurred near the dam, at the penstocks and in the powerhouse area.

The natural moisture content was mostly in the range of 40 to 80 percent, the higher values generally occuring in the lower portion of the soil profile, where a number of values of over 100 percent were recorded. The optimum moisture content in laboratory compaction tests with an energy rating of 20,000 ft-1b per cu. ft. was only about 27 percent, and the large difference between the natural and optimum moisture contents was one of the reasons for not using the saprolite as fill, resulting in adoption of quarried rock for the dam and canal embankments.

The specific gravity of soils at the dam and along the power canal ranged from 3.3 to 4.0, averaging about 3.7. However, the dry density was only about 1.5 to 1.5 $t/m^3$, compared to a maximum dry density over 1.8 $t/m^3$ in laboratory compaction. The low density resulted from a very high void ratio, ranging from about 1.2 to as high as 7.7 and averaging about 3.0. Many of the lowest unit weights and highest void ratios were determined on samples from the lower portions of the soil profile. The high void ratios were believed to be a result of intensive leaching, but may also be attributable to the presence of the mineral halloysite, which was identified in samples of comparable soils during an earlier exploration at the Soroako nickel plant site.

Shear strength and compressibility characteristics.

In contrast to highway or even railroad earthworks where local instability and settlements often can be economically corrected by maintenance, a 14 m wide reinforced concrete canal requires very stable earthworks to minimize the risk of rupturing the conduit. As complete as possible understanding of the shear strength and compressibility of the clay silt overburden was therefore necessary for the design of the deep cuts and high fills. As the exploration extended along the canal alignment and sufficient test results became available, it was found that the shear strength, particularly in terms of the effective stresses, was remarkably uniform despite variations in particle size distribution and Atterberg limits. The uniformity of strength data permitted the adoption of average strength parameters for the design of cuts and fills along the entire canal. Some local variations in the measured strength were taken into account in the design at individual locations. Adjustments were needed mostly at the penstock intake, where residual soils derived from a clay-stone formation occured, having a very similar effective stress parameters but a higher undrained shear strength and slower rates of consolidation. The following are some details of the strength and compressibility of the saprolite.

(a). <u>Undrained shear strength</u>. The upper part of Figure 11 shows the results of unconsolidated undrained triaxial and unconfined compression tests on undisturbed samples of the lateritic and saprolitic clays and clayey silts. The results marked with crossed circles were excluded from the least-square determination of the average curve. The samples were tested at the natural moisture content; the degree of saturation was usually on the order of 90 percent. The average strength line has a curved shape, the rate of increase in shear strength decreasing with an increase in normal stress. The strength of overburden was found to remain more or less constant with depth, despite the fact that unit weight was lower and the void ratio and moisture content were considerably higher at greater depth.

The average values of undrained shear

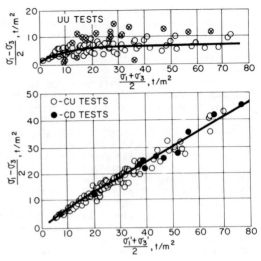

Fig.11. Undisturbed saprolite - shear strength results.

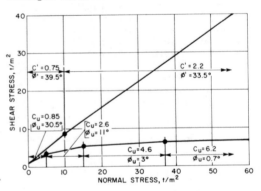

Fig.12. Undisturbed saprolite - selected shear strength parameters.

strength used in the design are shown on Figure 12 as the apparent cohesion intercept $c_u$ and the apparent angle of shearing resistance $\emptyset_u$.

Consistency of the test results and analysis of failure of some relatively low but steep cuts suggested that the selected undrained strength parameters were reasonable. Also, analysis of some very uniform, high natural slopes existing along the canal using these strength values resulted in factors of safety near unity; these slopes probably reached an equilibrium configuration by creep action. The undrained shear strength controlled the design of cut slopes above 12 m high, as well as the short term stability of fill foundations prior to any consolidation.

(b). <u>Effective stress parameters.</u> The lower portion of Figure 11 shows the results of consolidated undrained and drained triaxial tests expressed in terms of effective stresses. In order to permit direct comparison of the two types of tests, an energy correction was applied to allow for the external work done during drainage. All samples were saturated by back pressure which was maintained during the shearing stage of the tests. The effective strength results have a very high coefficient of correlation (99 percent), and the average strength line is slightly curved at lower normal pressures. Design strength parameters selected from this data are shown on Figure 12. The effective strength parameters controlled the stability of cuts up to 12 m high and the stability of embankment foundations during and at completion of the consolidation process.

(c). <u>Gain in strength with consolidation.</u> Figure 12 presents the unconsolidated-undrained and the fully drained strength curves for the lateritic and saprolitic overburden along the canal. When additional normal stresses are applied to the soil by the weight of the canal fill, excess pore water pressures are induced, but the strength gradually increases from the initial undrained to the fully drained condition as the pore pressures dissipate by consolidation. The magnitude of the induced pore pressures depends on Skempton's pore pressure coefficient A, while the rate of gain in strength depends on the coefficient of consolidation and on the length of the drainage path.

Pore pressure coefficient A was determined from the results of consolidated undrained triaxial tests, and was in the range of 0.4 to 0.8, indicating that high fills constructed at a rapid rate could induce considerable pore pressures in the foundation, endangering stability. However, as indicated by results of consolidation tests performed on undisturbed samples, the rate of dissipation of pore pressure and consequent rate of gain in shear strength were high. This made it practicable to control the rate of placement os as to obtain the gain in strength required to achieve the required factor of safety. For this reason, fills constructed at a controlled rate compatible with the rate of consolidation of the foundation were thought to present less of a stability problem than high cuts in the selection of the canal alignment.

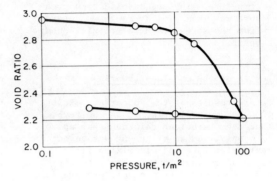

Fig.13. Undisturbed saprolite - void ratio - log pressure curve.

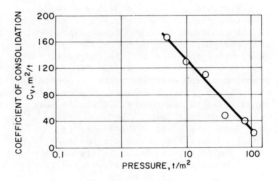

Fig.14. Undisturbed saprolite - coefficient of consolidation.

(d). <u>Consolidation properties.</u> The void ratio-log of pressure curve presented on Figure 13 is the average of 64 consolidation tests. It shows that the virgin compression index $C_c$ is 0.88. The compressibility is very high as could be expected for a soil with a void ratio averaging 3.0. Figure 14 shows the average value of the coefficient of consolidation for

62 consolidation tests. The coefficient of consolidation decreases markedly with increased pressure, as the void ratio is reduced by compression. "Preconsolidation" pressure determined from the consolidation tests varied from 10 to 40 t/m$^2$, averaging 27 t/m$^2$. This variation was believed to be caused by differing degrees of weathering of the material and intensity of the relict structure. The preconsolidation pressure was in a similar range as determined from the stress path in triaxial shear tests.

### Design of the canal earthworks

Before results of the subsurface exploration became available, it was believed that the overburden thickness was up to 12 m only, and an alignment was selected maximizing cuts and minimizing fills, based on maximum economy and schedule considerations. As the soil exploration progressed, it became apparent that overburden thickness could reach over 30 m. Stability analyses using undrained shear strength values determined from laboratory testing indicated that some of the existing slopes along the canal slignment were barely stable and could be subject to long-term creep. The stability of such slopes could not be economically improved, and could be endangered by excavation of deep cuts for the canal. In contrast, the test data showed that the consolidated strength of the residual soils is considerably higher than the existing, unconsolidated strength, so that fills can be more relied upon than deep cuts to be stable on a long-term basis. The long-term stability of fills is subject to fewer uncertainities than the stability of deep cuts. Morever, the rate of consolidation of the soils was found to be high, and computations showed that an adequate factor of safety could be achieved in most fills with a rate of construction corresponding to the actual mobilized capability of fill placement. A further consideration in favor of fills over cuts was the ease of placement of rock fill in the wet climate, when excavation and handling of the soft clayey silt presented serious problems.

For these reasons, the canal alignment was adjusted to minimize the height of cuts, at the expense of significantly increased volume of embankments. Because the wet climate precluded the use of the silty overburden soils having a high natural moisture content, all embankments were constructed of quarried rock. Every high fill was designed individually, so that the factor of safety during construction should not be less than 1.3 taking into account consolidation during construction.

The cut slope criteria developed on the basis of stability analyses allowed 1.5:1 slopes for cuts up to 12 m deep, and progressively flatter slopes for deeper cuts. The flatter overall slopes were achieved by inclusion of benches at vertical intervals not exceeding 12 m. Because some of the natural slopes along the canal are steep, the question was raised why relatively flat slopes are required for deep cuts when natural slopes of comparable or greater height occur in the area. A reason for this is illustrated on Figure 15, where the thickness of saprolite overburden determined in boreholes is plotted against the natural ground slope at the locations of the boreholes. It is seen that wherever the natural slope is 2.5:1 or steeper, the thickness of the saprolite does not exceed 9 m. This is consistent with the results of stability analysis. In clayey material whose strength does not vary significantly with depth, the factor of safety for a given slope angle decreases as the thickness of the clayey soil increases so that stable slopes will be progressively flatter as the soil deposit becomes deeper.

In order not to endanger the stability of the natural slopes in the vicinity of the canal, the width of the corridor stripped of vegetation was kept at a minimun, and strict control of excavation

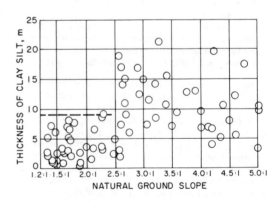

Fig.15. Slope angle vs. depth of saprolite at natural slope.

waste disposal was enforced, even though this required significantly increased waste haul distance and limitations to the amount of dozer work. Areas for waste disposal were selected in locations where any instability would not affect the canal, although the waste piles were graded to provide stability.

Settlement of canal embankments

In order to accelerate the consolidation, a surcharge exceeding the weight of the canal filled with water was applied typically for 30 days, which was sufficient to induce most of the settlement prior to the placement of the reinforced concrete structure of the canal. This duration of surcharge was indicated by laboratory tests and by measurements at an instrumented test fill, and was later confirmed by settlement monitoring of other canal embankments. The progress of consolidation was monitored by means of settlement plates in fill foundations and measurement of deflection in the inverts of large diameter CMP drainage culverts placed under the fills. The maximum measured settlement was 87 cm, with several of the fills settling more than 50 cm. Figure 16 shows the computed and actual settlements of one of the canal embankments constructed at an early stage, which was treated as a test fill. The computed settlement was obtained using the ISBILD finite element program developed at the University of California in Berkeley, employing hyperbolic strain-dependent modulus and Poisson's ratio parameters. The finite element mesh is shown on Figure 17, and the hyperbolic parameters are listed in Table 1.

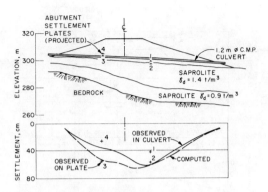

Fig.16. Test fill embankment - settlement profile.

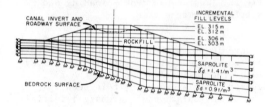

Fig.17. Finite element mesh used in settlement analysis of test embankment.

The parameters for the saprolite were initially obtained from drained triaxial tests but were adjusted to fit the average consolidation curves such as shown on Figure 13. The parameters for rock fill were obtained from technical literature for material of comparable type and gradation.

TABLE 1

Parameters Used in Finite Element Analysis of Settlement of Test Fill

| Parameter | Higher Density Saprolite | Lower Density Saprolite | Rock Fill |
|---|---|---|---|
| Unit Weight | 2.06 | 1.76 | 2.40 |
| Effective Cohesion, $t/m^2$ $C'$ | 2.20 | 2.20 | 0 |
| Effective Angle of Friction, $\phi'$ | 33.5 | 33.5 | 44 |
| Failure Ratio, $R_f$ | 0.75 | 0.75 | 0.64 |
| Modulus Number, K | 150 | 90 | 210 |
| Modulus Exponent, n | 0.67 | 0.54 | 0.51 |
| Unloading Modulus, $K_{ur}$ | 505 | 303 | 450 |
| Poisson's Ratio Parameters,* G | 0.272 | 0.272 | 2.250 |
| F | 0 | 0 | 0.09 |
| d | 4.5 | 4.5 | 4.5 |

* Where: G is the initial tangent Poisson's ratio at one atmosphere; F is the rate of change of initial tangent Poisson's ratio with confining pressure; d is the rate of change of initial tangent Poisson's ratio strain.

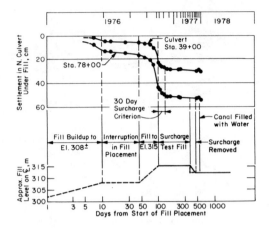

Fig.18. Test embankment - loading and settlement versus time.

Figure 18 shows the settlements measured at two of the survey points in the culvert below the fill, together with the fill construction sequence. It is seen that the 30 day duration of surcharge, initially selected on the basis of laboratory consolidation tests, was appropriate for the average subsurface conditions encountered along the power canal, since most of the final settlement developed in that period. The reasonably satisfactory agreement between the observed and computed settlements, as shown on Figure 16, provided a measure of confidence in the design of the canal embankments. Typically, on the downhill sides the higher embankments were designed with a 2:1 slope in the top 15m, followed by a 20m wide stabilizing berm with a 4:1 slope, terminating in another 2:1 slope. Steeper downhill slopes were adopted where overburden was thin or stronger than average. The uphill slopes of the canal embankments were generally 1.5:1. The stability of every embankment was analysed to provide a factor of safety of at least 1.3 during construction and at least 1.8 on long term basis. This relatively high latter value was adopted in anticipation of the large deformations of the highly compressible foundation, although laboratory testing has shown that the saprolite does not undergo significant reduction of effective strength at large strains.

POWERHOUSE AREA

While the Lamingko powerhouse is founded in sound peridotite bedrock, an approximately 50 m deep excavation was required to reach foundation level, including about 35 m of cut through sedimentary rocks overlying the peridotite. These strata consisted of siltstone and fissured claystone with layers of silty sandstone. When the excavation reached a depth of about 20 m, movement occured along an ancient slide which was not evident during site exploration. A plan of the excavation and the extent of the slide are shown on Figure 19. The downhill movement was initially monitored at a point located on the eastern edge of the slide area where a well defined shear plane intersected the ground surface. Observations made at this location from the time when movement was first noticed are shown on Figure 20. At first it appeared that the slide may be shallow and that the slope could be made stable by excavating to a flatter angle and improving drainage. However, when movement accelerated and reached one meter, excavation below the cut slope was stopped and a series of survey points were placed on the bench below the slope. Because these indicated that movement was taking place at the bench level, boreholes and trenches were put down in an attempt to determine the extent of the slide.

A shear zone was encountered in the trenches, and was also indicated by bent casing in the boreholes. Extensive heaving developed in an area up to 60 m from toe of excavation. Locations of the survey points, boreholes, trenches and slide features are shown on Figure 19. Figure 21 shows vectors of movements at three of the survey points. From the various observations, contours of downward and upward movements were determined,

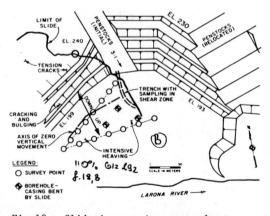

Fig.19. Slide in powerhouse excavation - plan.

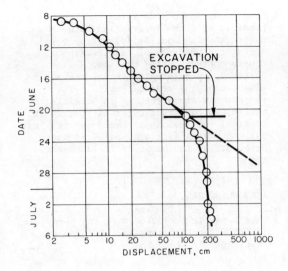

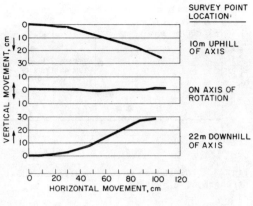

Fig.20. Slide in powerhouse excavation - movement along slide plane.

Fig.21. Slide in powerhouse excavation - typical trajectories of survey points.

permitting estimation of the approximate location of the axis of rotation of the sliding mass, as shown on Figures 19 and 22. It became apparent that the slide is deep seated, and that its toe follows the contact between the claystone and the peridotite, as shown on Figure 22. Consequently, the powerhouse was relocated about 200 m in downstream direction, at the expense of increased penstock length.

The slide area was stabilized by installation of horizontal drilled drains in the cut slopes, placement of drainage trenches filled with pervious material surrounded by Mirafi filter fabric and construction of a rockfill berm. For the design of the berm, a stability analysis was carried out using the subsurface profile shown on Figure 22. The probable location of the shear plane was estimated from the collected slide monitoring evidence, and shear strength parameters were obtained by back analysis of the slide and from triaxial tests on undisturbed samples of the claystone. These samples included some taken in the shear zone by jacking steel tubes into the side of the trench.

It is of interest that the strength determined from the analysis of the slide was in agreement with the strength indicated by laboratory testing, in terms of both total and effective stresses. The strength

data is shown on Figures 23 and 24. The resulting shear strength parameters for the fissured claystone were used in stability analyses of the cut slopes above the relocated powerhouse and for the design of the stabilizing berm placed in the slide area.

SURVEILLANCE PROGRAM

Regular surveillance of the dam, powerhouse and related facilities is essential to ascertain the continued integrity of the structures and to define areas requiring maintenance or repairs. A detailed

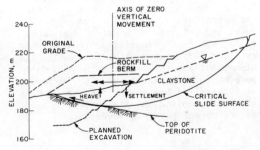

Fig.22. Slide in powerhouse excavation - stability analysis.

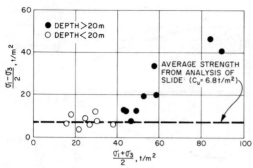

Fig.23. Powerhouse area - undrained shear strength of claystone.

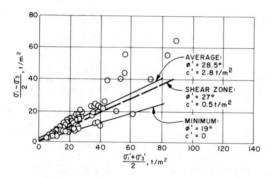

Fig.24. Powerhouse area - effective strength of claystone.

surveillance manual was prepared for use by the plant operations personnel. The monitoring system which was provided at Batubesi Dam consisted of piezometers in the sedimentary rocks at the left abutment and at the downstream toe, and displacement markers on the dam crest. For the power canal, marks were provided at the construction joints for monitoring total and relative movement of the canal walls, mostly in transition areas between cuts and fills, as well as in some of the higher cut and fill sections. At the powerhouse, piezometers and inclinometers were installed behind the deep excavation slopes to monitor the piezometric levels and to detect possible movements.

CONCLUSIONS

Successful completion of exploration, design and construction of the hydro-electric project under very adverse site condiitons in only 2½ years was made possible by close interaction between design studies and field experience gained during the progress of construction. Comprehensive testing provided design parameters consistent with the actual behavior of the residual soils. The project has shown that it is possible to design and construct major earthworks in tropical residual soils based on classical soil mechanics approach, adjusted where necessary on the basis of field observations.

ACKNOWLEDGEMENT

This paper is published with the permission of P. T. International Nickel Indonesia, subsidiary of Inco Ltd.

# Core material for Lahor Dam

SOERJONO
*State Electricity Corp., Jakarta, Indonesia*

## 1 INTRODUCTION

The Lahor Dam, which is situated about 20 km north of the Karangkates Dam, has a topographical characteristic that can be divided into two catagories :
1. below El. 250 with a steep bank slope of 1 by 0.8 - 1 by 0;
2. above El. 250 which is relatively flat.

The dam, which is of a rock fill type, has its crest at El. 278 and its foundation at El. 206.

For the supply of core material, an investigation was carried out an the soil and rock materials of which those for soil are taken from the areas around the dam, and for rock, the materials are taken from the Kalipare and Sukoanyar quarry sites (see figures 1 and 2).

In general, the characteristics of the soil properties around the dam site area are of the same kind as that during the construction (embankment) of the Karangkates dam. Consequently, under certain occasions, the methods applied during the construction of the Karangkates dam may well be used as reference.

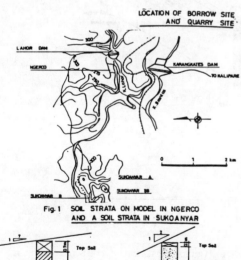

Fig.1 SOIL STRATA ON MODEL IN NGERCO AND A SOIL STRATA IN SUKOANYAR

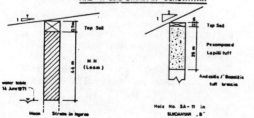

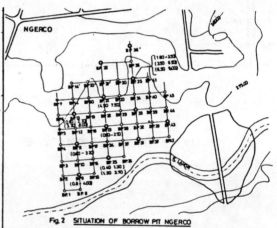

Fig.2 SITUATION OF BORROW PIT NGERCO

| BP. NO. | ELEVATION IN (m) | BP. NO. | ELEVATION IN (m) | BP. NO. | ELEVATION IN (m) | BP. NO. | ELEVATION IN (m) |
|---|---|---|---|---|---|---|---|
| 1 | 219.37 | 19 | 265.91 | 37 | 241.04 | 14' | 268.21 |
| 2 | 229.67 | 20 | 268.96 | 38 | 256.27 | 20' | 274.76 |
| 3 | 244.57 | 21 | 271.71 | 39 | 254.73 | 21' | -279.31 |
| 4 | 261.39 | 22 | 264.39 | 40 | 266.47 | 22' | 232.36 |
| 5 | 267.42 | 23 | 266.45 | 41 | 271.91 | 23' | 242.16 |
| 6 | 279.85 | 24 | 263.65 | 42 | 241.26 | 30' | 272.40 |
| 7 | 269.45 | 25 | 242.25 | 43 | 262.69 | 31' | 292.72 |
| 8 | 217.45 | 26 | 232.00 | 44 | 272.49 | 36' | 280.26 |
| 9 | 233.35 | 27 | 258.40 | 45 | 284.66 | | |
| 10 | 252.05 | 28 | 266.05 | | | | |
| 11 | 261.25 | 29 | 270.91 | | | | |
| 12 | 265.75 | 30 | 279.06 | | | | |
| 13 | -270.05 | 31 | 254.67 | | | | |
| 14 | 226.13 | 32 | 262.52 | | | | |
| 15 | 231.49 | 33 | 265.32 | | | | |
| 16 | 244.57 | 34 | 270.36 | | | | |
| 17 | 253.46 | 35 | 277.16 | | | | |
| 18 | 263.01 | 36 | 272.86 | | | | |

BP = BORROWPIT

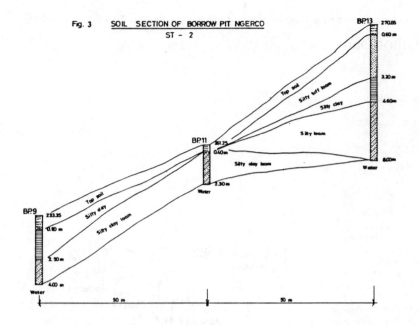

Fig. 3  SOIL SECTION OF BORROW PIT NGERCO ST - 2

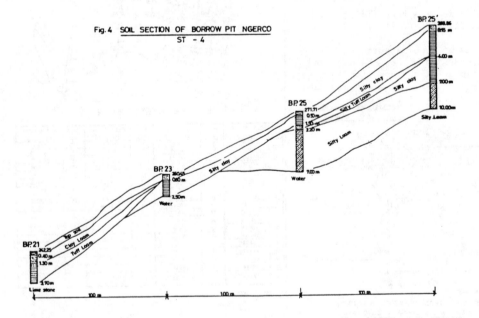

Fig. 4  SOIL SECTION OF BORROW PIT NGERCO ST - 4

The clayey silt type or the MH group of soil material (see figures 3 and 4) found at the borrow pits are not suitable for core embankment, whenever used in its natural condition due to the following matters :

1.0. High natural water content; so that during the compaction test the desired density of $\gamma_d > 1.1$ ton/m$^3$ at $w < 50\%$ (see figure 5) could not be reached. Another test with natural

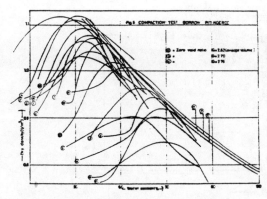

drying system also failed to give good results due the length of time required ( 8 hours) for a depth of 15 cm, although a ripper was provided for the test (see figure 6).

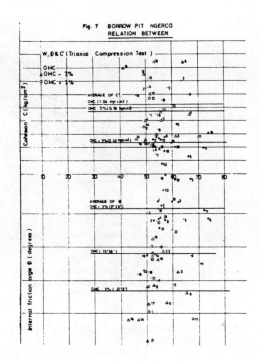

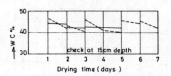

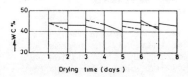

2.0. Due to the relatively high water content mentioned above, the soil properties were practically of the same kind.
At OMC (Optimum Moisture Content) only an average Ø (internal friction angle) of 15°58' was reached (see figure 7).

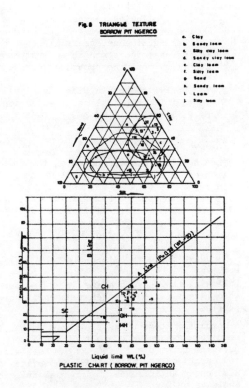

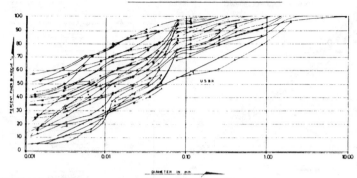

Fig 9 GRAIN SIZE DISTRIBUTION OF LOAM MATERIAL FOR BORROW PIT NGERCO

3.0. From its grain size distibution, the soil material is considered to be of bad quality and is not suitable for embankment material, although its plasticity index (Ip>15%) is fairly good for piping resistance and its permeability is of $n \times 10^{-6}$ cm/sec. (see figure 8 & 9).

According to the USBR standard, the above soil material belongs to the group of soil that can easily crack whenever undergoing a reduction of water content.
To cope with the above factors, the soil material is to be mixed with weathered rock for the following purposes :
1. To reduce the water content with less processing time necessary;
2. To increase the strength so that enough stability could be provided for safety factor;
3. To improve the grain size distribution so that its physical deficiences could be eliminated.

## 2 TEST ON MIXTURE PROPORTION

The investigation conducted on the combination of this mixture and that which is done with reference to what was done during the implementation of the Karangkates Dam, consists of tests with the following combinations :
1. 70/30      3. 50/50      5. 30/70
2. 60/40 *)   4. 40/60      6. 20/80
*) 60/40 means the mixture of 60% weathered rock and 40% soil by weight proportion.
Following the above combined mixture, further investigation is then made on its properties.

A compaction test using the CBR mold has resulted in a water content of w<45% and dry density $\gamma_d = \pm 1.60$ ton/m³ (see figure 10). Tests on its mechanical properties have also shown an increase of internal friction angle value $\emptyset$ to $\emptyset > 30°$.

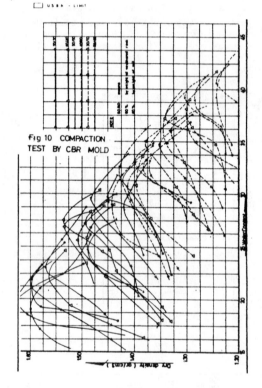

Fig 10 COMPACTION TEST BY CBR MOLD

With the success in decreasing the water content to lower than 45%, the problem of pore-pressure during the construction and work ability has been well taken care of since the soil can now be classified as of SC group.

Based on the possibility of economical implementation and the tolerable condition of water content, an intensive mechanical test will only be conducted to proportion of mixtures of 70 by 30 and 60 by 40, and to mixtures where the rock material is < $\frac{1}{3}$ of the total volume but can still maintain its soil characteristics.

The mechanical test for 20 by 80 mixture will be made only for checking its value, since soil material with a high consistency index of Ic > or natural water content Wf < PL will have favorable mechanical properties.

With reference to what was done during the implementation of Karangkates Dam, Jatiluhur and some other high dam projects, and as seen from the results of mechanical tests on the combination of the above mixtures, a conclusion can be made that in general the characteristics of the materials (as design value) are as follows :
1. mixture proportion = 70 by 30
2. specific gravity $G_s$ = 2.68
3. water content (OMC) w (%) = 35
4. dry density $d \cdot (t/m^3)$ = 1.27 (=1.34x0,95)
5. permeability k (cm/sec) = $10^{-6}$
6. shear strength $C(t/m^2)$ = 5.0 in wet condition
   = 3.0 in saturated condition
   $\tan \emptyset = 0.5 (\emptyset = 26,5°)$

with the consideration of the following conditions :
1. mixture of 70/30 will be fairly economical, since the rock volume will still be within the limit of soil characteristics stability;
2. water content should be somewhat higher than 30% - 37% range, so that the embankment could be still kept within the safety range (see figure 10).
3. dry density should be equal to that of point 2, namely a little bit lower than the dry density at water content 35 % to maintain safety during the embankment.
4. the lowest permeability during laboratory test should be n x $10^{-7}$ cm/sec., but considering that various core conditions and layer compactions can occur during the embankment, permeability of n x $10^{-6}$ cm/sec will be sufficient.
5. shear strength derived from the result of 70/30 test needs an adjustment on its cohesion and frictions as follows : the cohesion should be adjusted by 0.8 factor as against the result of 60/40 mixture, and the friction should be adjusted by 0.9 factor as against the 70/30 mixture.

3 THE SPECIFICATION OF CORE MATERIAL

To meet the requirements of the design of actual core embankment at the dam site, technical specification modified in accordance with the use of core-compaction machineries and other equipment are prepared.
The specifications cover among others :
1. the quality, especially that of organic matters for soil, and clay content of less than 7 % for weathered rock, of which those having a diameter larger than 10 cm should be put aside during the stock piling.
2. the stock piling, which should be made with the proportion of 30/70 could be carried out by arranging the thickness of each soil and weathered rock layers at 50 cm and 20 cm respectively.
3. the embankment at dam site using materials from stockpile with 30 cm thickness at each layer, and of 7 cm thickness for embankment at contact side.
4. the control of embankment quality during the implementation, especially of water content and the density of embankment material; where as the control of shear strength should be undertaken after the layer has reached each 3.0 meters of thickness.

In order to know that the equipment used afterwards could meet the requirements set by the design and whether it could be economically carried out, a kind of embankment test at a certain scale should be conducted.

The data required for the embankment test are among others :
1. compaction effect against the thickness of layer and the number of compaction of each layer so that the most efficient proportion could be reached for producing optimum density;
2. permeability of compacted core so as to reach the design value;
3. expected shear strength by the time of the embankment is carried out as compared to the results of laboratory test.

During its implementation the core embankment at the Lahor Dam site can be divided into 2 stages, namely :

First stage : stock piling of materials taken from borrow pit and quarry site;
Second stage : transportation to and embankment at the dam site, of which the materials are taken from the stockpile.

A stockpile consists of layers of embanked soil materials and weathered rock with a thickness of 50 cm/20 cm respectively (for mix proportion of 70 by 30).
In this way it is expected that from a stock pile reaching a volume of 6000- 8000 $m^3$ and height of about 10.0 meters at an equal rate of thickness for each layers so that an equal proportion of mixture at the embankment volume could be produced. The height of the stockpile is decided on the basis of the height of loader used; in this case a Koeh-

ring loader with an effective height of 8.0 meter and bucket capacity of 1.2 m³ is employed.

For the transportation of weathered rock material heavy dump trucks with 13.5 ton capacity are employed; whereas the soil material is taken from the borrow pit by way of employing a motor scraper of 6.0 m³ capacity.

Rock material used as mixture for soil taken from the borrow pit is transported from the Kalipare quarry site, about 10 km away, while the stock pile is made on the left bank. On the right bank, however, weathered rock material from the Sukoanyar quarry site is placed.

During the stockpile, the machineries used are among others :
 a. 1 Loader     -) at the quarry site.
    1 Bulldozer  -)
 b. 10 - 15 heavy dump trucks for transporting the rock materials, and also because of the far distance.
 c. 1 Bulldozer            -) for use at the
    3-4 motor scrapers-)    borrow pit :
    however, due to the short distance, however, the use of motor scraper will be much more effective.
 d. 1 Bulldozer for levelling
    1 Bulldozer with a rigger attached; for removing boulders larger than Ø 10 m.

The volume of stockpile, however, is made in accordance with the work volume during the dry season of the respective year plus that of spare volume of work for the coming dry season (± 1 - 2 months of core embankment work).

For core embankment at the dam site, the material is taken from the stock pile already provided.

The location of the stockpile is made as close as possible to the dam site so that the speed of the embankment could be effectively maintained.

Transportation is carried out by way of employment of heavy dump trucks of 13.5 ton capacity or motor scraper of 6.0 to 15 tons capacity (during the completion of stockpiling).

In the case the motor scraper is employed as means of transportation from the stock pile to the dam site, the bucket feeding for the motor scraper should be conducted by the loader (Power shovel).

This is to make sure that during the transportation of the material from the stockpile the soil and the rock could already be homogeneously mixed.

The machineries used during the core embankment consist among others of :

 a. 1 - 2 loaders  -)
    1 Bulldozer    -) for stockpiling;
    depending on the availability of the vehicles.
 b. 10 - 15 heavy dump  )
           trucks       ) For transportation to the dam
           or           ) site
    4 - 6 motor scrapers-)
 c. 1 - 2 bulldozer        -) for levelling
    2 Sheepfoot rollers    -) and compaction
    (drawn by bulldozers   ) of core
 d. 6 - 8 air tamper or    )
           tamping rammer -) especially for
    1 tire roller          -) joint compaction

4 RECOMMENDATIONS

Control of material quality based on the technical specification is performed during the stockpiling and during the core embankment at the dam site.

Data of quality control will be evaluated during the embankment and after its total completion.

This is necessary in order to know the divergence occuring against the specification.

In the mean time, observations on cases due to internal factors (pore pressure, etc.) and external factors (i.e. settlements and quakes) will always be conducted even after the whole completion of the embankment.

*Symposium on Problems and Practice of Dam Engineering / Bangkok / 1-15 December 1980*

# Stability of rock cavern for underground pumped-storage power stations

KEIICHI FUJITA
*Hazama-Gumi Ltd., Tokyo, Japan*

## 1 INTRODUCTION

In Japan, more than forty underground pumped-storage power stations, including those under construction at present, have been constructed to utilize the generated surplus electricity during night time.

Dimensions of these power stations range from 20 to 27 meters in width, 35 to 55 meters in height, and 61 to 163 meters in length. Underground power stations are mostly being excavated in relatively weak rock with many faults and joints, where the modulus of deformation ranges from 50,000 to 100,000 kg/cm$^2$ and should be located between upper and lower reservoirs (Fig.1).

During excavation, most of the horizontal displacement of vertical side walls amounts to between 5 to 40mm; sometimes, a collapse of a side wall occurs. Therefore, careful studies, on stability of caverns especially in case that the location of underground power station is close to the dam, should be conducted at each stage of investigation, planning, design and construction.

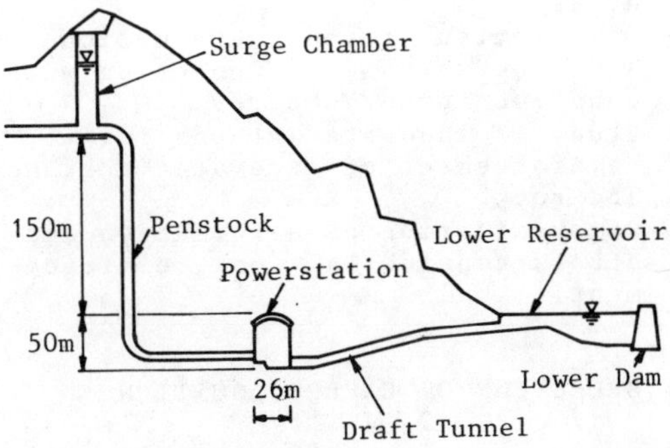


Fig.1. Typical cross section at Numazawa No.2

For the study of cavern stability during excavation M. Hayashi proposed the progressive stress analysis method considering non-linear visco-elastic and visco-plastic characteristics of rock. In many cases of underground power station excavation in Japan, this method is used for estimating the relaxed zones of surrounding rock, predicting the deformation of side walls and planning of side wall reinforcement, and this has been proven to be very efficient.

## 2 RELIABILITY OF ANALYSIS FOR CAVERN STABILITY

The method of analysis proposed by M. Hayashi has been applied to more than 20 sites of underground power stations to study their stability during the excavation. This method has been proven to be very reliable.

An example of analized results in estimating the relaxed zones of the surrounding rock and predicting side wall displacement during the excavations is shown in Fig. 2. In these calculations, representative rock characteristics are selected from among the complex rock strata formation, and analysis was carried out without taking into account the existence of cracks, seams, or of small caverns, openings in the vicinity, although the influence of these was considered.

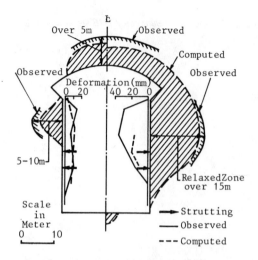

Fig.2. An example of comparison between computed & observed values on relaxed zone & horizontal displacement of side wall

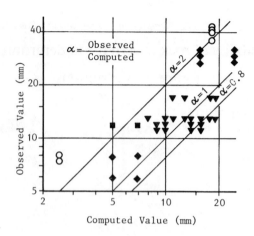

Fig.4. Comparison between computed & observed horizontal displacement of side wall

Fig. 3 shows the relation between the modulus of deformation and the horizontal displacement are in an inverse proportion to the modulus of deformation.

Fig. 4 shows that the observed values of side wall displacement amount to between 0,8 to 2 times of the prodicted values by M. Hayashi's method, therefore, they are similar. These predicted values have been computed before excavation took place.

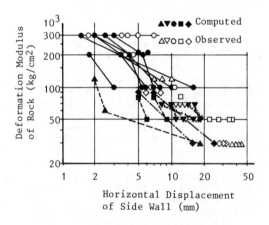

Fig.3. Deformation modulus of rock vs. horizontal displacement of side wall

The reason for the observed value being larger than the predicted one, is considered to be mainly due to the influence of the existing joints in rock.

## 3 STUDY PROCEDURES ON CAVERN STABILITY

The procedures are as follows:
- Selection of an approximate location of cavern by means of reconnaissance and geological survey.
- Selection of cavern location by means of excavating investigation adits and borings from it.
- Testing of rock in situ.
- Selection of representative rock properties through assessment of test results.
- Selection of construction procedures.
- Calculation of rock deformation and rock stress in connection with construction procedures.
- Study of adequate methods for reinforcement of excavated surface and rock.
- Re-examination of deformation and stress considering rock reinforcement.

## 4 SELECTION OF CAVERN LOCATION

In many cases, the location of cavern was changed more than ten times, so as to locate it in a rock of as good a quality as possible.

Important items when selecting a cavern location;
- A cavern should be located as far apart as possible from fault and fractured rock zones.
- The cavern should be located so that the longitudinal axis of it crosses at nearly right angle to most of the joint strikes.
- The direction of maximum initial principal stress should be kept as parallel as possible to side walls.

However, when fault and fractured rock zones can not be avoided, selection of cavern location is made as follows;
- The zones should cross at nearly right angle to longitudinal axis of power station.
- Large openings, such as draft tunnels, etc. located at fractured zones, should be avoided.
- End walls should not be located near fractured zones.

## 5 INVESTIGATION AND ROCK TESTS

Various kinds of in situ rock tests and laboratory tests are performed, using investigation adits and additional boreholes.

From these test results, geological structure of rock mass, classification of rock, location of faults, direction of joints and groutability or drainability of rock mass etc. are made clear. Geotechnical maps and drawings of cross-sections are prepared (Fig. 5).

## 6 ASSESSMENT AND INTERPRETATION OF GEOLOGICAL TEST DATA

Various rock properties are obtained as a result of aforementioned investigation and tests. From these data, representative values should be selected as input data for the FEM analysis.

While establishing an average of test data is simple, the selection of representative value is very difficult, because the values should be selected so that the estimated and measured values regarding the behaviour of cavern during construction are in agreement. In case the deformation of the side walls is excessively estimated, reinforcement and/or countermeasures are apt to be neglected because the observed values amount to less than the predicted values.

As the test data should be assessed and interpreted carefully based on each engineer's experiences, it is natural that the representative values should vary even when the same test data are used.

Table 1 shows the summary of deformation test results obtained at Numazawa Power Station No.2. In this table, the data of V-1 to V-5 show vertical loadings, and H-1 to H-5, horizontal loadings, and there is a great difference between two groups, in the rock properties except for the creep factor. The former shows a 2 to 2,5 times larger modulus than the latter. Based on the interpretation that this phenom-

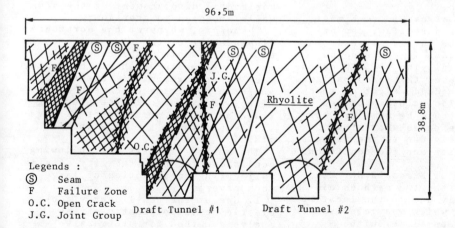

Fig.5. Geological conditions of the side wall at Numazawa No.2

TABLE 1. Summary of deformation test

Vertical

| Test No. | V-1 | V-2 | V-3 | V-4 | V-5 | Average |
|---|---|---|---|---|---|---|
| $D_o$ (×10³kg/cm²) | 46 | 120 | 182 | 92 | 64 | 100 |
| $E_t$ (×10³kg/cm²) | 94 | 166 | 194 | 176 | 161 | 158 |
| $E_s$ (×10³kg/cm²) | 53 | 120 | 175 | 153 | 63 | 113 |
| $\alpha$ | 0,13 | 0,07 | 0,06 | 0,30 | 0,03 | 0,12 |
| $\beta$ (day⁻¹) | 14 | 11 | 24 | 6 | 8 | 13 |

Horizontal

| Test No. | H-1 | H-2 | H-3 | H-4 | H-5 | Average |
|---|---|---|---|---|---|---|
| $D_o$ (×10³kg/cm²) | 47 | 42 | 69 | 22 | 18 | 40 |
| $E_t$ (×10³kg/cm²) | 78 | 93 | 86 | 51 | 59 | 74 |
| $E_s$ (×10³kg/cm²) | 56 | 50 | 77 | 27 | 20 | 46 |
| $\alpha$ | 0,13 | 0,09 | 0,18 | 0,08 | 0,05 | 0,10 |
| $\beta$ (day⁻¹) | 9 | 14 | 8 | 10 | 9 | 10 |

where,
$D_o$ : Modulus of deformation
$E_t$ : Tangent modulus of elasticity
$E_s$ : Secant modulus of elasticity
$\alpha, \beta$ : Creep factor

enon results mainly from the loading direction and joints position, $D_o=100 \times 10^3 \text{kg/cm}^2$ has been adopted as the representative modulus of deformation for the standard rock section at this site.

The modulus of elasticity was determined using the data of unloading cycle at the level of initial stress of surrounding rock. Furthermore, the estimation of displacement at each measuring location during construction was calculated by adjusting the values of $D_o$ and others, considering the size of joints, the direction of joints, the size of seams and the degree of fractures, etc.

The initial stress in rock mass, as well as its direction, can be measured by the in-situ stress released method. It can also be obtained by employing the FEM calculation which uses the land configuration and the dead weight of rock overburden. The values obtained by two methods should be similar, however there are cases where they differ extremely due to the effects of configuration and the conditions of joints.

In the case of Numazawa No. 2, the values of the two methods differed extremely due to the location of the cavern which was relatively close to a steep slope on the surface. This factor caused engineers to select varied representative values.

7 PLANNING OF SUPPORTS FOR THE SIDE WALLS

The maximum horizontal displacement of side walls ever observed at power stations amounted to about 45mm. However, considering the amount of estimated displacement from the time of excavation to the beginning of measurement, the maximum value would be about 60mm. In all the past experiences where the side wall either collapsed and/or fell under extremely dangerous conditions, the horizontal displacement of the side walls amounted to over 40mm.

Therefore, whenever the horizontal displacement of the side walls would exceed allowable displacement or 40 mm, countermeasures should be taken, such as arrangement of rockstruts and/or a reduction of cavern dimensions, as a change in the planning stage and as an arrangement of side wall supports in the construction stage.

The author proposes the following conditions for the stability of side wall during construction of large rock caverns.
1. Use of the non-linear visco-elastic or visco-plastic stress analysis method for progressive stress relaxation around excavated rock caverns, by M. Hayashi.

2. Allowable displacement of (40 -60) mm/S.F. for the side wall during excavation.
3. When anchor work is critical, the extent of the relaxed zone does not exceed 15 m.
4. Anchors should be prestressed, but the amount of anchor elongation after this shall be greater than the displacement of the side wall.

Here, safety factor (S.F.) for obtaining allowable displacement will be decided according to the accuracy of the calculation method, the extent of investigation of the rock and the reinforcing or supporting method of the wall; tentatively, the author recommends a value of 1,2 to 1,5.

Relaxed zones specified by the analysis indicate the zones where Poisson's ratio of rock exceeds 0,45, due to excavation of the cavern. Considering the possibility that there may be a danger of collapse in these zones or the zones where tensile stress may occur, appropriate supports should be planned for safety reasons. The common method applied for side wall supports is the prestressed rock anchors and/or shotcrete or temporary concrete lining.

The expected effects from the application of the anchors are:
- Resistance against sliding along the relaxed zones.
- Providing confined pressure zones around the cavern.

Criteria for the design of anchor for the side wall supports:
- Placing the anchor so that it would cover all the relaxed zones.
- Placing the bonded part of the anchor beyond the relaxed zones.
- Prestressing the anchor sufficiently so that it could resist the sliding force along the sliding plane which envelops the relaxed zone.
- Providing the elongation of the anchor larger than the amount of rock deformation at same section.
- Choosing materials which set rapidly after the installation of anchors, and which permanently remain in the same condition.

As an additional support, it is desirable to use shotcrete. For the purpose of preventing a fall of rock and loosening of the rock surface, the shotcrete should be mixed in order to obtain early-strength; also it should have the proper thickness.

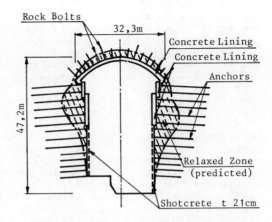

Fig.6. Supports of cavern for Numazawa No.2 (for standard section)

An example of the supports for the cavern at the standard section of Numazawa No. 2 are shown in Fig.6.

8 CONSTRUCTION METHOD CONSIDERING STABILITY

At Numazawa No.2, the rock properties were considered as being moderate compared with the existing power stations. However, there were unfavorable conditions in the side wall stability such as fractured seams in every section, crack openings (which sometimes amounted to several millimeters) as a result of blasting, draft tunnels crossing diagonally to the side wall, etc.

Procedures employed for stability of side walls during construction:
- Trying to achieve a smooth finish on the side walls, pre-splitting and smooth blasting was carried out in the area around the cavern in order to minimize the damage to the surrounding rock caused by blasting.
- To minimize the impact to the surrounding rock, the height of the bench for cavern excavation was set at 3 meters; blasting was excuted by a delay-blasting procedure, using a deci-second detonator, and the amount of powder per decisecond was limited to 20 kg. Based on the results of measurements, it has been proven that the damaged zones caused by blasting could be limited to an area less than 3 meters in depth.

- To decrease developing of relaxed zones of rock, shotcreting of the side walls should be done as soon as possible after blasting.
- At the time of drilling, grouting was applied with nearly no pressure to a hole for an anchor, where leakage of water through joints occured.
- Concrete lining was applied immediately after excavation to the area of the upper 4,5 meters of the total excavation height of 41 meters. This was done in order to stabilize the upper part of the side walls where tensile stress appears quite easily.
- Also, to assure the stability of the upper part of the cavern, the concrete lining was applied simultaneously in the upward direction during excavation of the remaining portion, when the excavation height of about 15 meters was completed.
- Tunnels, such as penstocks and draft tunnels connected to the cavern, had been completely constructed before reaching the stage of excavating the side walls, and reinforced with back-fill grouting, consolidation grouting etc.

These construction procedures were considered to be efficient in the relatively hard surrounding rock with many cracks.

## 9 PROCEDURE OF CHECK-UP OF STABILITY DURING CONSTRUCTION

As a first step, the check-up of stability during construction commences with a comparison between various predictions at the time of the planning and design stage and the measurements at the construction stage. If necessary, the modification of designs and changes of construction procedures are carried out, in order to assure safety during construction.

For this purpose, planning of observations and measurements is made prior to construction. The standard section as well as predicted troublesome sections are chosen as measuring points, and instruments are installed in these points before or immediately after the progress of excavation.

The geotechnical conditions, such as distribution of the geological structure, rock properties, seams, joints, etc. are recorded together with the progress of excavation of access and other tunnels, the arch and side walls. Also, the geotechnical conditions of the overall rock mass are studied and forecasted, even to unexcavated portions, and modifications of the geological maps and profiles should be done with the progress of excavation.

Data obtained from various measurements are immediately examined, in order to check whether the work is proceeding as safely as expected.

## 10 JUDGEMENT OF SAFETY BASED ON MEASUREMENTS

In order to evalute safety conditions during construction, it is necessary to compare the figures obtained from measuring the instruments installed in various places, such as in rock close to a side wall, with those from the FEM analysis. Based on the results, the behaviour around the cavern is examined and the judgement of safety is made.

The following items should be carried out at the time of measuring:
- Evaluation of accuracy or tolerance of predicted and observed values, prior to measurement.
- Confirmation of reliability of observed values by comparing the results obtained from one or more methods.
- Checking of the instruments whether they are operating normally.
- Checking of the observed values whether they represent the behaviour of the cavern and the surrounding rock.

Observed data are further examined as follows:
- Comparing the predicted values with measurements at each measuring points, and investigating the reasons whenever there is a difference.
- Finding the measuring point where the observed value shows an abrupt change with the time, without a specific reason.
- Finding the measuring sections which should be observed carefully.
- Finding any other dangerous sections, other than the measuring sections.
- Predicting the maximum or final value according to the tendency of observed values.

- Determining whether any counter-measures are required or not.

The judgement of the safety in the observation and/or the final stage must be made according to the abovementioned procedures. Especially, concerning troublesome section, it is recommended that observation should be made often by many engineers.

## 11 COMPARISON OF PREDICTION BY THREE ENGINEERS

At Numazawa No. 2. three engineers proposed their predicted values of horizontal displacement of the side walls during excavation applying the analytical calculation. Table 2 shows the comparison between their input data and the results. The differences of interpretation and assessment of the modulus of deformation, shearing strength, the direction of initial stress (the angle formed by the maximum principal stress and the perpendicular) and the damaged zone formed by blasting are shown in the table.

Consequently, the computed and predicted maximum horizontal

TABLE 2. Comparison of predictions proposed by three engineers for the standard section of Numazawa No.2

| Engineer | | A | B | C |
|---|---|---|---|---|
| **Input data chosen** | | | | |
| Initial stresses | $\sigma_2$ (kg/cm$^2$) | -55 | -50 | -51,2 |
| | $\sigma_1$ (kg/cm$^2$) | -30 | -30 | -31,4 |
| | $\theta$ (degree) | 27 | 62,3 | 62,3 |
| Unit weight | $\gamma_t$ (g/cm$^3$) | 2,6 | 2,5 | 2,5 |
| Modulus of deformation | $D_o$ (kg/cm$^2$) | 1x10$^5$ | 1x10$^5$ | 1x10$^5$ |
| for damaged zone | 0-1m | 0,4$D_o$ | 0,4$D_o$ | ——— |
| due to blasting | 1-3m | 0.7$D_o$ | 0,7$D_o$ | ——— |
| | 0-2m | ——— | ——— | 0,5$D_o$ |
| Poisson's ratio | $\nu_o$ | 0,25 | 0,25 | 0,25 |
| for damaged zone | 0-1m | 0,4 | 0,4 | ——— |
| due to blasting | 1-3m | 0,35 | 0,35 | ——— |
| | 0-2m | ——— | ——— | 0,375 |
| Creep factor | $\alpha$ | 0,1 | 0,1 | 0,11 |
| delay ratio | $\beta$ (day$^{-1}$) | 10 | 10 | 11 |
| Shear strength | $\tau_{RO}$ (kg/cm$^2$) | 10 | 14 | 13,9 |
| Shear stress ratio | $\sigma_t/\tau_{RO}$ | 0,1 | 0,1 | 0,15 |
| Days of excavation | | 430 | 430 | 500 |
| Number of steps for excavation | | 6 | 6 | 5 |
| **Predicted values** | | | | |
| Thickness of | Hill side (m) | 7-8 | 7 | 16 |
| relaxed zones | River side (m) | 7-8 | 7 | 10 |
| Hor. displacement | Hill side (mm) | 13 | 16 | 25 |
| of side walls | River side (mm) | 14 | 18 | 22 |
| **Observed values** | | | | |
| Horizontal displacement (mm) | | 13 - 14 | | |

where,
$\sigma_2$, $\sigma_1$ : Minimum, maximum principal stress
$\theta$ : Angle between direction of initial principal stress to perpendicular line in degree
$\tau_{RO}$ : Initial shear strength in case that normal stress equals zero
$\sigma_t$ : Normal stress in case that shear stress equals zero

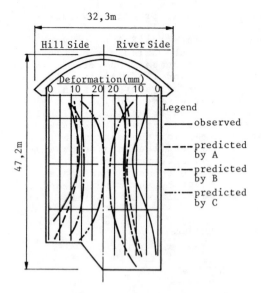

Fig.7. Comparison between predicted and observed values on horizontal displacement of side walls (Standard section Numazawa No.2)

displacements (absolute values) of side walls amounted to 13-13mm, 16-18mm, 22-25mm, respectively. The relative displacement between the surface and the 25m beyond surface was observed as being at 12-13mm, but considering the amount of estimated displacement from the excavation to the beginning of measurement, the maximum horizontal displacement of side walls would be 13-14mm (Fig. 7).

12 REFERENCES

Hayashi, M., Kitahara, Y., Hibino, S. 1969, Timedependent Stress Analysis in Underground Structure in Visco-plastic Rock Masses, Proc. of International Symposium on the Determination of Stress in Rock Masses, Lisbon.

Hayashi, M., Hibino, S. 1970, Visco-plastic Analysis on Progressive Relation of Underground Works, Proc. of the 2nd Congress of the International Society for Rock Mechanics, Beograd, No. 4-25 (Vol. II): 565-575.

Fujita, K., Ueda, K. and et al. 1977, An Empirical Proposal on Stability of Rock Cavern Wall during Construction, Storage in Excavated Rock Caverns, Rockstore '77, Proc. of 1st. Int. Symposium (Stockholm) Vol. 2, Pergamon Press: 309-314.

Takahashi, Y., Takeda, K., Fujita, K., Yoichi, H. 1980, Stability of Vertical Walls during Excavation of Underground Power Stations, International Symposium "The Safety of Underground Works", Brussels, Brussels International Conference Center: 29-34.

Section D:
Country reports

*Symposium on Problems and Practice of Dam Engineering / Bangkok / 1-15 December 1980*

# Dam engineering in Pakistan

AMJAD AGHA
*National Engineering Services (Pakistan) Ltd., Lahore*

SYNOPSIS. Pakistan has one of the largest and most intricate irrigation systems in the world with about 64,000 kilometers of major canals, 20 barrages and 24 important dams, two of which are amongst the largest in the world.

This country report briefly traces the history of irrigation development from the ancient inundation canals to the modern diversion barrages and dams. As a result of the geo-political developments of 1947, Pakistan lost the water supplies of its three eastern rivers. This led to the implementation of the gigantic Indus Basin Project including Tarbela and Mangla, two of world's largest earthfill dams, so that water supplies could be restored to the network of irrigation canals. A brief resume is given of some of the important dams already constructed and those under construction alongwith their salient features and typical embankment sections. The future development schemes to meet the requirement of water and power after Tarbela have been discussed, emphasizing the necessity of Kalabagh dam as the immediate need of the country. About 180 potential dam sites have also been identified, some of which are in various stages of planning.

Certain problems of dam construction are peculiar to Pakistan. These are, costly nature of surface water storages with unsuitable topography, very high floods, phenomenal sedimentation, difficult foundations and seismically unstable areas. Such problems had to be confronted at an unprecedented scale both at Tarbela and Mangla where many aspects were on or beyond the boundaries of present experience. Some of these problems have been discussed in this report.

The construction of dams in last two decades has had a marked social and economic impact on the development of Pakistan, particularly in the hydro-power and agriculture sectors which form the backbone of the national economy. Benefits derived from irrigation and power have been briefly dealt.

1 DEVELOPMENT OF IRRIGATION

1.1 General

Pakistan is bounded by Iran on the west, Afghanistan on the north-west and north, China on the north-east, India on the east and south-east and Arabian sea on the south-west. The total area of the country is 1,050,000 sq. kms. (405,700 sq. miles); out of this 662,000 sq. kms. (255,700 sq. miles) comprises rugged mountains, narrow valleys and foothills. The remaining 388,000 sq. kms (150,000 sq. miles) consist of sandy desert and flat alluvial plains, of which some 30 million hectares or 303,030 sq. kms. is estimated to be arable, this includes 20.25 million hectares of the Indus Plains (Append. B).-

The massive glaciers emerging from the snow clad Himalayas have developed one of the most fascinating river systems of the world in this foreland comprising the mighty Indus and its six tributaries; Kabul, Jhelum, Chenab, Ravi, Sutlej and Beas.

Pakistan lies in a semi-arid zone, most of the Indus plain is deficient in rainfall, annual mean being generally under 250 mm. The river flows, though perennial, are variable within a year and from year to year. This presents a serious challenge to the further development of irrigated agriculture in the area.

Agriculture is the largest single sector of Pakistan's economy. In the semi-arid environments prevailing over a greater part of the country, the availability of adequate and timely irrigation water supplies is the key to progressive agriculture. In order to appreciate the developments of Dam construction in Pakistan, it is necessary to briefly review the development of the irrigation system, and the impact of the geopolitical changes brought about by the Indus Water Treaty of 1960.

1.2 Ancient Irrigation Practice

Historians believe that about 5000 years ago, two great civilizations of the world, Mohenjo-Daro on the lower Indus and Harappa on the Ravi, flourished on irrigated agriculture. The ancient system of irrigation, which practice continued for several centuries, involved breaching the banks or 'natural levies', and providing shallow channels to direct the floods. In this way it was possible to bring water to fields in low-lying portions of flood plains. This could be done, however, only during high flow periods, and so, 'inundation irrigation' as it is called, was limited in the Indus Basin to summer season. During the winter season the water levels were too low for the purpose.

The inundation irrigation was, at best, a precarious means of agriculture, which required a great deal of maintenance. By 1872 A D. the inundation canals in Punjab alone aggregated some 4000 kms.

(2500 miles) in length and irrigated more than 40 thousand hectares (1 million acres). In lower Indus by year 1900, such canals were providing irrigation water to about 1.21 million hectares (3 million acres).

1.3 Mid-Nineteenth Century

The irrigation system which exists today has been built up over the course of more than a century by means of a series of engineering works, many of them unprecedented in the world at the time they were undertaken. With the increasing demand of water the first perennial canal is known to have been built by emperor Jehangir (reigned 1605-27) which was taken from the river Ravi. However, the development of the modern complicated and extensive canal system took place during the British era. To serve the expanding population around mid-nineteenth century large scale diversion of water from the main rivers was accomplished with a network of canals of varying sizes and later by means of barrages. In 1947 when Pakistan became independent, the average annual river diversions were about 8.27 billion cu. m. (65 million acre feet). During this period need for water storage dams was not felt as the diversion dams or barrages aided by the perennial rivers were adequately meeting the irrigation requirements.

1.4 Mid-Twentieth Century

After Independence the Government of Pakistan continued the expansion and development of agriculture. By 1962, with the completion of Guddu Barrage on the Indus, another 3.6 million hectares (9 million acres) were added to the culturable commanded area under the canals. So in a period of 15 years (1947-1962) the irrigation system was not only successfully maintained and operated, but also expanded by 30 to 40 percent.

The pattern of irrigation development was aimed predominantly to formulate isolated schemes to meet the needs of localized areas. With the completion of the Guddu Barrage on the Indus in 1962, almost all the dependable or utilizable natural flows in the river were exhausted. Further development could not proceed on old lines, i.e. building a weir or barrage for river diversions and a canal system for irrigation. In fact, a turning point in the water resources planning for the Indus Basin was reached. It was realized that further development has to be based on unified and integrated planning for the optimum utilization of water and power resources of Pakistan at minimum cost.

1.5 Indus Basin Replacement Plan

The mid-twentieth century was a period of great political upheavals resulting into partition of the sub-continent. These geopolitical changes were a major event in the history of irrigation development and greatly influenced the future construction of water storage schemes. Prior to partition the waters of the Indus river system were primarily used in areas of Pakistan, with relatively small quantities used in areas now included in India. However, the head of the waters of the major tributaries of the Indus River, i.e. Sutlej, Beas, Ravi, Jhelum and Chenab lie in India. Taking advantage of this, India started diverting the waters of these rivers to its own areas after partition.

Pakistan considered it as usurpation of her historical water use rights. The dispute continued over years but was resolved in 1960 by concluding a treaty between the two countries through the good offices of the International Bank for Reconstruction and Development (IBRD). According to this treaty, India obtained the exclusive rights over the three eastern rivers, the Sutlej, Beas and Ravi. This reduced the flows available to the Indus Plains in Pakistan from about 215.8 B.cu.m. (175 MAF) to 175.1 B.cu.m. (142 MAF). Pakistan in turn was provided with necessary funds for transferring the water from the western rivers, the Chenab, Jhelum and the Indus main stem, to make up for the loss of water, it historically used from the eastern rivers. This resulted in the formulation of a huge project called Indus Basin Replacement Plan. Soon after the signing of the Indus Water Treaty, the Government of Pakistan appointed the Water and Power Development Authority (WAPDA), as its agent for implementation of the Settlement Plan. Two major storage dams, Mangla Dam on the Jhelum river with a usable capacity of 6.6 B.cu.m. (5.34 MAF) and Tarbela Dam on the Indus river at Tarbela with a usable capacity of 11.5 B.cu.m. (9.3 MAF), six new barrages or diversion dams; eight new inter-river link canals of 198 to 623 $m^3$/sec. (7,000 to 22,000 cfs) capacity, aggregating to a total of 3630 $m^3$/sec. (128,100 cfs) and remodelling of a barrage at Balloki on the Ravi river and three existing link canals and several irrigation systems (Append. C). The total inter-river transfer capacity including the new and old links, now stand at 5.2 B.cu.m. (4.2 MAF) per month.

2 DAMS IN PAKISTAN

The history of Dam construction in Pakistan is rather short. The perennial rivers served the irrigation need. Drinking water supply was taken care by tapping the vast underground water reservoir. In fact before independence, there were only three dams in the whole country with none on the major rivers. Two of them in the water scarce area of Baluchistan, the Khushdil Khan Dam - 1890 and Spin Kariz - 1945 and Namal Dam - 1913 in Mianwali district of Punjab.

The growing needs of irrigation and power and the rapidly increasing drinking water supply scarcity in some heavily populated towns, made it necessary to construct dams. Later when India stopped water supplies to the network of canals irrigating about 3.2 million hectares (8 million acres) of land, it became imperative to build large storages and link canals to restore water to the affected canal system. This resulted in the construction of two gigantic dams Mangla and Tarbela, part of the Indus Basin replacement works mentioned earlier. Apart from the replacement works a number of relatively smaller schemes of irrigation dams were also taken up during 1960 - 80 as given in Append. A.

Two of the rapidly expanding towns of Islamabad and Karachi also required water storage to meet the water supply requirements. This made it necessary to construct the Hub Dam near Karachi and Rawal, Simly and Khanpur Dam near Islamabad which are basically water supply schemes but would also cater for some irrigation requirement of the area.

The construction of the Mangla Dam and Tarbela Dam were two such projects which involved some of the world's best brains in dam building. It would not be an exaggeration to state that work on these projects was being done on the frontiers of knowledge. The construction of large complex dams involved various aspects which were unprecedent in the world and hence there are many lessons to be learnt, some of which are discussed later.

Construction of Dams in Pakistan was initiated in 1955 when it was facing an acute power shortage. To meet its power requirements, Pakistan in 1947 was buying about 12,000 KW of electricity from India. On supplies being stopped by India, an alternative source had to be immediately found. Therefore, the construction of the Warsak Dam on the Kabul river near Peshawar was undertaken.

During the last 25 years there have been a number of dams built in various parts of Pakistan, out of which the salient features of 24 dams are given in Append.A.

In a basically agricultural country with river flows having large seasonal variations, prudent use of its water is of paramount importance. Though we have many problems relating our potential sites for dams discussed later, but nevertheless dam construction for future is an important aspect linked with the country's development. For future construction of dams to meet the growing demands of irrigation, power and water supply, various potential sites have been identified which are given in Append. L. The more important post-Tarbela schemes are discussed in this report. Some of the important constructed dams and a few under construction are discussed below:

## 2.1 Completed Dams

**Warsak Dam:** Warsak Dam is located on Kabul River about 41 km. downstream of the border with Afghanistan and about 30 km. northwest of Peshawar. It is a multipurpose project for irrigation and generation of power. The minimum residual capacity (after sedimentation) is estimated as 12.3 million cu. m. (10,000 acre feet). The dam is a concrete gravity structure 180 meter long at its crest, and 76 meter high above lowest foundation level. The power plant at Warsak Dam presently has 160 MW installed capacity and operates under an average effective head of 63 meters. The power station houses four generators of 40 MW each. Turbines are fed through a concrete lined power tunnel branching off into six steel lined penstocks of 5.5 meter diameter each. Two more units of 40 MW capacity each are under construction. A circular horse shoe section tunnel provides 14 cu. m/sec (500 cusecs) for irrigation supplies to Kohat Canal.

**Rawal Dam:** Rawal Dam completed in 1962, has been built across Korang River, about 14 km from Rawalpindi. Apart from irrigating an area of about 4,860 hectares in the vicinity of Rawalpindi, the Rawal Dam reservoir provides the much needed potable water supply of 90 million liters (20 million gallons) daily to Rawalpindi and Islamabad. The project has also created an artificial lake near the Federal Capital which has developed into a very attractive tourist resort.

The dam is masonry type, 213 meter (700 ft) long, 34.5 meter (113 ft) high from the river bed and has a storage capacity of 58.6 million cu. m. (47,500 acre feet). It was the first large masonry dam built in Pakistan, designed and constructed by Pakistani engineers and contractors without any external help.

The rocks exposed in the area consist of alternate beds of sandstone and claystone of Murree Formation of Miocene age.

Tanda Dam: Tanda Dam Project is located about 9.6 km. (6 miles) from Kohat city in North West Frontier Province. The project consists of an earthen embankment which is 35 m (115 feet) high and 670 m (2,198 feet) long. The reservoir which has a gross capacity of 99 million cu.m. (80,260 acre feet) is fed by the flood waters from a small barrage on Kohat Toi through a tunnel and a lined canal.

To provide the irrigation supply, which is the main purpose of the project, a pressure tunnel has been constructed which drains the water into the canal system to irrigate about 12,950 hectares (32,000 acres) of land. The canal system constructed for this purpose is 90 km (56 miles) long.

Mangla Dam: Mangla Dam completed in June 1967, was built across the Jhelum river. Mangla Dam is a multipurpose project and a part of the Indus Basin Settlement Plan. The project includes Mangla Dam across Jhelum River, Sukian Dyke on the south-eastern periphery reservoir and Jari Dam across Jari nullah. All these dams are rolled fill embankments with central impervious core. General layout and typical section of main Dam is shown in Append. E. The broad features of these dams are as follows:

The project has ultimate power generating capacity of 1,000 MW of which 600 MW is already in operation and 2 more units of 100 MW each are under construction and expected to be commissioned by middle of 1981.

The reservoir has a gross storage capacity of 7.2 B.cu.m. (5.88 MAF) and the live storage of 6.6 B.cu.m (5.34 MAF). The life of the reservoir has been estimated more than 70 years. In order to check and reduce deposition of sediments in the reservoir, a long term watershed management project in the catchment areas extending over an area of 8,466 sq.km. (3,269 sq. miles) is under execution. The bedrock in the Mangla area belongs to Siwalik Formation of Tertiary Period, consisting of alternating beds of weakly cemented sandstones, siltstones and over-consolidated fissured clay.

Over 1000 instruments comprising electrical and hydraulic piezometers, total pressure cells, slope indicators, plumb-lines and settlement gauges have been installed in the foundation bedrock, body of earthfill, cut slopes and below concrete structures. In addition to these instruments, seepage through the drainage networks provided under the Embankments and Main Spillway is also measured and analysed chemically for

| Name | Max. Height | | Crest Length | | Fill | |
|---|---|---|---|---|---|---|
| | (m) | (ft) | (m) | (ft) | (M.C.m) | (M.C.yds) |
| Mangla Dam (Main and Intake Embankment) | 115.8 | 380 | 3,140 | 10,300 | 65.0 | 85.0 |
| Sukian Dyke | 44.0 | 144 | 5,152 | 16,900 | 9.5 | 12.5 |
| Jari Dam and Rim Works | 83.5 | 274 | 4,421 | 14,500 | 32.5 | 42.5 |

There are also two spillways - the Main Spillway and the Emergency Spillway. The Main Spillway and Emergency Spillway are capable of discharging 24,635 cu.m/sec (870,000 cfs) and 6,513 cu.m/sec (230,000 cfs) of water respectively.

soluble and insoluble salts at some locations.

Chashma Barrage: A low dam called Chashma Barrage is located on river Indus about 56 km. (35 miles) downstream of Jinnah Barrage and Kalabagh town. The main concrete structure is 1084 meters (3555 ft) long with a flood discharge capacity of 26,900 cu.m/sec. (950,000 cusecs). The Barrage has a storage capacity of 1.07 B.cu.m (0.87 MAF) which is used during winter months. Supplies from the river Indus including releases from Tarbela Reservoir are diverted by Chashma Barrage into Chashma-Jhelum Link Canal for ultimate use in the area to be served by the Trimmu-Sidhnai-Mailsi-Bahawal Link Canals System and Haveli and Rangpur Canals. The Chashma reservoir is also used for efficient re-regulation of Tarbela and Warsak releases.

Tarbela Dam: The multipurpose Tarbela Dam Project has been constructed on the Indus river about 80 km. (50 miles) north west of Rawalpindi-Islamabad. Tarbela Dam is one of the two storage reservoirs included in the Indus Basin Settlement Plan. The reservoir created by the dam has a gross storage capacity of 13.7 B.cu.m. (11.1 MAF) and a net usable capacity of 11.5 B.cu.m. (9.3 MAF). In addition, project has also an ultimate power generating capacity of 2,100 MW. A significant feature of Tarbela Dam will be its future role in creating silt free off-channel storages of about 49.3 B.cu.m. (40 MAF) in the adjoining side valleys of Haro and Soan Basins discussed in the chapter on future developments.

The principal element of the project is the Main Embankment Dam 2,743 meters (9000 ft) long with a maximum height of 143 meters (470 ft). The Main Embankment involved 105 million cu.m. of fill (138 million cu.yd), which makes it the largest earthfill dam in the world. There are also two Auxiliary Dams on the left bank of the river 710 meters (2,340 ft) and 261 meters (860 ft) long. Auxiliary Dam No.1 involved 15 million cu.m (20 million cu.yd) and Auxiliary Dam No.2 involved 1.5 million cu.m. (2 million cu.yd) of fill material. In addition 16 million cu.m. (21 million cu.yd) of fill was also used for upstream impervious blanket. The Main Dam rests on a foundation of alluvial gravel having a maximum depth of about 229 meters (750 ft). An impervious blanket, continuous with the core of the dam covers the river bed for more than 1.5 km (one mile) to restrict the underflow through the permeable gravel foundations. General layout of the project and typical section of Main Embankment Dam is shown in Append. F. A line of drainage wells is provided at the downstream of the dam to collect the seepage which finds its way under the blanket and dam.

There are two gate-controlled Spillways on the left bank which discharge into a large channel excavated along the natural side valley joining the main river channel downstream of the dam. The total spillways capacity is about 42,475 cu.m/sec (1,500,000 cusecs).

Four tunnels varying in length from 730 to 820 meters (2,400 to 2,700 ft) have been constructed through the right abutment rocks. Upstream of the main gates at mid-tunnel the four tunnels are concrete-lined with a common diameter of 13.7 meter (45 ft). Downstream they are steel-lined with a diameter of 13.3 meters (43.5 ft) for tunnels 1, 2 and 3 and 11 meters (36 ft) for tunnel 4. The steel linings of tunnels 3 and 4 extend as penstocks, each of which then divide in a massive Y-branch into two rectangular conduits; the flow from each is controlled by 4.9 x 7.3 meters (16 ft x 24 ft) radial gate. The discharge arrangements on tunnels 3 and 4 are designed to pass through stilling basins a total of 3,851 cu.m/sec (136,000 cfs) at low reservoir and correspondingly larger discharges at higher heads. During the last phase of construction of Dam the two tunnels 1, 2 were used for the diversion of river. The tunnels 1, 2 and 3 will be used for power generations while tunnel 4 will be used for irrigation releases. The power installation will ultimately, consist of 12 units of 175 MW capacity each for a total of 2,100 MW. Each of the three tunnels devoted to power will serve a group of four units. Four units at tunnel 1 are already in operation while four units at tunnel 2 are under construction.

A fifth tunnel on the left bank has also been constructed between the two spillways to release irrigation water at low level. This tunnel is 910 meter long and 11.0 meters diameter, and is also steel lined in the downstream portion. There were two main features in this tunnel, different from right bank tunnels. Firstly, the energy dissipation at outlet of this tunnel is through a flip bucket, while the tunnels 3 & 4 on right bank had a stilling basin. It was learnt through the successful operation of this tunnel that flip bucket worked as a much better system for energy dissipation. Another special feature of this tunnel was that it was constructed while the reservoir was being filled. Therefore a comprehensive drainage umbrella was designed over the tunnel to ensure its timely completion.

Hub Dam Project: Hub Dam has been constructed on Hub River about 56 km. northeast of Karachi. The project is to irrigate about 12,545 hectares (31,000 acres) of land in Karachi and Lasbela districts in addition to supplying 513 million liters (114 million gallons) water per day for domestic and industrial purposes.

The main features of project include Earthen Embankment 8,007 m (26,264 ft) long and with a maximum height of 46 m (151 ft) (Append. G), a spillway of free overfall and an irrigation outlet 1.82 m (6 ft) dia, 141.4 m (464 ft) long with a design discharge of 14.15 cu.m/sec. (500 cfs). The gross storage of the reservoir which extends on the surface area of 98.5 sq.km (38 sq. miles) is 1,139.8 million cu.m (924,000 acre ft).

2.2 Dams Under Construction

There are at present three dams under construction. Khanpur Dam and Simly Dam apart from the purpose of irrigation, would be meeting the water supply demands of rapidly growing Federal Capital region of Rawalpindi and Islamabad. The third, Bolan Dam near Quetta is basically an irrigation project which is being rebuilt as the original dam failed in 1976.

Khanpur Dam: Khanpur Dam is under construction across the river Haro, a left bank tributary of River Indus, about 14 km. (9 miles) from the town of Taxila. The project is proposed to irrigate about 14,974 hectares (37,000 acres) of land in Hazara and Campbellpur districts in addition to the provision of supply of 590 million liters (131 million gallons) of water daily to capital city of Islamabad, Rawalpindi, Wah Cantt. and Taxila for domestic and industrial uses.

The gross storage capacity of the reservoir is 132 million cu.m. (107,000 acre ft) with a live storage of 112 million cu. m. (91,000 acre ft). The principal feature of the project include a Main Embankment, a Spillway, three Saddle Embankments, a diversion tunnel and a system of irrigation canals. The main dam is 51 m (167 ft) high, 471 m (1,546 ft) long, earth and rock fill type with an impervious upstream blanket (Append. H). The three numbers saddle embankments with maximum height of 16.5 m (54 ft) and total crest length of 1,270 m (4,166 ft) also involve 657,800 cu.m. (23.2 million cu.ft) of earthfill. The spillway constructed has a design discharge capacity of 4,700 cu.m/sec (166,000 cfs). The spillway is chute type with Ogee crest and flip bucket and provided with 5 Nos. radial gates. A tunnel 3.05 m (10 ft) diameter, horse shoe shape and 518 m long has been constructed to divert river water during the final closure of the river channel. An irrigation outlet of cut and cover conduit type has been provided with a design discharge of 18.5 cu.m/sec. (650 cfs). A system of irrigation canals has also been constructed to convey water for irrigation and other purposes.

The rocks exposed in the project area consist of alternate beds of limestone and shales of Eocene age.

Simly Dam: Simly Dam is being constructed about 39 km north-east of capital city of Islamabad across Soan River which originates from Murree Hills. The gorge at damsite controls a drainage area of 153 sq.km (59 sq. miles) and drains about

80.2 million cu.m. (65,000 acre ft) of water per year. The essential purpose of the project is to supply water to fast developing capital of Islamabad. The reservoir will store the flood as well as the perennial flows of the Soan river. Water will be conveyed to Islmabad through twin conduction pipes after treating it by a filtration plant installed near the dam. When completed the dam would provide 108 million liters (24 million gallons) per day of drinking water to Islamabad. The reservoir created by the structure will have a gross storage capacity of 35.5 million cu.m. (28,750 acre ft) and a net live storage of 24.7 million cu.m (20,000 acre ft). The principal element of the project is an embankment, 308 m (1,010 ft) long with a maximum height of 76 m (250 ft). It will have a central impervious clay core with upstream and downstream rockfill shells (Append.J).

An overflow type concrete spillway is located at the left abutment. The flow over the weir will be carried to the lower channel by a steep chute, and would be discharged into the river through a double stilling basin arrangement. A 1.83 m (6 ft) diameter tunnel and pipeline outlet system will be located on the left abutment between the spillway and the dam, which will be provided for the release of water.

Rock formations at the damsite comprise of massive sandstone interbedded with claystone and siltstone beds. The rocks dip very steeply on the downstream and are intersected by three prominent joint systems.

Bolan Dam: Bolan Dam was originally constructed in 1958 across Bolan River about 60 km from Sibbi in Baluchistan. The dam remained operational till September 1976 when it failed due to overtopping. Most of the embankment dam has been breached and eroded away. The main reason for this failure was the non-existence of an adequate spillway. The Government of Baluchistan is now reconstructing the dam to rehabilitate and improve upon a facility which was meeting the irrigation requirement of the area.

The gross storage proposed to be created by the Bolan Dam will be 92.5 million cu.m. (75,000 acre ft), with a live storage of 64.75 million cu.m. (52,500 acre ft.) The Bolan Dam entails the construction of a main embankment 20 meter (66 ft) high and 510 meter (1,673 ft) long across Bolan river gorge, and a saddle embankment to fill up a 135 m (443 ft) long low lying gap in the reservoir periphery. The main body of these embankments will consist of the compacted homogeneous fill of sandy silt and clayey silt. The layout and typical section of main embankment is shown in Append. K.

An open cut spillway with an overflow section of 210 m (690 ft) width has been provided on the right flank. The spillway structure has been designed to cater for a discharge of 2,124 cu.m/sec (75,000 cfs), which results from an extreme flood of 13 hours duration and a peak of 2,832 cu.m/sec (100,000 cfs).

The rocks at the damsite and spillway consist of alternate beds of claystone/siltstone and sandstone which belong to Sibbi Group ranging in age from Oligocene to Pliocene. The rocks are generally moderately jointed to blocky and joints are generally filled with gypsum or clay.

2.3 Dam Monitoring

In order to ensure the safety of structures a Dams Monitoring Organization (D.M.O) under WAPDA has been set up by the Federal Government. Monitoring of dams and reservoirs operated and maintained by WAPDA carries high priority for purpose of safety. Behavioural monitoring is achieved through analysis and interpretation of piezometric, seepage, movement and seismic data. Long term studies are carried out to analyse stability of structures and their foundations. Measures are evolved for enhancing the service life of the structures. Investigation programmes are framed to study and analyse programmes. Remedial measures are evolved and their execution is supervised by an experienced staff. Studies are also conducted for rehabilitation and increasing the height of dams. D.M.O. also arranges independent inspections through a panel of engineers and geologists where required to tackle complex problems.

The responsibilities of dams monitoring organization extend to:
1. Analysis and interpretation of surveillance data including pore pressures, seepage and structural movements etc.
2. Study of seismic records and review of design parameters.
3. Diagnosis of any problems and advice and assistance in implementation of remedial and rehabilitation work.
4. Review of instrument observation schedule and data collection.
5. Arranging and rendering assistance for independent periodic inspections.
6. Making available consolidated data description and background histories of specific problems and longterm studies of various phenomena.

Dam Monitoring Organization of WAPDA has established a programme of yearly as well as periodic inspections of the major dams, including those which are under construction.

## 3 FUTURE DEVELOPMENT SCHEMES AND POTENTIAL RESERVOIR SITES

### 3.1 Future Developments

Development of surface water resources for purpose of irrigation and power is of primary concern to Pakistan, whose economy is based essentially on agricultural prosperity and ultimate self-sufficiency. The Master Planning Organization of WAPDA prepares a comprehensive plan for coordinated development and utilization of water and power resources of Pakistan on a unified and multi-purpose basis. It provides the direction for subsequent development activities in the water and power sectors in the light of changing socio-economic conditions, requirements and priorities.

The implementation of the two billion U.S. Dollars Indus Basin Replacement works have ensured the irrigation supplies to those parts of Pakistan which have in the past depended on the flows in Ravi, Beas and Sutlej. The total measured discharge entering the Indus Plains in Pakistan and India averages about 215 B.cu.m (175 MAF) per year. At present, about 206 B.cu.m (167 MAF) enters Pakistan. This includes the 30.8 B.cu.m (25 MAF) of the Ravi and the Sutlej which can be completely diverted. Thus the average annual flow assured for Pakistan remains to be 175 B.cu.m (142 MAF) from Indus, Kabul, Jhelum and Chenab (Append. D).

The average annual river diversions which were 80.2 B.cu.m (65 MAF) in 1947, increased to 105 B.cu.m (85 MAF) in 1970 through Indus Basin Barrages and Link Canals. After the completion of Tarbela Dam they increased to 117.2 B.cu.m (95 MAF).

Considering the requirements of irrigation water after the completion of Tarbela, there still remains a short fall in the winter season. This short fall is likely to increase to about 10 B.cu.m (8 MAF) by the year 1990. Therefore by 1990 a storage of about 10 B.cu.m would be required. In Pakistan more than 80% of the population is dependent on agriculture. The population of the country is likely to touch 200 million mark by the year 2000. This expected population increase would put a tremendous strain on country's food and fibre resources. To cope with the situation there is need for developing more storage of water to meet the requirements in the scarce water season.

According to the forecast for electric power the requirements are to rise from 6,197 MW in 1980 to 7,937 MW required in 1990, which has also to be generated to keep pace with the development. Pakistan has ample natural gas resources, which have been previously used extensively as a fuel for thermal power stations. However, with the rising costs of fossil fuels, the hydroelectric plants are expected to be much more economical, therefore more emphasis is anticipated, in the future, on construction of dams for hydroelectric power. The list of the future development projects is given below, followed by a brief discussion on some of these projects:
  Indus River - Main Stem Storage
1. Skardu Dam, 2. Bunji/Chilas Dam
3. Khaplu Dam and 4. Kalabagh Dam

Indus River - Off-Channel Storages:
1. Ghariala Dam, 2. Sanjwal Akhori Dam
3. Dhok Pathan Dam, 4. Dhok Abbaki
Pumped Storage and 5. Thal Reservoir

Jhelum River: 1. Raised Mangla Dam and 2. Kohala Project

Kunhar River: Kunhar Storage Projects

Lower Swat River: 1. Munda Dam, 2. Kalangai Dam, 3. Khazana Dam, 4. Ambahar Dam and 5. Bazargai Dam

3.2 Description of Post Tarbela Schemes

For large scale storage development on the Indus after Tarbela there appear to be four main alternatives; Kalabagh, Off-Channel Storage, Upper Indus Sites and Thal Offstream Storage Scheme. The alternative schemes are discussed below:

I. Indus River Storages

a. Storage Along the Main Stem

Out of the various sites identified the most promising storages on the Indus Main Stem are Skardu and Kalabagh Dam Projects.

i. Skardu Dam Project

A promising reservoir site in the Upper Indus Basin is at Skardu at an elevation of about 2,135 meters (7000 ft) above mean sea level. Access to the area is extremely difficult.

On the basis of discharge measurements it has been estimated that the annual discharge of the river at damsite is in the order of 43 B.cu.m. (35 MAF). No records of sediment load of the river at Skardu are available. Two heights of the dam have been considered, one of 80 meters (260 ft) to impound 6.5 B.cu.m. (5.2 MAF) of water, the other of 95 meters (310 ft) to impound 10 B.cu.m. (8.0 MAF).

In view of the above, dam at or near Skardu is generally attractive and the development of the site is potentially promising for power, at least for part of the year. Access and severe temperature conditions will continue to present a formidable problem and considerable further investigations will be required before any specific proposals can be formulated.

ii. Kalabagh Dam Project

The Kalabagh Dam Project envisages the creation of a storage reservoir on the Indus main stem and its tributary Soan by construction of an earthfill dam across river Indus just below confluence of these two rivers. The feasibility report of the Kalabagh Dam Project has already been prepared, and the project planning and detailed design work is about to commence. The major features of the project as envisaged in the feasibility report are main dam, auxiliary dam, sluiceway, power facilities and auxiliary spillway. The dam is proposed to be 81 meters (265 ft) high zoned embankment, which will create reservoir with gross capacity of 11.5 B.cu.m. (9.3 MAF), firm and secondary power capacity each of 880 MW, a total of 1,760 MW.

b. Off Channel Storage

The possibility of diverting water from Tarbela to storage in either the Soan or the Haro river valleys on the left side of the Indus has for some time been seen as additional alteration of Tarbela site. It has been suggested that storage capacity of the order of 37 B.cu.m (30 MAF) might be built on these rivers, to be filled almost entirely by directed Indus flows. The four potential sites are discussed briefly below:

i. Ghariala Dam

Ghariala Dam site is located on the Haro River a few kilometers upstream of its outfall into Indus River. With an earthfill dam about 114 meters (375 ft) high it can create a reservoir of about 10 B.cu.m. (8 MAF) live capacity. Most of the water to fill the reservoir would be diverted from the upper level of Tarbela reservoir through a conveyance system extending across the divide between the Siran arm of Tarbela and Jobikas tributary of the Haro. The conveyance system about 8 km. (5 miles) long would have a design capacity of 2,150 cu.m/sec (76,000 cusecs). There are some adverse foundation problems connected with the site. The submergence of Campbellpur town is a major factor to be reckoned with, before this project is further considered.

ii. Sanjwal-Akhori Dam

This is an alternative to Low Ghariala. Reservoir would be formed by two dams one near Sanjwal on the Haro River and the other at Akhori on the Nandnakas, a tributary of the Haro. The storage created behind the two dams would have a live capacity of 4 B.cu.m. (3.3 MAF). It would also require a water conveyance system having the same location and arrangement of structures as for Ghariala Dam Project but a flow of 906 cu.m./sec (32,000 cusecs) would be required from Tarbela during the filling season.

iii. Dhok Pathan and Dhok Abaki Dams

Dhok Pathan and Dhok Abaki sites close together on the Soan River would have dams similar to Ghariala and Sanjwal-Akhori. Dhok Pathan is probably much better site for storage of water diverted by gravity from Tarbela, whereas Dhok Abaki, 11 kms (7 miles) downstream, might be preferable as a pumped storage project in connection with the reservoir at Kalabagh. The dam envisaged for Dhok Pathan would be an earth and rockfill structure some 84 meters (275 ft) high with a crest length of about 3,660 meter (12,000 ft) containing 29 million cu.m. (38 million cu.yds.) of fill. The reservoir would provide a usable live storage capacity of about 9.2 B.cu.m. (7.5 MAF).

The Dhok Abaki scheme, in conjunction with Kalabagh, would provide about the same storage as Dhok Pathan and it would be cheaper, as it will not involve the long conveyance system.

Water would be conveyed from Tarbela to the Dhok Pathan reservoir by a 112 km (70 miles) long conveyance system requiring canals with a combined capacity of 2,150 cu.m/sec (76,000 cusecs). A supplementary dam, 72 meters (235 ft) high would have to be built at Bahtar where conveyance system crosses the Nandnakas. Maintenance cost on the long conveyance system would be high because it would be used only for some three months in the year. Only reconnaissance level studies have so far been carried out. Detailed feasibility is planned at a later date.

iv. Thal Reservoir

This scheme, consisting of an enormous shallow reservoir of some 26 B.cu.m. (21 MAF) gross storage capacity formed by a long dyke around an area of poor agricultural potential on the left bank of the Indus in the upper part of Thal Doab. Though the live storage capacity of the reservoir could be as much as 26 B. cu.m. (21 MAF), the actual yield of the reservoir would be substantially less due to high seepage and evaporation losses. A smaller scheme of storing about 2.8 B. cu.m. (2.3 MAF) is also being considered for this site. The project is in a preliminary planning stage and detailed feasibility is still to be carried out.

II. Jhelum River

1. Raised Mangla Project

Mangla Dam as completed has gross storage capacity of 8.2. B.cu.m (5.88 MAF) and a live storage capacity of 6.6 B. cu.m (5.34 MAF). Provision has been made for raising the maximum reservoir level by 15 meter (50 ft). This will increase the live storage capacity by about 4.3 B.cu.m (3.5 MAF). Raising the reservoir level by 15 meter (50 ft) would result in an increase of about 600 MW in the firm capability of power plant and provide additional generation of about 2000 million Kwh per annum. The project is still in preliminary planning stage.

2. Kohala Project

The project is located on river Jhelum upstream of Muzaffarabad in Azad Kashmir. The river Jhelum makes a big loop around Domel and it is proposed to connect the two limbs of the river loop through eleven miles long tunnel, thereby utilizing the 195 meters (920 ft) head of water for power generation. Theoretically, maximum power generation capability of the project will be 3760 MW during high flow in April, falling to a minimum of 306 MW in December.

III. Kunhar River

The project is located in Kaghan Valley some 64 km (40 miles) north of Muzaffarabad. The project would consist of 2 reservoirs namely the Naran Dam and the Suki Kinyari Dam. The project has generating capacity of about 500 MW. It has, however, very little storage potential. The scheme is in reconnaissance stage.

IV. Swat River

Lower Swat Gorge Projects

Development possibilities in the lower Swat Gorge hold sufficient promises for storages of surplus flows for irrigation and generation of hydro-electric power. After making an allowance for anticipated Swat River Irrigation

Development it was estimated that an average annual quantum of about 2.5 to 3.0 B.cu.m (2.00 - 2.5 MAF) could be available for storages in the Lower Swat Gorge. The potential sites in the Lower Swat Gorge are High Munda, Kalangai Dam, Khazana Dam, Ambahar and Bazargai Dam sites. The various alternatives proposed are High Munda 1.8 B.cu.m (1.5 MAF), Kalangai Dam 7.4 B.cu.m (6.0 MAF), Low Kalangai 4.3 B.cu.m (3.5 MAF), Khazana Dam 3.0 B.cu.m (2.5 MAF), Ambahar 9.2 B.cu.m (7.5 MAF) and Low Munda 0.037 B.cu.m (0.03 MAF) and Bazargai 9.2 B.cu.m (7.5 MAF).

### 3.3 Immediate Need

In order to fulfil the additional water and power requirements, we should have in service by the year 1990, additional surface water storage to provide a minimum capacity of 8.6 to 10 B.cu.m (7 to 8 MAF) and additional power capacity of 1078 MW by 1985. Off channel storage dams are not being considered at present, these would probably become more important in the future when the Tarbela reservoir would be silted up. Skardu Dam cannot be constructed at present because of the accessibility on the site. Reservoirs in Swat Valley would be too small. Therefore, the obvious choice for a second stage storage lies with the Kalabagh Dam Project. This project would not only provide a live storage of 9.5 B.cu.m (7.7 MAF) but would also generate 1760 MW of electric energy. Dam on the Indus at Kalabagh would be the second major storage on the main stem.

Additional 11.5 B.cu.m (9.3 MAF) water storage available would considerably help to boost the agricultural production particularly during the Rabi season during which the river flows are very low. The project would also generate a substantial amount of electricity; 880 MW of firm power and 880 MW of secondary power which will constitute a most significant, effective and economical means of overcoming the power shortage. The menace of floods would also be greatly moderated by the Kalabagh reservoir.

The principal features of the Kalabagh zoned earthen embankment dam would be as follows:

| | |
|---|---|
| Height of main dam above average river bed | 81 m (265 ft) |
| Number of power units | 8 |
| Capacity of each power unit | 220 MW |
| Firm Power | 880 MW |
| Secondary Power | 880 MW |
| Capacity at full Reservoir level | 11.56 B.cu.m (9.375 MAF) |
| Live Storage | 9.58 B.cu.m (7.771 MAF) |

### 3.4 Other Potential Dam Sites

Numerous other potential dam sites have been identified in various regions of Pakistan given in Append. L.

There is a general scarcity of water in the Baluchistan province of Pakistan. A number of multipurpose projects are in the various stages of planning and investigation. This includes the Mirani Dam for Water supply to the Coastal area and three schemes near Sibi; Baber Kach in the Kachi plain for irrigation, power and flood control, Tali Tangi and Beji also in Kachi Plain for irrigation and flood control. Another scheme is the Akra Kaur Dam project situated near Gwadar in the south of Baluchistan which is in the feasibility stage.

There is a tremendous amount of potential for hydel power generation in the northern mountainous area particularly along the Indus. The areas are at present sparsely populated and there is problem of accessibility in the rugged mountainous regions. Therefore construction of dams in these remote areas and the transmission of power over long route is for the present extremely uneconomical. However, there are a number of other small dam schemes which are in the planning and investigation stage.

## 4. PROBLEMS OF DAM ENGINEERING IN PAKISTAN

Some of the problems faced in construction of large dams in Pakistan were unprecedented and threatened the very completion of the projects. Many aspects of Tarbela were on, or beyond the boundaries of present experiences which sometimes required untried techniques in solving the problems associated with the features. However, they have been successfully tackled and solutions found. Today it can be said with satisfaction that a substantial contribution has been made to the dam building technology.

Dams have very significantly contributed towards the economy of Pakistan and have been found to be quite feasible projects. However, there are certain difficulties in the construction of dams, peculiar to Pakistan, due to which not many dams were built in the country in the past. In this chapter some of the problems of dam engineering in Pakistan are being discussed.

## 4.1 Unsuitable Topography

There are a number of factors accounting for the costly nature of surface water storage in Pakistan, which is a land of alluvial plains with hills and mountainous regions on the north and north westerly areas. Major power and irrigation needs are in the plains whereas most of the ideal damsites are present in the sparsely populated mountainous regions. Sites with potential for construction of dams, having suitable storage areas are usually located in broad valleys. The valley topography is generally such that the length and height of the dams has to be unusually large relative to the storage capacity created. Presented below are some of the main statistics regarding the two major projects of Tarbela and Mangla in order to assess the magnitude of construction work and capacity of storage created:

Principal Features - Mangla and Tarbela Projects

|  |  | Mangla | Tarbela |
|---|---|---|---|
| Maximum Initial Live Storage | (B.cu.m) | 6.6 | 11.5 |
| Length of Reservoir | (km) | 64 | 80 |
| Site Excavation | (M.cu.m) | 42 | 73 |
| Total Earth and Rockfill | (M.cu.m) | 107 | 136 |
| Main Dam: Earth and Rockfill | (M.cu.m) | 65 | 121 |
| Maximum Height | (meters) | 116 | 143 |
| Crest Length | (meters) | 3140 | 2743 |

It is because of unsuitable topography that long rim embankments had to be built for Mangla which include 580 meter long intake embankment, 5,152 meter long Sukian Dyke and 4,421 meter long Jari and Kakra embankments. Tarbela Dam as well required the construction of two auxiliary embankments on its left rim.

The tunnels at Tarbela had to be squeezed in the right abutment because of unfavourable topography and thus leaving insufficient space in the outlet area for adequate geometry of energy dissipation structures.

Similarly for Hub Dam near Karachi it was because of unsuitable topography that 8,007 meter long embankment had to be constructed for a relatively small storage of 1,139 million cu.m.

## 4.2 Floods

Apart from the size of the valleys, another major factor accounting for the high costs of surface water storage is the large amount of spillway capacity that has to be provided. High floods also have to be catered for during construction of dams. Therefore, the river diversion during construction constitutes a substantial part in the cost of construction of dams in Pakistan. The cost of spillway represents a a very significant item of total cost; for instance at Tarbela and Mangla the spillway constitutes about 20 percent of the main civil engineering contract. Floods and spillway capacities at Mangla and Tarbela are tabulated below:

|  | Mangla Jhelum | Tarbela Indus |
|---|---|---|
| Maximum Flood of Record cu.m/sec (cfs) | 28,320 (1,000,000) | 24,780 (875,000) |
| Date | Aug. 1929 | Aug. 1929 |
| Design Floods cu.m/sec (cfs) | 73,630 (2,600,000) | 60,230 (2,127,000) |
| Spillway Capacity total cu.m/sec (cfs) | 31,148 (1,100,000) | 42,475 (1,500,000) |
| Main Spillway cu.m/sec (cfs) | 24,635 (870,000) | 17,410 (615,000) |
| Auxiliary Spillway cu.m/sec (cfs) | 6,513 (230,000) | 22,510 (795,000) |

At Tarbela the main component of the design flood is the maximum probable monsoon storm, but substantial allowances had also to be made for late snowmelt runoff from the Himalayas and for the possibility of a natural dam break upstream. Natural dams have formed fairly frequently in the past on the Upper Indus and its tributaries, due to glacial movement of avalanches. The flood of August 1929 on the Indus, the highest ever recorded, occurred in conjunction with the breaking of a natural dam on the Shyok River.

In the case of under construction Khanpur Dam on Haro river, the spillway capacity provided is for a discharge of 4,700 cu.m/sec (166,000 cusecs). Though the record for about twenty past years shows that the discharge has never exceeded 1130 cu.m/sec (40,000 cusecs). Catchment area and rainfall characteristics are prone to produce flash flood in the Haro.

Floods in themselves are a great menace in Pakistan. We have no suitable storage site on Sutlej, Ravi and Chenab where floods can be controlled. On Jhelum there is a small 2 MAF super storage capacity available at Mangla which is barely sufficient to shave off the flood peaks. On Indus at Tarbela there is very little super storage capacity. The high cost of construction of these dams does not allow to build super storage capacity for the purpose of averting floods only. Floods have, therefore, caused great havoc in the country periodically, and other solutions are being planned to combat them.

## 4.3 Sedimentation

Another noteworthy factor accounting for the high cost of water storage schemes on the Indus river system is the siltation that is expected to take place in the reservoir and consequently rather rapid depletion of live storage capacity. For instance, average annual sediment load of the Indus at Tarbela is estimated at 350 million short tons and almost all of it is borne by the summer flood flows. The sediment load can reach 10 million tons a day. Not much is known about its origin, but it appears that some of it is due to current glacial action and avalanches in the upper reaches of the river and its tributaries while some is due to landslides and river scouring in the vast piles of debris built up in the lower parts of the Upper Indus by past glacial movement and silt deposition. The Jhelum also has a sizeable sediment load, although much smaller (about 110 million short tons per year) than the Indus. However, the sediment load on the Jhelum appears to be the result of human activities to a much greater extent than that on the Indus, and it has been estimated that about 30 percent of it could be eliminated by conservation measures and erosion control. Sediment on the Indus, being apparently much more the result of geological forces, may be harder to reduce.

Thus planning to date has had to be carried out on the assumption that an average of about 110 million tons of sediment will be deposited in Mangla Reservoir each year and about 350 million tons in Tarbela Reservoir. The heavy silt load of the water does, of course, have other implications besides those for the life of storage dams. It will cause problems of abrasion, for instance, of uncertain severity, on the spillways, tunnels and turbine blades, and this will require added maintenance attention. These factors, the designs adopted to minimize problems of abrasion and the limited life that is anticipated for storage projects on the main stem of the Indus, do add to the cost of stored water.

Warsak dam was silted up in the first few years. Thereafter the silt laden water started causing excessive damage to the rotary blades of the turbines, thus considerably increasing the maintenance cost.

## 4.4 Foundations

Foundation conditions encountered in most of the dams constructed in Pakistan have posed considerable problems. Broadly speaking the foundation problems can be attributed to three main geological phenomena.

The outer Himalayas and the Potwar plateau where most of damsite are located in Pakistan, are mainly formed by the soft Siwalik rocks composed of overconsolidated clays and sandstones with lenses of gravels. These weak rocks, as foundations for large structures, have caused complex problems which are difficult to assess as discussed later. Secondly the rivers have cut deep gorges and backfilled them with alluvium. The hetrogeneity of these alluvial deposits and especially the open work gravels within them have caused problems of unpredictable underseepage even after thorough investigations. Thirdly, the region of Pakistan has undergone immense amount of tectonic disturbance which has continued to recent geologic times. This factor has transformed hard igneous and metamorphic rocks of sites like Tarbela into a mass of highly discontinuous and pulverised rocks, causing unprecedented problems. The adverse geologic conditions encountered at most of the damsites have resulted in extensive investigations and costly foundation treatment measures. Some of these problems on major dams are discussed below:

a) Foundations at Mangla:

Mangla Dam is built on the soft young sedimentary rocks of Siwaliks age.

Problem of Sheared Clay: The rock formation are mainly of alternating beds of sandstones and clays with subsidiary beds of siltstone. Gravel beds occur in the upper part of the formation which out-crop at the eastern end of Sukian Dam and at Jari. The clays have been well consolidated by the weight of overlying strata but they are fissured and the surface of some of the fissures are slickensided.

In 1963 when foundations of the dams were exposed, a large number of undisturbed block samples were taken from these areas and an extensive programme of sampling and testing was carried out. It was discovered that there are sheared zones in the clays, and that the shear strengths of the clay bed-rocks were lower than had been found from laboratory tests or assumed for the contract design. The design parameters had to be revised and the $\emptyset$ values for bed rock clays were reduced from $32^\circ$ to $28^\circ$ for main Dam and $28^\circ$ to $20^\circ$ for Jari Dam.

The sheared zones are not readily apparent, and are apt to be overlooked, unless one is specifically looking for them. The existence of sheared clays and consequent remedial measures required, increased the cost of the project by over 12 million dollars.

Swelling Clays: Sukian Dyke Mangla is underlain by alternating beds of sandstones and clays. Strike of foundation beds is at varying angles to the dam axis and the sandstones are exposed in a number of downstream nullahs. This set up has produced a number of seepage exits in downstream areas. The seepage monitoring has indicated an increase in the quantity of seepage through the foundations. This is partly attributed to the swelling clays existing in the foundations. The swelling and shrinkage of the clays in relation to reservoir changes has caused the development of fissures in clays near contact zones with sandstones, which has caused the rise in seepage. This problem is being extensively studied and remedial measures are undertaken where required. Swelling has been measured as a maximum net rise in embankment crest of the order of 7.6 cm against an expected settlement of about 30 cm had there been no swelling.

Potential for Liquefaction: The proposed Mirpur dyke was to be built in a low lying saddle of Mangla reservoir rim. It had to be abandoned, because of instability in earthquake and instead a more costly solution had to be found. The silt strata forming the foundation of the proposed dyke were prone to liquefaction under the expected earthquake loading of 0.15g, therefore, the construction had to be dropped and instead two dams totalling 4,420 meter length had to be built at Jari.

b) Foundation Problems at Tarbela:

Various foundation problems were encountered at Tarbela, some of which are discussed below:

Sinkholes in Upstream Blanket: The dam rests on highly pervious alluvium average of about 180 meter thick. A vertical cutoff wall was impractical and use had to be made of an upstream horizontal blanket. Tarbela Dam is more than twice the height of the next largest embankment dam employing an upstream blanket and first to be constructed on such foundations with extensive associations of open work gravel and sand. This led to the problem of development of sinkholes on filling of reservoir, which were revealed when the reservoir was depleted in 1974 due to problems in one of the Tunnels. An elaborate treatment of sinkholes was carried out, and the blanket was also thickened in the areas of sinkhole formation. The repairs of the blanket has continued even after refilling of the reservoir. Further development of sinkholes was monitored by side-scan sonar equipment. Treatment of sinkholes was done by dumping material from bottom-dump barges. Between 1977-78 about 375 sinkholes have been treated in this manner.

The problem of sinkholes in the blanket is more or less resolved, although surveillance and treatment of blanket would probably continue for some years to come.

Seepage through Right Abutment and Embankment Foundation:

Seepage control under adverse geological conditions pose difficult problems. The right abutment at Tarbela has excessive structural discontinuities filled with marly material. The situation is further complicated by the presence of soluble rocks (gypsum), friable rocks (sugary limestone) and caverns. Under these conditions, the grout curtain has been only partially effective and high drainage flows have been encountered with a large content of dissolved material. This condition has resulted in providing additional drainage adits in this abutment.

The foundations of Tarbela are on alluvial deposits of cobble gravel and fine sand with extensive association of open works. This made the foundations highly pervious. As discussed earlier, a positive cut-off was impracticable, therefore, a blanket was provided in an area of about 5 sq. km. upstream of dam. The under seepage was controlled by a system of relief wells at the downstream toe of dam.

The volume of underseepage at Tarbela was much larger than estimated in the design stage. On filling of the reservoir in 1974, it was realized that some of the relief wells were discharging 70 l/s instead of the designed capacity of 28 l/s. Consequently 92 additional wells were installed before 1975 filling, bringing the total wells to 203 to adequately control the underseepage. Over the period from 1974 to date the seepage has gradually reduced from 6400 l/s (226 cusecs) to 3653 l/s (129 cusecs) in 1980. Auxiliary spillway is founded on karstic limestone. An upstream grout curtain in approach channel and a downstream drainage curtain under headworks was provided to control under seepage. Migration of fines from rock joints and gouge material from rock fissures has given problem of piping into the drains of the drainage curtain. All drains are now being replaced by filtered new drains. The upstream grout curtain had also to be strengthened and approach channel provided with concrete cover.

4.5 Seismicity

Another important problem faced in building dams in Pakistan is the hazards of earthquakes. In recent years the environment safety demands have become far more stringent. Dams have to be safeguarded against risk of surface faulting, strong ground shaking, rock slides in reservoir triggered by large shocks etc.

The geographical position of Pakistan happens to be within the syntaxial bends of Himalayan orogenic belt. The belt is well known for its seismic instability having been the locus of four earthquakes exceeding magnitude 8.3 in the past 75 years, two of which are among the greatest ever recorded.

During known history, this part of the world has experienced some disasterous earthquakes, causing a colossal loss of life and property. To mention a few, the historic city of Taxila in north west Pakistan was ruined by an earthquake in about 25 A.D which led to basic changes in the architectural design of the new city. In the current century the destructive earthquakes of Baluchistan and particularly the Quetta earthquake of 1935 are well known, in which 30,000 lives were lost. More recent events were that at Pattan in December 1974 killing about 5,000 people.

The northern and western mountainous areas of Pakistan are still undergoing through a phase of seismic activity. Destructive earthquakes have occurred in this region in the past and are likely to continue in the future. Both Mangla, Tarbela and most of our future sites including Kalabagh are in zones of recent seismic activity.

The proposed Kalabagh dam site also happens to lie in the vicinity of an active fault. This requires the structure to be designed on a high seismic factor, which is likely to increase its cost.

Increasing attention is being paid for realistic evaluation of seismicity. Tarbela Dam project is being extensively monitored by a micro seismic network installed to cover the tectonic features in the critical range of the project. The dam has been designed to withstand an acceleration of

0.15g and has later been checked for an acceleration of 0.25g. Recent seismic studies indicate more stringent values. A dynamic analysis is now underway to check its stability against the designed earthquake. Microseismic networks have been set up for certain other important projects and is also being proposed for Kalabagh project. Therefore, in the years to come Pakistan would have an extensive network covering most of the region so that the hazard from earthquake to our dams and other important structures could be evaluated with greater assurance for the safe design of structures.

## 5. ECONOMIC CONSIDERATIONS

The construction of dams in Pakistan has imparted a tremendous impact on the development of the country which has been briefly dealt in this chapter. For the efficient operation of reservoirs and related facilities in line with modern management techniques Government of Pakistan has established a Water Resource Management Directorate (WRMD) under WAPDA. The basic assignment of this outfit is to plan the reservoir operations through evaluation of the seasonal operational criteria with the object of maximising the benefits to the water and power sectors of the national economy. As a follow-up WRMD has to collect and maintain the record of pertinent reservoir operations and furnish appropriate water accounts to various users. In addition, thorough analyses of the data is carried out to evaluate their effect on the 'Operational Criteria' in respect of reservoir releases, operating levels and power production.

i) Increase in the Irrigation and Hydroelectric Power Supplies:

The construction of dams in the last two decades have played a vital role in building the country's economy. Two major dams of Tarbela and Mangla have been able to provide the water lost because of the diversion of Eastern rivers. The controlled water can be utilized at will to match with the cropping season, which has given the farmer more confidence who can be now sure of his required water supplies. Hydel power available from Warsak, Mangla and Tarbela are meeting 70% of the country's power requirement. The hydel power capacity is being increased by addition of more units to be ultimately able to supply 1770 MW of electricity by 1985.

The judicious use of water has significantly changed the water supply pattern to the area within the commands of Mangla and Tarbela. Though essential commitment of the storage for replacement so far has come in the way of significant increase in the total water supply, distribution pattern of water over the cropping seasons has definitely improved by applying the required quantity of irrigation water. The farmer is prepared to invest for better seeds, fertilizers, mechanization and modern methods of farming, thus having a favourable impact on the agricultural production of the country.

Benefits from Mangla Dam: Mangla Dam has been in operation since 1967. Benefits from storage releases have been at the rate of R$ 65.00 per acre-foot (1233 cu. m) only from power generation at the rate of R$ 0.10 upto June, 1978. Some idea of the magnitude of the benefits derived from its operation can be had from the TABLE 'A' appearing on the next page.

Benefits from Tarbela Dam: The development benefits of Tarbela to the Pakistan economy have been most noticeable in the power sector where WAPDA reports that it has already contributed over 9 billion killowatt hours worth about U.S. $ 375 million in terms of savings in imported oil at contemporary oil prices during the period. With the present installed capacity of 700 MW the average annual benefits in terms of imported oil savings are estimated to be around U.S. $ 325 million per annum, and U.S. $ 760 million per annum when the total rated capacity reaches the planned 2100 MW. These savings are based on the average cost of fuel for thermal generation in northern Pakistan of 6.5 U.S. cents per kwh calculated from January 1980 prices.

In addition there are further savings to the economy measured by the cost of installing equivalent thermal generating capacity, conservatively estimated (on the basis of a recent cost study in another country) at U.S. $ 400 million.

The development benefits of Tarbela in the agricultural sector would increase manifold with modifications in the water supply system.

## TABLE 'A'

| YEAR | STORAGE RELEASES | | | GENERATION | | CUMMULATIVE |
|---|---|---|---|---|---|---|
| | B.cu.m | (MAF) | Rs in Million | Million KWH | Rs in Million | Rs in Million |
| 1966-67 | 0.3 | (0.24) | 15.64 | - | - | 15.64 |
| 1967-68 | 5.65 | (4.58) | 297.83 | 1180.30 | 59.02 | 356.85 |
| 1968-69 | 6.0 | (4.85) | 315.72 | 1478.59 | 73.93 | 389.65 |
| 1969-70 | 6.65 | (5.39) | 350.59 | 1668.24 | 83.42 | 434.01 |
| 1970-71 | 5.95 | (4.82) | 313.69 | 2131.05 | 106.55 | 420.24 |
| 1971-72 | 4.45 | (3.61) | 234.79 | 2437.88 | 121.88 | 356.88 |
| 1972-73 | 6.62 | (5.37) | 348.96 | 2979.91 | 148.99 | 497.96 |
| 1973-74 | 6.67 | (5.41) | 351.87 | 2936.67 | 146.83 | 498.70 |
| 1974-75 | 4.18 | (3.39) | 220.51 | 3238.60 | 161.93 | 382.44 |
| 1975-76 | 5.82 | (4.72) | 307.06 | 4393.41 | 274.85 | 581.91 |
| 1976-77 | 6.36 | (5.16) | 335.37 | 4216.23 | 421.62 | 756.99 |
| 1977-78 | 5.23 | (4.24) | 272.92 | 3155.69 | 315.57 | 591.49 |
| Total | 61.88 | (51.81) | 3367.95 | 29816.45 | 1914.59 | 5282.54 |

Regardless of the improvements in irrigation supplies being undertaken, the capacity for undertaking cropped acreage expansions and improvements has been provided through the water storage and flood attenuation capability of Tarbela.

The economic value of the increase in water availability may be roughly estimated by taking the total quantity of stored water released during the year multiplied by a national unit value. The real value of stored water will depend on the crops being grown. At a value of $15-$45 per acre foot, 9 MAF per annum would provide a benefit of $135-$405 million per annum. Another approach has been to estimate the potential for increased grain production from use of stored water from Tarbela. If this is taken at 1,000,000 tons per annum the saving to Pakistan's economy would be about $200 million per annum in grain imports at present value.

Thus the total potential development benefits of Tarbela today may be of the order of U.S. $525 million per annum. So far optimum benefits from Tarbela have not been derived, because of costly repair work which had become necessary due to various mishaps which occurred in the initial operation. However, the Tarbela Dam Project, despite various set backs, has played its pivotal role in terms of irrigation and power, and is expected to provide greater benefits in the future economy of Pakistan.

ii) Hydropower Supplies:

The consumption of power in Pakistan has grown at a very rapid pace. At the time of independence in 1947, the total capacity was 31.3 MW including import of 10 MW from India. In 1959 when WAPDA took over the electricity department the capacity was only 119 MW, which has by 1978 grown to 2642.2 MW recording more than 21 fold increase since 1959. During the same period the hydel power generation has gone up from 66.7 MW in 1959 to 1567.8 MW in 1978 which represents a 23 fold increase over a period of 19 years.

It is mainly through the construction of Warsak, Mangla and Tarbela Dams that WAPDA is today obtaining 74% of its total capacity from hydel generation. The generation capacity at Warsak, Mangla and Tarbela is being further augmented through installation of more units.

6. SOCIAL IMPACTS

The phase of dam building activity in the last decade and investment of billions of dollars in a developing country is bound to have a great social impact.

As dealt earlier the construction of dams was inevitable for the economy of Pakistan from the point of view of irrigation and power. These two aspects are of vital importance on which is dependent the socio-economic structure of the country. The benefits derived from the dams construction has been dealt briefly above, some social aspects which need to be mentioned include the resettlement aspect, transfer of technology, development of skilled labour, industry, tourism and fisheries.

Resettlement of Displaced Persons: Great sacrifices were made by the people who had to leave their homes in the cause of nation to bring prosperity to millions of their fellow country-men. About 150,000 people had to be resettled in areas of Punjab and Sind. From Mangla about 5136 families have been given possession of 84,000 acres of land. A number of families who had no original land holdings were settled along the periphery of the reservoir. In Tarbela out of 16,000 families, 6400 were eligible for alternative agricultural land, for which about 60 thousand acres were produced in Punjab and Sind.

Transfer of Technology: The construction of large dams afforded a unique opportunity to the Pakistani engineers and workers to be associated with some of the best engineers and technicians in the world. During the construction of these projects, transfer of technology took place which resulted in grooming of scores of engineers, geologists and research scientists.

The largest assemblage of construction equipment in the world produced thousands of skilled labour which are today a source of foreign exchange earnings, working in the middle east and elsewhere.

Tourism: The lakes created by the large reservoirs are a great tourist attraction. Attention is being paid to the development of aquatic sports like boating, yatching, sailing, fishing etc. In years to come this is likely to boost the tourist industry.

Fisheries: Development of fish culture in the reservoirs created by the dams is being rapidly undertaken. This is providing the much required protein food in the region as well as employment to the local population.

7. CLIENTS, CONSULTANTS AND CONTRACTORS

All dams in Pakistan are either owned by the Federal Government or Provincial Governments, therefore the government and its subsidiary agencies are the main clients for the construction of dams. The biggest client is the Water and Power Development Authority. WAPDA is the premier organization of the country, which handles, on behalf of the Government of Pakistan, the planning and executing of schemes in the field of Irrigation and Power. Other client for dam construction include the Small Dams Organization and the Irrigation Departments.

In the past, these government agencies used to design as well as construct such projects themselves. However, in the recent past, the pattern is changing and now for most projects, independent consulting firms are appointed and the construction work is also awarded to contracting companies.

Pakistan has had a rich experience in the development of water resources and related Barrages and Dams. The practice of irrigated agriculture in the region is perhaps as old as the history of civilization. Today with almost 65,000 km of major canals and water courses and a wide array of dams, barrages, regulators, syphons and other hydraulic structures, Pakistan has one of the largest and most intricate irrigation systems of the world. The local engineering talent in Pakistan is therefore highly experienced in the field of design and construction of dams, barrages and irrigation network. In the past the design work was mostly done by the design offices within the Public Work Departments, and the profession of engineering consultancy is relatively new. Consequently for the design of Indus Basin Projects, foreign consultancy firms had to be employed, however, Pakistani engineers worked very closely with these firms. During the last 10-15 years a number of local consultancy firms have been established, and the well recognized consultancy firms now number about 100. The largest and the leading consulting firm in the country is the National Engineering Services (Pakistan) Limited (NESPAK) which was established by the Government of Pakistan in 1973, by pooling technical expertise available within the country in various departments. NESPAK has on its full time staff over 650 highly experienced engineers, geologists, economists and other professionals. NESPAK is working on most of the major projects in Pakistan including various dam projects in UAE, Oman, Somalia, Tanzania and Nigeria.

The other leading consultancy firm highly experienced in dam engineering is Associated Consulting Engineers (ACE) which was established in 1958 and has about 200 professionals working with it in Pakistan as well as in its branches in Saudi Arabia, Malaysia, Iran, Libya and UAE. ACE has designed three major dams in Pakistan i.e. Rawal, Hub and Khanpur. NESPAK and ACE have jointly established a subsidiary company in Nigeria alongwith a Nigerian partner; this company is particularly undertaking a number of dam projects in Nigeria. So far this joint venture has completed the design of 5 dams and water supply schemes, and is working on three other such projects.

In Pakistan now, most development projects are being designed by the local consultancy firms. On some very major projects, joint ventures of local and foreign firms are encouraged. Pakistani consulting firms are also providing services in other developing countries in the middle and far east. A proposal has been discussed by the developing countries in various forums, that in order to achieve self sufficiency in technical capability, the developing nations should pool up their technical resources. Pakistan is a strong supporter of this point of view, and has been discussing with some other developing countries to set up joint ventures in consultancy and contracting fields.

As mentioned earlier the last two decades in Pakistan saw a major activity in dam construction. Projects such as Tarbela and Mangla which were amongst the largest contracts in dam construction, attracted the major construction companies of the world. During this period a number of local firms of contractors also came up. The leading Pakistani contracting organization in the dam construction is the Mechanized Construction of Pakistan Ltd. (MCP), which has constructed Tanda Dam, and Hub Dam and are now constructing the Simly and Khanpur Dams. The company has a substantial fleet of earth moving equipment, technical staff and operators. MCP is also constructing a large irrigation project in Iraq. National Construction Company of Pakistan has been associated in the construction of irrigation tunnel at Tarbela and is presently handling apart from various other projects, civil engineering works of Mangla Power House Extension Project. Other companies engaged in dam building in Pakistan include Mark International, Hashtam Khan and Company, Nazir & Company, and Izhar Ltd. Apart from the major Civil Contracting organizations, construction of dams is also carried out by certain Clients themselves. The Small Dam Organization, Provincial Irrigation Department, the Agricultural Development Authority and the Pakistan Army Crops of Engineers have also built a number of dams. For construction of large dams in Pakistan, international tenders are invited and firms from all over the world participate in these tenders.

There has been a restraint in the development of local construction industry for heavy construction work. The restraint is due to the fact that heavy construction equipment and its spare parts have to be imported in Pakistan. Foreign exchange resources have not allowed sufficient imports in this respect, and the construction machinery in the past was purchased on project to project basis. However, now efforts are being made by the Government of Pakistan to develop a well established heavy construction industry, by encouraging the local contractors to undertake large construction projects.

.-.-.-.-.

APPENDIX A

SOME OF THE IMPORTANT DAMS IN PAKISTAN

| Dam Site | Location | River | Status | Installed Capacity (MW) | Engineering by/Consultant | Contractor/ Constructor | Particulars | |
|---|---|---|---|---|---|---|---|---|
| Namal | Mianwali | Golar Nullah | Completed 1913 | | PWD Punjab Govt. | S.B. Lehna Singh | Type<br>Height<br>Crest Length<br>Volume Content of Dam<br>Reservoir Capacity | Gravity (masonry)<br>25.9 m.<br>46.6 m.<br>—<br>27.6 million cu.m |
| Spin Karaiz | Quetta | Nar & Murdar | Completed 1945 | | Military Engineering Services | — | Type<br>Height<br>Crest Length<br>Volume Content of Dam<br>Reservoir Capacity | Earthfill<br>21.3 m.<br>759.0 m.<br>805,000 cu.m.<br>6.8 million cu.m. |
| Warsak | Peshawar | Kabul | Completed 1960 | 240 | H.G. Acres & Co. Canada | Augus Robertson & Co. Canada | Type<br>Height<br>Crest Length<br>Volume Content of Dam<br>Reservoir Capacity | Gravity (concrete)<br>41.0 m.<br>198.1 m.<br>369,000 cu.m.<br>24.7 million cu.m. |
| Rawal | Rawalpindi | Kurang | Completed 1960 | | Associated Consulting Engineers | Mir Aslam Khan | Type<br>Height<br>Crest Length<br>Volume Content of Dam<br>Reservoir Capacity | Gravity (masonry)<br>34.5 m<br>213.4 m.<br>40,500 cu.m.<br>58.6 million cu.m. |
| Baran | Bannu | Baran | Completed 1961 | | Irrigation Department Pakistan | Irrigation Department Pakistan | Type<br>Height<br>Crest Length<br>Volume Content of Dam<br>Reservoir Capacity | Earthfill<br>36.6 m.<br>1090.6 m.<br>2,722,000 cu.m.<br>120.9 million.cu.m. |
| Wali Tangi | Quetta | Wali Tangi | Completed 1961 | | Military Engineering Services | Pak Army | Type<br>Height<br>Crest Length<br>Volume Content of Dam<br>Reservoir Capacity | Rock fill<br>22.9 m.<br>61.0 m.<br>25,000 cu.m.<br>0.51 million cu.m. |
| Misriot | Rawalpindi | T/Soan | Completed 1963 | | West Pakistan Agricultural Development Corporation | West Pakistan Agricultural Development Corporation | Type<br>Height<br>Crest.Length<br>Volume Content of Dam<br>Reservoir Capacity | Concrete<br>10.0 m.<br>113.0 m.<br>3,400 cu.m.<br>0.6 million cu.m. |

APPENDIX A (Contd...)

| Dam Site | Location | River | Status | Installed Capacity (MW) | Engineering by/Consultant | Contractor/Constructor | Particulars | |
|---|---|---|---|---|---|---|---|---|
| Sipiale | Fatehjang | T/Haro | Completed 1964 | | West Pakistan Agricultural Development Corporation | West Pakistan Agricultural Development Corporation | Type<br>Height<br>Crest Length<br>Volume Content of Dam<br>Reservoir Capacity | : Concrete<br>: 12.8 m.<br>: 55.0 m.<br>: 3,500 cu.m.<br>: 0.7 million cu.m. |
| Kahl | Haripur | T/Haro | Completed 1965 | | West Pakistan Agricultural Development Corporation | West Pakistan Agricultural Development Corporation | Type<br>Height<br>Crest Length<br>Volume Content of Dam<br>Reservoir Capacity | : Earthfill<br>: 22.9 m.<br>: 34.7 m.<br>: 86,400 cu.m.<br>: 1.0 million cu.m. |
| Tanaza | Fateh Jang | T/Sil | Completed 1965 | | West Pakistan Agricultural Development Corporation | West Pakistan Agricultural Development Corporation | Type<br>Height<br>Crest Length<br>Volume Content of Dam<br>Reservoir Capacity | : Concrete<br>: 9.4 m.<br>: 24.4 m.<br>: 22,900 cu.m.<br>: 0.1 million cu.m. |
| Dhurnal | Campbellpur | T/Gabhir | Completed 1967 | | West Pakistan Agricultural Development Corporation | West Pakistan Agricultural Development Corporation | Type<br>Height<br>Crest Length<br>Volume Content of Dam<br>Reservoir Capacity | : Earthfill<br>: 19.2 m.<br>: 18.2 m.<br>: 137,000 cu.m.<br>: 1.7 million cu.m. |
| Tanda | Kohat | Kohat Toi | Completed 1967 | | Water and Power Development Authority | Mechanized Construction of Pakistan | Type<br>Height<br>Crest Length<br>Volume Content of Dam<br>Reservoir Capacity | : Earthfill<br>: 35.0 m.<br>: 670.0 m.<br>: 2,290,000 cu.m.<br>: 99.0 million cu.m. |
| Mangla | Jhelum | Jhelum | Completed 1967 | 600 | Binnie & Partners U.K. | Guy F. Atkinson USA | Type<br>Height<br>Crest Length<br>Volume Content of Dam<br>Reservoir Capacity | : Earthfill<br>: 115.8 m.<br>: 3,140 m.<br>: 110,096,600 cu.m.<br>: 7154.3 million cu.m. |
| Mang | Haripur | T/Haro | Completed 1970 | | West Pakistan Agricultural Development Corporation | West Pakistan Agricultural Development Corporation | Type<br>Height<br>Crest Length<br>Volume Content of Dam<br>Reservoir Capacity | : Concrete<br>: 15.5 m.<br>: 33.5 m.<br>: 24,700 cu.m.<br>: 0.7 million cu.m. |
| Gurabh | Talagang | T/Soan | Completed 1971 | | West Pakistan Agricultural Development Corporation | West Pakistan Agricultural Development Corporation | Type<br>Height<br>Crest Length<br>Volume Content of Dam<br>Reservoir Capacity | : Earthfill<br>: 19.8 m.<br>: 31.0 m.<br>: 84,900 cu.m.<br>: 0.8 million cu.m. |

| Name | Location | River | Status | Consultants | Contractors | Parameter | Value |
|---|---|---|---|---|---|---|---|
| Dungi | Gujar Khan | T/Kansi | Completed 1971 | West Pakistan Agricultural Development Corporation | West Pakistan Agricultural Development Corporation | Type<br>Height<br>Crest Length<br>Volume Content of Dam<br>Reservoir Capacity | Concrete<br>18.0 m.<br>18.2 m.<br>22,400 cu.m.<br>2.2 million cu.m. |
| Chichali | Kalabagh | T/Indus | Completed 1971 | West Pakistan Agricultural Development Corporation | West Pakistan Agricultural Development Corporation | Type<br>Height<br>Crest Length<br>Volume Content of Dam<br>Reservoir Capacity | Concrete<br>37.5 m.<br>91.0 m.<br>40,600 cu.m.<br>13.0 million cu.m. |
| Chashma | Mianwali | Indus | Completed 1971 | Coode & Partners, U.K. | Dumez France | Type<br>Height<br>Crest Length<br>Volume Content of Dam<br>Reservoir Capacity | Gravity Concrete<br>9.8 m.<br>1083.2 m.<br>485,500 cu.m.<br>999.0 million cu.m. |
| Tarbela | Haripur | Indus | Completed 1976 | TAMS, USA | Tarbela Joint Venture (Impregilo) | Type<br>Height<br>Crest Length<br>Volume Content of Dam<br>Reservoir Capacity | Rockfill & earthfill<br>143 m.<br>2743 m.<br>126,208,200 cu.m.<br>13,691.9 million c.u.m. |
| Kahuta | Rawalpindi | Ling | Completed 1979 | National Engineering Services (Pakistan) Limited 'NESPAK' | Izhar Ltd. | Type<br>Height<br>Crest Length<br>Volume Content of Dam<br>Reservoir Capacity | Concrete & earthfill<br>22 m.<br>211.8 m.<br>Earth 23,721 cu.m.<br>Conc.28,326 cu.m.<br>0.3 million cu.m. |
| Hub | Karachi | Hub | Completed 1979 | Associated Consulting Engineers | Mechanized Construction of Pakistan | Type<br>Height<br>Crest Length<br>Volume Content of Dam<br>Reservoir Capacity | Earthfill<br>46.0 m.<br>8,007.3 m.<br>8,779,200 cu.m.<br>1,139.8 million cu.m. |
| Khanpur | Taxila | Haro | Under Construction | Associated Consulting Engineers | Mechanized Construction of Pakistan | Type<br>Height<br>Crest Length<br>Volume Content of Dam<br>Reservoir Capacity | Earthfill<br>50.9 m.<br>471.3 m.<br>4,407,442 cu.m.<br>131.9 million cu.m. |
| Simly | Rawalpindi | Soan | Under Construction | Water and Power Development Authority | Mechanized Construction of Pakistan | Type<br>Height<br>Crest Length<br>Volume Content of Dam<br>Reservoir Capacity | Earthfill<br>76.22 m.<br>307.93 m.<br>1,977,867 cu.m.<br>35.5 million cu.m. |
| Bolan | Dhadar | Bolan | Under Construction | National Engineering Services (Pakistan) Limited 'NESPAK' | | Type<br>Height<br>Crest Length<br>Volume Content of Dam<br>Reservoir Capacity | Earthfill<br>20.6 m.<br>510.0 m.<br>560,000 cu.m.<br>92.50 million cu.m. |

Notes:
1. T- denotes tributary
2. Height of dam is given as height above ground level
3. Reservoir capacity indicates the gross capacity

COMPILED BY NATIONAL ENGINEERING SERVICES (PAKISTAN) LIMITED

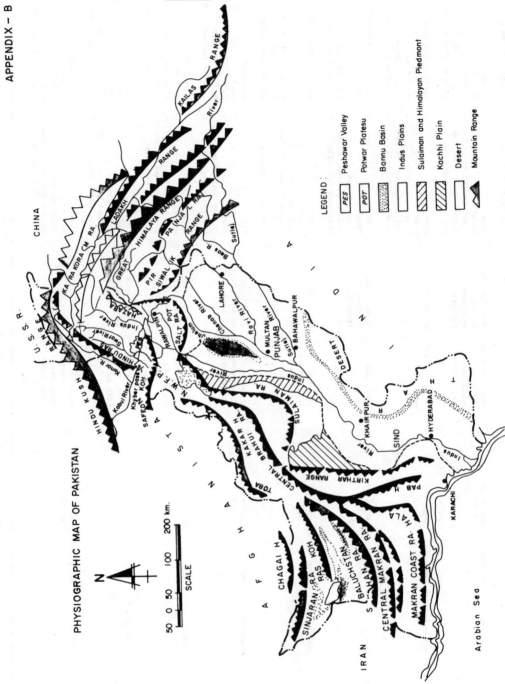

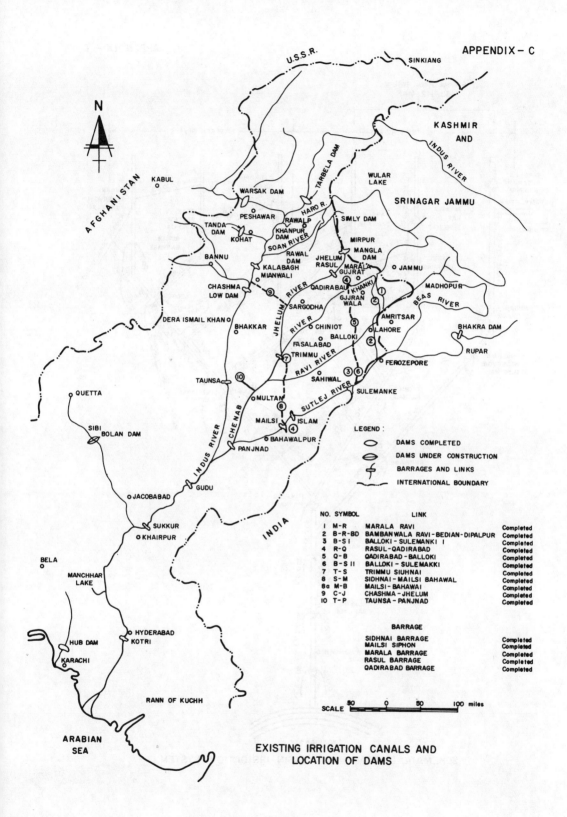

APPENDIX - D

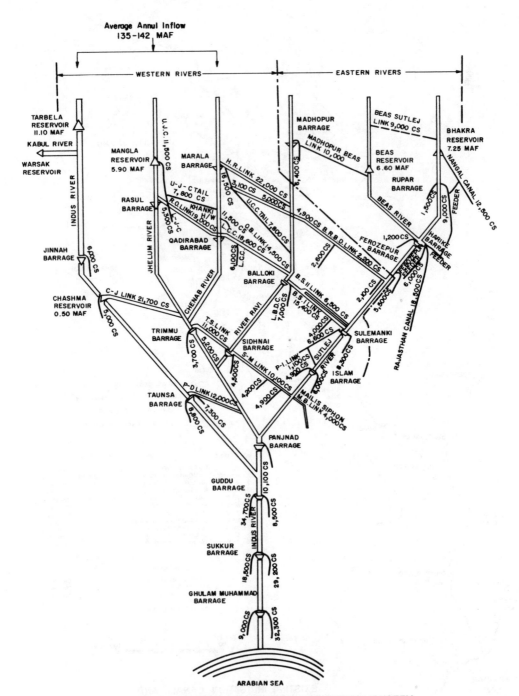

SCHEMATIC DIAGRAM INDUS BASIN IRRIGATION SYSTEM

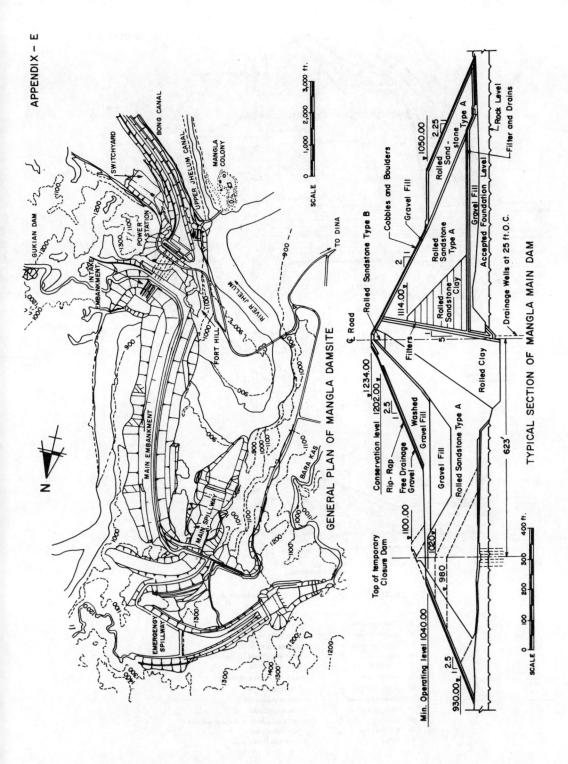

APPENDIX - F

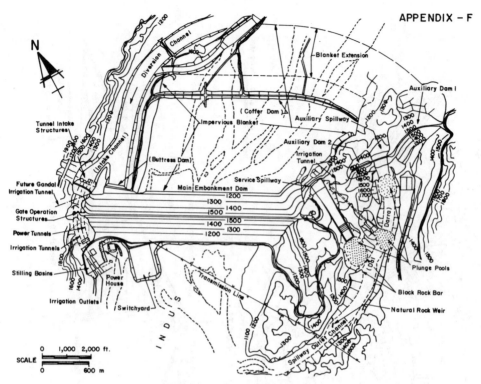

GENERAL PLAN OF TARBELA DAM

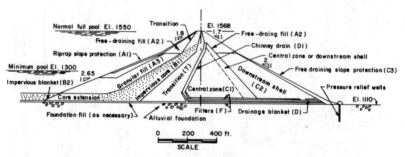

TYPICAL CROSS-SECTION THROUGH MAIN EMBANKMENT DAM

| ZONE | | MATERIAL |
|---|---|---|
| A1 | Slope protection | – Heavy rock |
| A2 | Upstream shell | – Free draining fill |
| A3 | Upstream shell | – Coarse granular fill |
| B1 | Core | – Impervious blend of gravels, sand and silt |
| T | Transition zone | – Fine granular fill |
| C1 | Central zone | – Semi-pervious granular fill |
| D1 | Vertical drain | – Permeable gravel |
| C2 | Downstream shell | – Coarse granular fill |
| C3 | Slope protection | – Free draining fill |
| B2 | Blanket | – Impervious earth or decomposed rock |
| D | Drainage zone | – Permeable gravel |
| F | Filter | – Fine gravel and sand |

APPENDIX – G

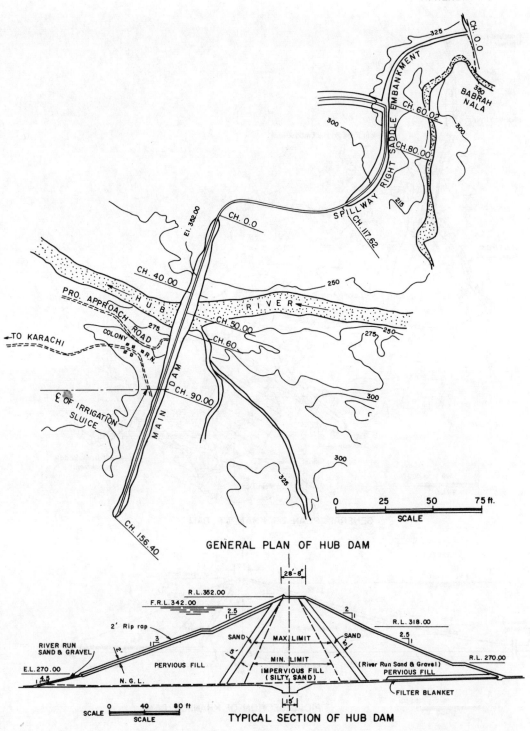

APPENDIX - H

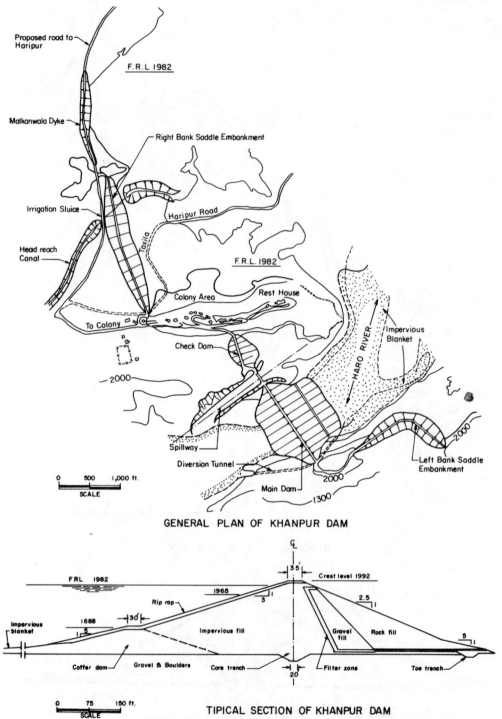

GENERAL PLAN OF KHANPUR DAM

TIPICAL SECTION OF KHANPUR DAM

APPENDIX - J

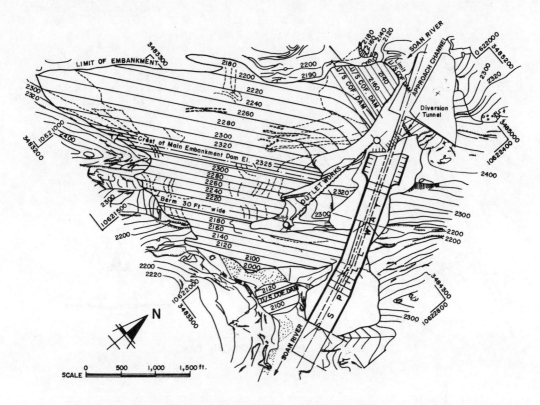

GENERAL PLAN OF SIMLY DAM

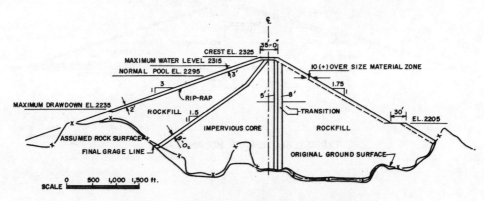

TYPICAL SECTION OF SIMLY DAM

APPENDIX - K

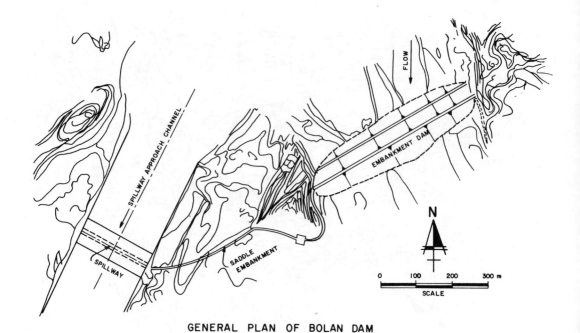

GENERAL PLAN OF BOLAN DAM

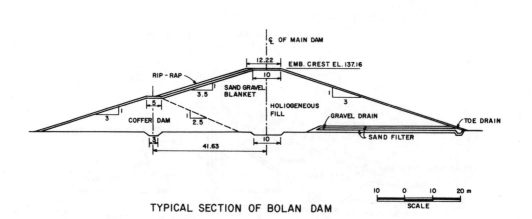

TYPICAL SECTION OF BOLAN DAM

APPENDIX L

## POTENTIAL DAM SITES IN PAKISTAN

REGEND: TYPE: E - EARTHFILL: R - ROCKFILL: G - GRAVITY: A - ARCH
PURPOSE: I - IRRIGATION: P - POWER: M - MULTIPURPOSE:
W - WATER SUPPLY: R - RIVER REGULATION:
F - FLOOD CONTROL: S - SEDIMENT CONTROL
PRESENT STATE: O - IN OPERATION: C - UNDER CONSTRUCTION:
P - IN PLANNING; F - FOR FUTURE:
S - SUPERSEDED OR ABANDONED

| NAME OF DAM | CHARACTERISTICS | | | | PURPOSE | POWER CAPACITY (MW) | | NOTES | PRESENT STATE |
|---|---|---|---|---|---|---|---|---|---|
| | TYPE | HEIGHT (FT.) | LENGTH (FT.) | GROSS CAPACITY OF RESERVOIR (MAF) | | INITIAL OR INSTALLED | ULTIMATE | | |
| Adam Kot | | | | | | | | See Khajuri Kach | P |
| Ahnai Tangi (or Ahnai Kili) | | | | | | | | Superseded by Hinnis Tangi | S |
| Akhori | E | 250 | 15,800 | 3.6 | I | | | Superseded by Gariala | S |
| Aktar Kili Ambahar | R | 920 | 850 | 7.9 | I P | 1,270 | | " by Khajuri Kach | S |
| Anambar | E | 80 | 2,600 | 0.055 | F | | | | |
| Attock | E | | | 30 | I P R | | | Supersedded by Kalabagh | S |
| Babar Kach I | R | 180 | 400 | 0.715 | I P F | 15 | | | |
| Babar Kach II | R | 120 | | | | | | Abandoned | P |
| Badin Zai | | | | | | | | Superseded by Khajuri Kach | S |
| Bahtar | E | 235 | 7,500 | 0.9 | I | | | | |
| Bakhuwala Bandagai | E | 80 | 2,500 | | I | | | | P |
| Banda Saidu | G | 210 | 500 | 0.003 | P I | | | | F |
| Banda Tanda | E | 115 | 2,340 | 0.078 | I | | | | |
| Bara | | | | | | | | See Tanda Dam | C |
| Bara Tanda | | | | | | | | See Miri Khel | |
| Barahotar | | 150 | | 0.007 | P W | | | See Baran Superseded by Chaniot | S |
| Basund | | 271 | 750 | 0.075 | I P | | | | |
| Bazargai | E R | 960 | | 8.0 | M | 1,140 | | | F |
| Beji Diversion | R | 120 | 750 | | F | | | | P |
| Bhaun | E R | 206 | 2,050 | 0.026 | I | | | | P |
| Boya Post | | | | | | | | | |
| Bunha I | | 300 | | 0.28 | | | | See Tochi | |
| Bunha II | | | | 2.8 | | | | | |
| Bunji | | 250 | | 4.6 | P | 2,000 | | | F |
| Burj Zam | E | 168 | | 0.171 | I P | 1.8 | | | P |
| Butta | | | | | | | | | |
| Chaniot | G | 176 | 675 | 0.0095 | P W | 1,536 | | | P |
| Chaphar Rift | | 185 | | 0.079 | I W | 1 | | | F |
| Charah | G | 175 | 832 | 0.067 | I W | | | | P |
| Chaudhwan Zam | E | 262 | | 0.150 | I P | 2.24 | 5.1 | | P |
| Chilas | | | | | P | | | | F |
| Chiniot | | | | 1.4 | I | | | | P |
| Chitral | | | | | P | | | | F |
| Choti | | | | | | | | | |
| Chutiatan | | | | | P | 12 | | | F |
| Dabar | | | | | | | | | |

343

POTENTIAL DAM SITES IN PAKISTAN                                            APPENDIX L (Cont'd)

| NAME OF DAM | TYPE | HEIGHT (FT.) | LENGTH (FT.) | GROSS CAPACITY OF RESERVOIR (MAF) | PURPOSE | POWER CAPACITY (MW) INITIAL OR ULTI-INSTALLED MATE | NOTE | PRESENT STATE |
|---|---|---|---|---|---|---|---|---|
| Dadar | G | 260 | 1,620 | 0.028 | I P W | 4 | | F |
| Dagarai | | 230 | | 0.10 | I | | | |
| Dhamtour | | 225 | 780 | 0.027 | I P W | 15 | | F |
| Dara Tang | | 130 | 8,500 | | | | | F |
| Daraban Zam | | 100/175 | | 0.03/0.15 | | | See Burj Zam | P |
| Darazinda | | | | | | | Superseded by Burj Zam | S |
| Darwat | | 110 | | 0.073 | I | | Infeasible | |
| Data Khel | | 200 | | 0.35 | I | | See Tochi | |
| Dau | | | | | | | | |
| Dhaabi | | 77 | | 0.012 | I | | | |
| Dhaimgrah | | | | | P | | | F |
| Dharyal Weir | | 42 | 920 | | I | | | |
| Dhok Abbaki | E R | 295 | 24,000 | 9.0 | P | 105 | | P |
| Dhok Ham | G | 131 | 700 | 0.013 | I | | | P |
| Dhok Mila (Power Plant) | | | | | P | 1,200 | | P |
| Dhok Pathan | E R | 275 | 12,000 | 8.5 | I | | | P |
| Dhok Sial | G | 72 | 360 | 0.012 | I | | | P |
| Dhrabi | E | 77 | 985 | 0.02 | I | | | P |
| Domanda | | | | | | | See Chaudhwan Zam | |
| Drosh | | 300 | 1,700 | | P | | | F |
| Dulhal | | 70 | 3,000 | | | | | |
| Fort Sandeman | | | | | | | See Khaiwri Kach | |
| Gahirat | | 100 | | | P | | | F |
| Gaj (Gaja Nai) | | 300 | 120 | 0.150 | I F | | | P |
| Gambila | | | | | I | | | P |
| Gandalat Tang Weir | | 10 | | | | | | |
| Gariala | E | 375 | 40,000 | 8.2 | I | 110 | | P |
| Ghatti Bridge | | | | | | | | |
| Ghaziabad | | | | | | | Infeasible | |
| Giddar Pur | E | 52 | 1,124 | 0.004 | I | | | |
| Gomal Zam | | | | | | | See Khajuri Kach | |
| Gomal Zam Weir (Murtaza) | | | | | | | See Mian Nur | |
| Gul Kach | G | 195 | | 0.50 | I F | | | F |
| Haranbar | | | | | | | | |
| Havelian | E | 173 | 6,600 | 0.018 | I | | | |
| Hinnis Tangi | | 236 | | 0.068 | I P | 0.015 | | F |
| Jaro | G | 350 | | | | | | |
| Jathaput | | | | 2.7 | | | Superseded by Mangla | S |
| Jhelum | | | | | | | | |
| Kaham | | | | | | | | |
| Kalabagh | E R | 260 | 4,150 | 9.375 | I P | 1760 | | P |
| Kalam | | 480 | | 0.363 | I P | 110 | | F |
| Kalangai | G/E R | 580 | | 6.5 | I P | 750 | | F |
| Kamalan Kach | R | 200 | | | I F | | | F |
| Kanshi | | 270 | | 1.1 | S | | | P |
| Khairi Murat | E | 220 | | 0.021 | I | | | P |
| Khajuri Kach | A/G | 500 | 630 | 2.15 | I P | 127 | | P |
| Khajuri Post | | 150 | | 0.130 | I | | | |
| Khapulu | | 600 | | 10 | I P F R | 600 | | P F |
| Kharikan Kas | E | 230 | 9,750 | 0.13 | | | | |
| Khazana | E R | 520 | | 3.0 | M | 170 | | F |
| Khirgi Weir | | | | | P | | | F |
| Khushhalgarh | | | | | M | | Superseded by Kalabagh | S |
| Khwaja Khizar | E/R | 100 | 400 | 0.14 | I | | See Yozara | |
| Kirpalian | | | | | I P | | Superseded by Tarbela | S |
| Kesu | | | | | P | | | F |

344

POTENTIAL DAM SITES IN PAKISTAN                                          APPENDIX L (cont'd)

| NAME OF DAM | TYPE | CHARACTERISTICS HEIGHT (FT.) | LENGTH (FT.) | GROSS CAPACITY OF RESERVOIR (MAF) | PURPOSE | POWER CAPACITY (MW) INITIAL OR ULTI- INSTALLED MATE | | NOTE | PRESENT STATE |
|---|---|---|---|---|---|---|---|---|---|
| Kot Fateh | E | 80 | 8,000 | 0.016 | I | | | | P |
| Kotkai | | | | | M | | | Superseded by Tarbela | S |
| Kohala | | | | | | | | | |
| Kotli | E | 320 | | 0.3 | I P S | 220 | 3760 | | P |
| Kud | | | | 0.047 | I F | | | | |
| Kunbat | | | | | P | | | | F |
| Kurram Garhi | | - | | | I P | 4 | | | O |
| Kurram Tangi | E | 300 | | 1.50 | I F P | | | | P |
| Ladoo | | 80 | | | F | | | | |
| Lohi Bhir | | | | | | | | | |
| Lohar Gali | G | 530 | | 0.8 | I P | | | | P |
| Lower Taba Kas | | 170 | 4,000 | | | | | | |
| Main Swat | | 500 | | | I P | | | | F |
| Makhad | | 280 | | 6.0 | I P | | 700 | Superseded by Kalabagh | S |
| Mastuj-Lutkho | | 200 | | | P | | | | F |
| Miannur | E | 77 | 3,460 | 0.089 | I F | | | | P |
| Mile 46 | | | | | | | | Superseded by Domanda | S |
| Mina Bazar | | 90 | 600 | | | | | Superseded by Khajuri Kach | S |
| Mirabandi | | 375 | | 2.1 | I | | | | |
| Miri Khel | | | | | | | | | F |
| Mirkhani | | 400 | | 0.8 | | | | | |
| Morgah Weir | G | 49 | | 0.116 | I | | | | P |
| Munda | E R | 660 | | 2.0 | I P | 370 | 760 | | P |
| Murtaza Weir | | | | | I P | 8 | | Superseded by Mian Nur | S |
| Nari Bolan | | | | | | | | | |
| Nari Bolan | E | 68 | 1,750 | 0.325 | I F | | | | C |
| Naran | G | 410 | 1,360 | 0.28 | I P | 50 | | | P |
| Naulung | R | 185 | | 0.306 | I | | | | P |
| Nawan | | 650 | | 2.6 | | | | Superseded by Mangla | S |
| Nazarai | | | | | | | | Superseded by Khajuri Kach | S |
| Nili Kach | E | 77 | | 0.018 | I P | | | | |
| Pak-Afghan | | | | | I P R | | | | F |
| Panjar | | | | 3.0 | P | | 1,500 | | |
| Papin | E R | 100 | 300 | 0.053 | I | | | | P |
| Paras | | | | | | | | Infeasible | |
| Pashtkhand (Raiko) | | 190 | | 0.244 | | | | Superseded by Maulung | |
| Pishi | | | | | | | | | |
| Porali | | | | | | | | | |
| Rajdhani | | E | 325 | | 0.86 | I P | | 40 | Superseded by Mangla | S |
| Rasul | | | | 10.0 | I P | 300 | | Superseded by Mangla | S |
| Rohtas | | 25 | | 1.90 | I P | | 60 | | P |
| Saggar Kas | E | 230 | 9,500 | 0.77 | | | | | P |
| Sanjwal | R | 165 | 5,800 | 0.177 | | | 2,250 | Superseded by Gariala | S |
| Sawawan | | | | | | | | | |
| Sehwan Barrage | | | 3,500 | 0.8(2.7) | I | | | | P |
| Shadi Kar | | | | | | | | | |
| Shah Bilawal | E | 73 | 1,300 | 0.021 | I | | | | P |
| Shah Pur | | | | | | | | | |
| Shakista Nala | E | 187 | 11,900 | 6.25 | | | | | P |
| Sheikh Haider Zam | E | 188 | | 0.0687 | | | | | |
| Sheikh Mela | | | | | | | | Superseded by Domanda | S |
| Shinki Post | | 250 | | 0.23 | I | | | | F |
| Singur | | | | | P | | | | C |
| Skardu | R | 310 | 3,700 | 8.0 | R | | | | P |

POTENTIAL DAM SITES IN PAKISTAN

APPENDIX L (cont'd)

| NAME OF DAM | CHARACTERISTICS | | | | PURPOSE | POWER CAPACITY (MW) INITIAL OR ULTI-INSTALLED MATE | NOTE | PRESENT STATE |
|---|---|---|---|---|---|---|---|---|
| | TYPE | HEIGHT (FT.) | LENGTH (FT.) | GROSS CAPACITY OF RESERVOIR (MAF) | | | | |
| Spli Toi | | | | | | | Superseded by Hinnis Tangi | S |
| Suki Kinari (Power Plant) | | | | | P | 500 | | P |
| Surgul-Chambai | | | | | | | Superseded by Tanda | S |
| Takari Kili | | | | | | | See Khajuri Kach | |
| Talli Tangi | G | 195 | 150 | 0.1385 | I P | | | P |
| Thalian | | | | | | | | |
| Thapla | | 210 | | 0.27 | I P | | Superseded by Tarbela | S |
| Tochi | | | | | | | See Data Kjel | |
| Torder | | | | | P | | | P |
| Total Nala | E | 148 | 14,500 | 6.25 | I P | | | F |
| Tung | | | | | | | | |
| Tungi | | | | | I | | | P |
| Turan China | | | | | | | Superseded by Hinnis Tangi | S |
| Upper Sukleji | | | | | | | See Gandalat Tang | |
| Upper Taba Kas | | 60 | 2,000 | | | | See Gandalat Tang | |
| Wadala | | | | | | | | |
| Wucha Sesta | E | | | | I | | Superseded by Domanda | F |
| Yozara | | | | | | | | P |
| Zhair Narai China | E/R | 100 | 400 | 0.14 | I | | Superseded by Khajuri Kach | S |

# Dam engineering in Philippines

EDUARDO P. ABESAMIS
*National Power Corp., Manila, Philippines*

## INTRODUCTION

The growing need to reduce dependence on costly oil-based power, increase agricultural production, minimize flood losses and provide domestic water focused the attention of the Philippine Government to the importance of control and regulation of the country's streams. Dams are essential in fulfilling these needs and are an integral part in water-resource development programs. Among the various public and private groups actively engaged in dam construction, the following are the government's implementing authorities in this field: (a) National Power Corporation (NPC); (b) National Irrigation Administration (NIA); (c) Metropolitan Waterworks and Sewerage Authority (MWSA); and, (d) Local Water Utilities Administration (LWUA).

## DAM ENGINEERING IN THE COUNTRY

The Philippines has demonstrated a significant momentum in developing its water resources as expressed by dam construction. Over the past decades, dam building has been largely concentrated on hydro power development but increasing attention is being given by the government to multipurpose programs which include irrigation, flood control and water supply schemes.

Among the more than twelve dams in existence, there are four which exceed 100 meters in height. At the time of its construction in 1950, the Ambuklao Dam in Bokod, Benguet, was one of the highest rockfill dams in the world, the structure rising 129 meters above the river bed. Its Philippine record, however, has since been surpassed by the Angat Dam in Bulacan with its 131 meters high earth and rockfill structure completed in 1967. Of the dams with the greatest volume content, the largest is the Pantabangan Dam on the Pampanga River in Nueva Ecija. The dam is composed of two-zoned earthfill structures which meet at a ridge in the center. The combined structure contains a total of 12.3 million cubic meters of earth and rock. The majority of the dams already completed are of rockfill type with earth and clay cores. All concrete dams are of gravity type with the highest being only 32 meters, used for industrial water supply. The trend in dam construction in the Philippines favors earth and rockfill owing to the availability of the needed material in most river basins plus the simple nature of preparation and processing required of earth and rocks. This observation is being confirmed by the fact that of the four dams currently under construction, three are of rockfill type, while the fourth one is a concrete gravity dam. The dams also tend to be low and wide on account of very few deep valleys in the potentially promising rivers. Three dams now being built are below 50 meters in height.

A review of the water resources in the country shows there are many sites which are yet to be developed. Possible hydropower sites alone reveal that millions of barrels of oil could be saved daily if they are harnessed. With the present energy crisis, the list of water-power schemes under planning is growing in number. Over twenty dams are in the planning stage in the country, most of them over the 100-meter category are under the various stages of investigation, with hydroelectric schemes becoming more emphasized.

REVIEW OF EXISTING PHILIPPINE DAMS

The Caliraya Dam

This structure is part of the Caliraya Hydroelectric scheme located in the town of Lumban, Laguna some 100 km southeast of Manila. It was the first major project undertaken by the National Power Corporation. The power plant contains four generating units with a total installed capacity of 32 MW. The first unit was put into operation in 1945 while the fourth unit was commissioned in 1950.

The Caliraya development actually includes two rolled earthfill dams. A 30 m-high dam on nearby Lumot River diverts water through a circular connecting waterway 2 m in diameter and 218 m long. The Caliraya Dam impounds water in its 86 million cubic meter reservoir which supplies turbine water to the powerhouse via a 2.3 m penstock, 740 m to a point where it branches into four manifolds of 1.0 m diameter each. Flood discharges are handled by three spillways; the service spillways near Caliraya and Lumot dams are of the drop inlet types while the third, for emergency, is a saddle-type spillway, 160 m along its axis, which also doubles as a roadway for vehicular traffic.

The Caliraya River dam is founded on layers of hard basalt rock for most of its base while the abutments rest on stable clay material. The upstream face slopes three to one down to elevation 265 where it is level and is paved with a 0.15 m thick concrete throughout its face. The downstream side contains three levels of 4.0 m berms and is generally sloped at 1.5:1 with its toe containing boulders and rocks. There is a thick central impervious core composed of rolled earth which forms the bulk of the construction material. The physical features of the Caliraya dam are shown on Figure 1 and Exhibit A and can be summarized as follows:

| Type | Earthfill | |
|---|---|---|
| Height | 42 | m. |
| Base Width | 240 | m. |
| Crest Length | 426 | m. |
| Crest Width | 10 | m. |
| Embankment Volume | 0.65 | MCM |

The Ambuklao Dam

Featuring one of the highest rockfill dams in the world during its construction in 1950, the Ambuklao Hydroelectric Plant is located some 36 km northeast of Baguio City, the summer capital of the Philippines. The hydro station has a capacity of 75,000 kw and provides a yearly output of $400 \times 10^6$ kwh of electrical energy to the Luzon Grid.

The Ambuklao dam on Agno River is sited across a narro gorge between two ridges. This location was chosen because of a large natural reservoir available, a favorable head for power and the spillway construction directly over the mountain slope on the left ridge. Earlier plans by Harza Engineering called for a homogeneous rockfill structure but materials obtained near the damsite were found to be too fine for the purpose. It was finally determined that a zoned rockfill dam would be the best alternative, the materials being available on-site. The dam, as built, stands 129 m high with a crest length of 452 m and its base measures 444 m at the maximum section (Figure 2). The embankment volume is 5.82 million cu. m. of earth and rock.

The foundation of the dam is of diorite, metamorphosed andesite and sedimentary rocks. No Major problems were encountered during its construction.

Other features of the development include a reservoir of 258 million cu. m. capacity, a 127 m-wide spillway with eight openings controlled by tainter gates, each measuring 12.5 x 12.5 m which can handle a design flood of 7,300 cms. An intake tower delivers water to the underground power station via three horseshoe-shaped power tunnels.

The Binga Dam

The second hydraulic power scheme along the Agno River, the Binga Plant is located 19 km downstream of Ambuklao. This IBRD-financed project was undertaken by NPC in 1956 to meet the increasing power demand obtaining then in the area which was induced by the operation of the Ambuklao Plant. Four years after, the Binga Hydroelectric Plant was commissioned to generate 100,000 kw of electricity. Design and construction drawings were undertaken by the consulting firm of Tippets-Abbet-McCarthy-Stratton (TAMS) while project implementation was done by the Philippine Engineering Syndicate, Inc.

The Binga Dam is an earth and rockfill structure, shown in fig. 3 with an inclined impervious rolled earth core and rockfill on both faces. It is 107 m. in height and the crest length is 215 m. at elevation 586 m. The top of the dam, 8 m wide, serves as a vehicular roadway and the dam itself contains 2,000,000 cu.m. of earth and rock. Preliminary investigation showed that rocks at the damsite are of metamorphic series with some fractures, hence, extensive grouting was undertaken during the dam construction to preclude a possible seepage problem posed by the sheared zones. The concrete spillway on the left abutment is 94.5 m wide and flood flows of 5,200 cms are controlled by six hoist-driven gates 12.5 m wide by 12 m high. During its initial operation, scouring was noticed downstream of the bucket section. Big-sized rocks and boulders were dumped into the area to prevent further damage on the outfall area.

The Binga reservoir which stretches 7 km long and 1 km at its widest section is capable of storing 48,200,000 cubic meters of water and provides ample opportunities for recreational activities like boating, swimming and fishing. Because of this, the man-made lake has become a favorite tourist attraction in this part of Luzon.

The Angat Dam

The construction of this multi-purpose dam in Norzagaray, province of Bulacan, has paid dividends in power generation, water supply and irrigation. One of the two diversion tunnels required during the construction of the Angat Dam was left in service to maintain the flow of water to the MWSA-owned Ipo Dam, 7.5 km downstream, which is Metro-Manila's principal source of domestic water supply. At the outlet of this tunnel is a secondary hydro station which, together with the main station, have a combined installed capacity of 212 MW. The reservoir also provides year-round irrigation water to some 30,000 hectares of agricultural lands. It forms a large semi-elliptical loop 35 km. long and contains a billion cubic meters of water from its 568 sq. km. watershed.

With its 131 m. high zoned earthfill dam, Angat currently holds the Philippine record of the highest dam in its class (fig. 4). The 10-m wide crest extends 568 m at elevation 220 m while its base is 550 m at the widest breadth. The upstream face slopes 1:1.4 down to elevation 203 m before it widens, 1:2.5, to the river bed. There is a roadway on the downstream side, which slopes 1:1.4. With dumped rockfill on both sides and an inclined impervious core, the dam contains a total volume of 7,000,000 cubic meters. Principal rock in the damsite consist of interbedded series of meta- morphosed lavas and sedimentary rocks intruded by andesite. In 1965, a foundation problem developed at the left abutment of the dam. Cracks were noted during the excavation, clean-up and preparatory filling operations. Studies were made and corrective measures were implemented among them, an adit along the clay core portion at the base and running up the ridge was excavated. This served as grouting gallery during the succeeding operations. A 3-meter deep trench was dug, ridded of fractured rocks, gunited and backfilled with concrete. Curtain grouting consisting of three lines extending 45 m. below the invert of the adit was performed. From the gallery, further foundation treatment was undertaken simultaneously with the dam filling; all these were executed under close supervision. Sufficient supply of clay material for the impervious core of the dam and dikes, sand for filters, gravel and rocks were available in the vicinity of the project.

The ogee-crested spillway structure is designed for a flood discharge of 7,500 cms and accommodates three 12.5 m x 15 m radial tainter gates. At the end of the concrete chute is a flip bucket which deflects water 100 meters downstream to avoid scouring of the toe.

The pit-type main hydro station contains four generating units of 50 megawatt-capacity each, while the auxiliary station houses two units generating a total of 12 MW. With their combined output of 212,000 kw, the Angat Hydroelectric development is today the largest hydro installation in the country.

The Pantabangan Dam

Among the twelve dams already built, Pantabangan Dam in Nueva Ecija is the most significant multipurpose development. As well as having a 100 MW generating capacity, the scheme has a flood cushion of 305 MCM to control floods which used to cause serious damages in the downstream flood plains and provide year-round dependable supply of irrigation water to 83,760 hectares of fertile crop lands. The Pantabangan Reservoir, with a volume of $3.0 \times 10^9$ cubic meters in a watershed of 853 sq. kms. is the largest in the Philippines and is becoming an important asset as one of the country's busy tourist and recreational centers.

The new dam is composed of two-zoned earthfill structures which meet at a ridge in the center. The combined structures have a total crest length of 1,615 m, a height above the river bed of 107 m at the maximum section and a maximum base width of 535 m. They contain a total of 12,300,000 cubic meters of earth and rock. Much of the material for the core came from excavation for the foundations, spillway and channel.

The dams impound separate reservoirs which are joined by a connecting channel 200 m. long and 50 m. wide, which was excavated as part of the project. Water is released thru the left abutment of the main structure. Used for diversion during construction, one of the tunnels was converted for a penstock for the powerhouse on the downstream side of the dam. The other will continue to function as an outlet. The concrete spillway, with three gated channels controlled by 8 x 15 m. radial gates and one ungated channel, is located at the left abutment of Aya Dam. Flood discharges from the Pampanga River will flow through the connecting channel into the Aya reservoir and thence to the spillway which is designed to handle a flood of 4,200 cms and the entire volume of turbine water in the event of complete shutdown. Typical Section of the dam is shown on Figure 5.

Hydro Resources, a group of local contractors who pooled their forces, proved its mettle by completing the Pantabangan Dam well ahead of schedule. This factor is expected to make the country largely independent of foreign contractors for this type of development.

DAMS UNDER CONSTRUCTION

The four dams currently under construction are: (a) Magat Dam (NIA-NPC); (b) Agus IV (NPC); (c) Agus VII (NPC); and, (d) Pulangi IV (NPC), Figures 6,7 and 8.

Among them, Magat Dam is the largest and is the only one discussed herein.

The Magat Multipurpose Project is expected to contribute great economic benefits to the country following the experience of Pantabangan. It will provide a dependable supply of irrigation water to 104,600 hectares of farmland and is expected to increase rice production by some 474,338 metric tons yearly. The power phase of the project, which involves a hydro station with 540 MW capacity, will add 1,103 GWH of electric energy annually to the system. When completed, the dam will put an end to the chronic flood damages occurring in the area. The new dam will also create a large artificial lake (44.6 sq. km.) ideal for recreation such as swimming, boating and water skiing, and is likewise intended for fish conservation and water supply to neighboring municipalities of Isabela Province. The Magat Multipurpose scheme is scheduled for completion by 1983.

The Project involves construction of dams in two sections - the Magat and Baligatan Dams. The central feature, the Magat Section, is 6 km upstream of an existing irrigation diversion dam, the Maris. Plans call for a composite-section Magat dam with a concrete portion 783 m. long, which incorporates the spillway; a right wing and left wing embankment, 1,199 m and 943 m long, respectively. The highest section is 114 meters while the base is 102 meters wide at the most. With a crest width of 12 m, the dam provides for a roadway throughout its 2,925 m length, including a bridge over the concrete spillway. The zoned rockfill components have an upstream slope of three to one and the downstream face slopes 2.5 to one. Estimated volume of fill for the embankments total 12,000,000 cubic meters.

The overflow chute spillway on center of the concrete section contains eight openings, all of which are controlled by radial gates. The openings are designed to pass 30,400 of flood waters. On this section is also located the power intake structure on the left side, which is 90 m wide, to serve six turbines at the powerhouse immediately downstream at the toe. The concrete dam will require a total of 700,000 cubic meters of concrete. The drawings on Figure 6 show the details of the proposed Magat and Baligatan Dams.

PROPOSED DEVELOPMENT

Over twenty dams, most of them higher than 100 m and of rockfill types, have been planned for construction, with priority still given to hydro power. A new expanding purpose for which dams are to be built, aside from hydro-power and water supply, are recreation and environmental enhancement associated with new community developments. The latter results from larger reservoir requirements caused by higher water demand, hence necessitating relocation of some communities. The consequent effect is to upgrade the natural scenery and existing sight seeing spots around the area for added benefits.

Notable among the many dams planned to be built are San Roque Dam in the Agno River in Pangasinan, the Gened-Abulug and Chico Dams in Kalinga-Apayao. The feasibility of major hydro-electric developments in these three sites have been confirmed. The Abulug development will feature the Philippines' first attempt on double-curvature arch dam construction.

While water exploitation must be pursued to sustain the nation's economy, the government is in no hurry to develop such sites. There are at present more restrictions on site use than there were decades ago. Preservation of the "wild rivers" has become a popular concern and objections raised against dam constructions by the people affected are often related to their centuries-old customs and traditions. Because of this, the government is taking time to evaluate alternatives and compromise.

An enumeration of the various dams either scheduled for construction or in the feasibility and reconnaissance stages are listed in Annex A.

SOCIAL AND ECONOMIC CONSIDERATIONS

Among the major benefits that can be derived from dam construction, the following stand out prominently: (a) Power Generation. The country's industrialization program would continue to depend on hydro and sources of power other than oil. Of the present total installed capacity of 4,035 MW, or an annual generated energy capability of 15,815 GWH, only 24% is attributable to hydro. If the current power expansion program is vigorously pursued, this percentage will have improved by 1990 in which hydropower is expected to reach 48% of total capacity and 36% of energy generated annually. This will mean higher energy and capacity value, increased industrial production and, therefore, stabilization of our economy; (b) Increased agricultural production. The Philippines is a rice-producing country. During the past decades, our irrigation systems drew their supply of water mostly out of releases from run-of-river powerplants and diversion from small rivers. Low diversion dams could not conserve river inflows during wet seasons. The general effect was that only negligible agricultural areas could be served during dry months. With the integration of irrigation into our on-going multi-purpose schemes, a year-round supply of water is assured and the farms could produce at least two crops of irrigated rice per year. Availability of regulated water would also permit farmers to adopt new high-yielding rice varieties for maximizing farm income; (c) Mitigation of flood losses. Typhoons are frequent during the wet seasons and the "rice granaries" are often seriously affected. Rice crops and properties are devasted resulting in low annual production. With better flood protection afforded by our regulation dams, benefits could be significant; and, (d) Provision of domestic, commercial and industrial water supply.

Aside from the direct benefits, dam projects would also generate important indirect benefits, some of which are: (a) increase employment as a result of manpower requirements by dam contractors, or creating employment opportunities for a large number of rural families whose standard of living would be greatly improved; (b) increased property values and personal income will materially increase tax revenues; (c) transportation system expansion to accommodate transport of agricultural production in the area; and, (d) promote tourism by means of man-made lakes suitable for recreational activities such as boating, swimming and fishing.

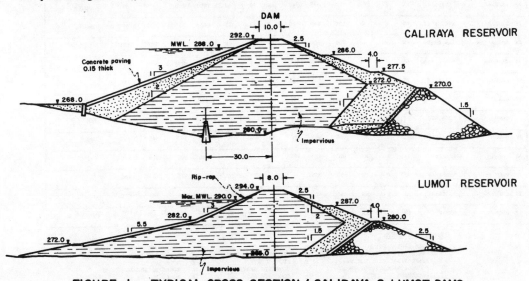

FIGURE 1    TYPICAL CROSS SECTION / CALIRAYA & LUMOT DAMS

# Annex A

STATISTICAL DATA OF PHILIPPINE DAMS

| | Name of Dam/Project | Location | Year Comp'd | Purpose | Type | Dam Ht,M | Crest LGT. M. | Max. Base Wd.M. | Top Wd. M. | Vol. of dam MCM. | Spillway Width M. | Spillway Cap. MCM | Res. Vol. MCM. | DR'GE Area Sq.Km. | Own'r | Consultant | Contractor |
|---|---|---|---|---|---|---|---|---|---|---|---|---|---|---|---|---|---|
| 1 | CALIRAYA | LAGUNA | 1945 | P | E | 42 | 426 | 240 | 10 | .65 | 4.2(s) | 260 | 78 | 92 | NPC | U.S. ARMY ENGRS. | SG1 |
| 2 | MA. CRISTINA (AGUS VI) | LANAO N. | 1953 | P | R | 13 | 67 | 36 | 9 | .25 | 35 | 35 | ~2 | 1891 | NPC | | PHESCO |
| 3 | AMBUKLAO | BENGUET | 1956 | P | R | 129 | 452 | 444 | 12 | 6 | 127 | 7300 | 258 | 690 | NPC | HARZA | GFA |
| 4 | BINGA | BENGUET | 1960 | P | ER | 107 | 215 | 400 | 8 | 2 | 95 | 5200 | 48 | 936 | NPC | TAMS-EDCOP | PESI |
| 5 | ANGAT | BULACAN | 1967 | P,I,W,F | ER | 131 | 568 | 550 | 10 | 7 | 36 | 7500 | 850 | 568 | NPC | HARZA-EDCOP | PESI |
| 6 | PANTABANGAN | NVA.ECI.JA. | 1977 | P,I,W,F | ZE | 107 | 1615 | 535 | 12 | 12 | 270 | 4200 | 2996 | 853 | NPC-NIA | ECI-EDCOP | HYDRO |
| 7 | AGUS II | LANAO S. | 1979 | P | E | 29 | 140 | 133 | - | 1 | 113 | 2400 | 2 | 1711 | NPC | SCFRELEC | PHESCO |
| 8 | MAGAT | ISABELA | UC | P,I,E,ZR,W,F | ZE 08 | 114 | 2925 | 280 | 12 | 13 | 276 | 30400 | 1464 | 4143 | NPC NIA | E S E D | HYDRO |
| 9 | AGUS VII | LANAO N. | UC | P | CG | 48 | 174 | 40 | 4 | .15 | 32 | 2000 | 1 | 1883 | NPC | E L C | CDCP |
| 10 | AGUS IV | LANAO N. | UC | P | ZR,CG | 32 | 1478 | 200 | 6 | 4 | 25 | 1400 | 25 | 1873 | NPC | LAHMEYER | HANIL |
| 11 | AGUS III | LANAO N. | FS (1978) | P | R | 55 | 180 | 450 | 10 | 1 | 42 | 2770 | 2 | 1739 | NPC | ELC-EDCOP | |
| 12 | SAN ROQUE | PANGASINAN | FS (1979) | P,I,W,F | ZE | 210 | 1130 | 800 | 20 | 43 | 90 | 15600 | 680 | 1250 | NPC | E L C | |
| 13 | GENED-ABULOG | KAL-APAYAO | FS (1979) | P,I,W,F | CA | 175 | 472 | 45 | 12 | 2 | 96 | 15000 | 1300 | 1661 | NPC | NEWJEC | |
| 14 | CHICO II | MT. PROV. | FS | P | CA | 160 | 400 | 530 | 9 | 2 | 41 | 4500 | 560 | 720 | NPC | LAHMEYER | |
| 15 | CHICO IV | KAL-APAYAO | FS | P | R | 155 | 880 | 657 | 6 | 18 | 84 | 8000 | 430 | 1410 | NPC | LAHMEYER | |
| 16 | AGOS I | QUEZON | FS | P | R | 140 | 960 | 725 | 15 | 21 | 150 | 17000 | 540 | 867 | NPC | NKOIE | |
| 17 | DIDUYON | QUIRINO | FS | P | CG | 95 | 375 | 88 | 8 | .70 | 105 | 8900 | 555 | 477 | NPC | NEWJEC | |
| 18 | BAGO I | NEGROS OC. | FS | P | R | 115 | 450 | 523 | 6 | 5 | 70 | 1706 | 195 | 387 | NPC | SECO | |
| 19 | BAGO III | NEGROS OC. | FS | P | R | 60 | 410 | 321 | 6 | 3 | 70 | 1856 | 33 | 421 | NPC | SECO | |
| 20 | PULANGI III | BUKIDNON | FS | P | ER | 120 | 1150 | 606 | 6 | 3 | | | 2280 | 1320 | NPC | MEISDOR-EDCOP | |
| 21 | PULANGI IV | BUKIDNON | FS (1978) | P | ER | 26 | 1070 | 85 | 7.5 | 1 | 28 | 6900 | 67 | | NPC | SCFRELEC | |
| 22 | ILAGAN | ISABELA | PS | P | R | 160 | 625 | 740 | 6 | 20 | 36 | 7500 | 830 | 875 | NPC | LAHMEYER | |
| 23 | DAGKAN | NVA. ECIJA | RS | P | R | 150 | 640 | 520 | 8 | 17 | 55 | 6400 | 365 | 565 | NPC | NEWJEC | |
| 24 | AMBURAYAN | LA UNION | RS | P | R | 200 | 900 | 800 | 10 | 28 | 66 | 10200 | 590 | 615 | NPC | | |
| 25 | TANUDAN | KAL-APAYAO | RS | P | R | 200 | 420 | 800 | 8 | 14 | 66 | 5100 | 480 | 296 | NPC | | |
| 26 | ABUAN | ISABELA | RS | P | R | 140 | 400 | 720 | 9 | 9 | 44 | 5600 | 450 | 458 | NPC | | |
| 27 | DIBAGAT | KAL-APAYAO | RS | P | R | 130 | 320 | 593 | 6 | 5 | 77 | 11600 | 290 | 763 | NPC | NEWJEC | |
| 28 | TABU | BENGUET | PS | P | CA | 113 | 330 | 630 | 8 | | 60 | | | 1070 | NPC | ELC | |
| 29 | CADENO | NVA.VISCAYA | RS | P | R | 150 | 640 | 700 | 10 | 17 | 55 | 6400 | 365 | 565 | NPC | NEWJEC | |
| 30 | AODULU | KAL-APAYAO | RS | P | R | 108 | 220 | 594 | 8 | | | 7100 | 400 | 510 | NPC | NEWJEC | |
| 31 | ULOT | E. SAMAR | RS | P | ER | 50 | 200 | | | 20 | | | 236 | 903 | NPC | | |
| 32 | CATUBIG | M. SAMAR | RS | P | ER | 68 | 400 | 220 | | 28 | | | 176 | 283 | NPC | | |
| 33 | JALAUR | PANAY | FS | P,I,W | ER | 150 | 300 | 580 | 12 | 4 | 20 | 1240 | 442 | 1065 | NK-NIA | ECI | |

LEGEND ON SYMBOLS:

| | | | | | |
|---|---|---|---|---|---|
| UC | UNDER CONSTRUCTION | P | POWER | E | EARTHFILL |
| FS | FEASIBILITY STUDY PHASE | W | WATER SUPPLY | ER | EARTH AND ROCK |
| PS | PRE-FEASIBILITY STUDY PHASE | F | FLOOD CONTROL | R | ROCKFILL |
| RS | RECONNAISSANCE STUDY PHASE | CG | CONCRETE GRAVITY | ZE | ZONED EARTHFILL |
| | I  IRRIGATION | CA | CONCRETE ARCH | ZR | ZONED ROCKFILL |

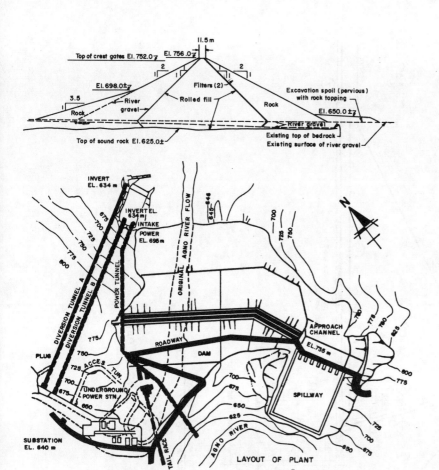

FIGURE 2 AMBUKLAO HYDROELECTRIC PROJECT

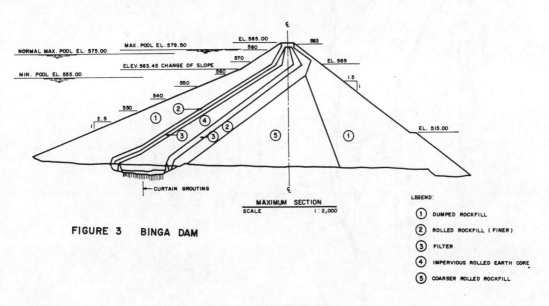

FIGURE 3 BINGA DAM

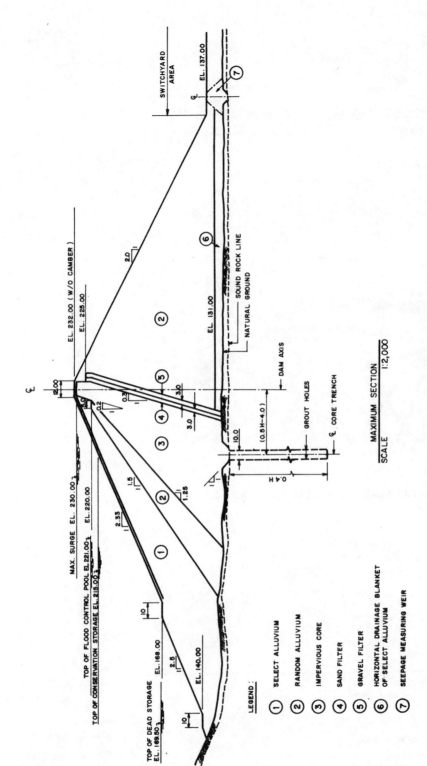

FIGURE 5   PANTABANGAN DAM

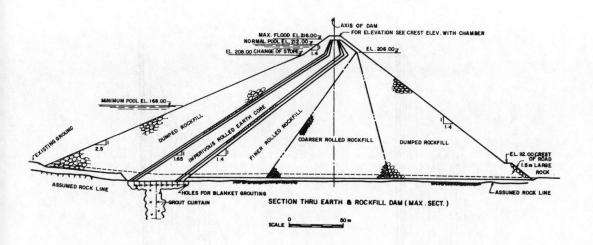

FIGURE 4   ANGAT DAM

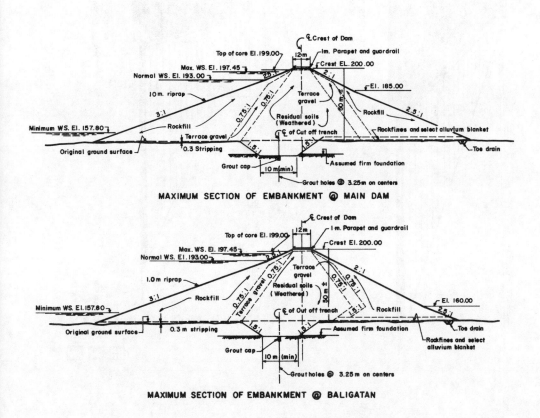

FIGURE 6   MAGAT MAIN DAM & BALIGATAN DAM

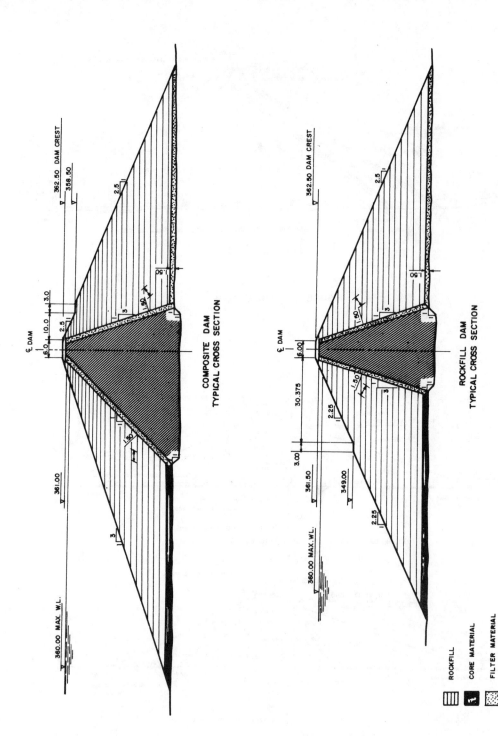

FIGURE 7 AGUS IV

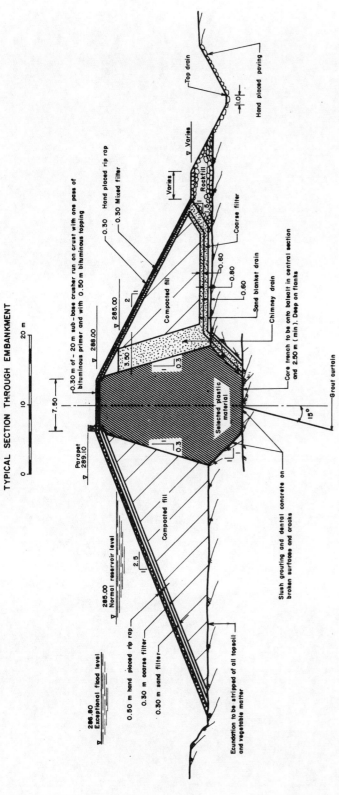

FIGURE 8    PULANGI IV EMBANKMENT

*Symposium on Problems and Practice of Dam Engineering / Bangkok / 1-15 December 1980*

# Dam engineering in Taiwan

DAVID S.L.CHU
*Taiwan Power Co., Taipei*

## INTRODUCTION

Taiwan is a province of the Republic of China, locating astride the tropic of Cancer, approximately 150 kilometers off the coast of China mainland. The island, tobacco leaf in shape, is about 394 kilometers long and 144 kilometers wide, with an area of about 36,000 square kilometers or 13,900 square miles.

## 1. Topography

Topography of the island is rugged. The Central Range running generally northsouth, divides the island into two parts, mountainous east and western plain. The highest peak in this range is Yushan at an elevation of 3,997 meters. There are 30 peaks over 3,000 meters in elevation. The remainder of the Central Range is generally about 2,500 meters. Approximately 70 percent of the area of Taiwan is composed of mountains and hills, the remaining 30 percent being mostly alluvial plains.

## 2. Rainfall and Streamflow

Taiwan is situated in the midst of East Asia monsoon system. In winter, the cool moist northeast winds prevail. In summer, the air currents from the southwest are warm and moist.

The Central Range intercepts these monsoons and provides a decided orographic effect to precipitation. The winter northeast winds provide significant amount of moisture to the northern end of the island and leave the rest of the island in the rain shadow. When wind direction is reversed in summer, the mountains intercept southwestern winds to provide summer precipitation to the southwestern section of Taiwan. Additional precipitation results from thunderstorms and typhoons which sweep Taiwan in the months from June to September and bring about 60 percent of annual rainfall.

Annual precipitation ranges from slightly less than 1,500 milimeters in the western lowlands to more than 6,000 milimeters at the northern tip. Plain areas generally receive from 1,300 to 2,500 milimeters; mountain areas receive from 2,500 to 3,000 milimeters. The annual isohyetal map is shown in Figure 1. Generally speaking, the wet season of the whole island is from June to October, while the dry season is from November to next May. As a result of short and steep rivers, the streamflow fluctuates considerably. Floods of magnitude over a thousand times of minimum flow were recorded in many streams. During typhoon the flood peak may be extremely high. Due to the landslide and difficult soil conservation condition in upper mountains the sediment contents of several streams in middle and downstream areas are very high.

## 3. River Basins

In Taiwan, there are more than 60 rivers. The majority of which originate from the Central Range at varying elevations and run down to the coast. Because of the location of the range, the rivers on the west side are generally flatter than those on the east side of the island. All rivers are short and rather flashy. Nineteen rivers are classified as the major ones. Location of basins is shown in Figure 2.

## 4. Dams in Taiwan

In Taiwan, the mountains are numerous and

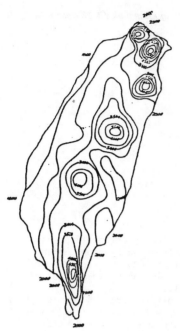

Fig. 1. Annual Isohyetals in Taiwan

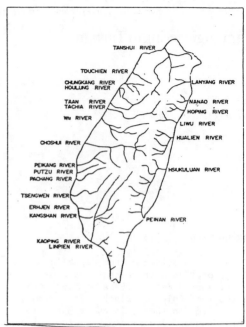

Fig. 2. Location of Major Rivers in Taiwan

steep; the rivers are short and rapid; the geological structure is weak and the rainfall distribution is uneven. It is frequently hit by typhoons, floods and earthquakes. It is apparent that without water resources development, flood and drought could hardly be prevented.

For the development of water resources, the Government of the Republic of China has endeavoured in the construction of dams and reservoirs in the island to reduce flood damage, to irrigate farmland, to generate power and to supply water for domestic and industrial uses. As shown in Table 1, there are forty seven dams were completed, two under construction, four under design. Eight most significant dams will be described in the report.

EXISTING DAMS

1. Techi Dam

The Tachia River lies in the central part of Taiwan and originates from the Central Mountain Range. It has a total length of 124 kilometers with a total catchment area of 1,272 square kilometers. Techi Dam is located on the upstream of Tachia River and controls a drainage area of 514 square kilometers.

The Techi Dam creates a reservoir which is the key to the development of the Tachia River system. The regulation flow is released to run a series of four powerstations already built. Also it has the effects of improving irrigation of downstream farmland, controlling floods, water supply and recreation. The basic project layout involves the following:

The dam is a thin arch, double curvature structure with a maximum structural height of 180 meters, an excavation volume of about 445,800 cubic meters and a concrete volume of 456,000 cubic meters. The dam crest is at elevation 1411 meters, with crest length of about 290 meters. The thickness of the dam varies from four meters at the top to 20 meters at the base.

The reservoir created by the dam has a maximum normal water level of 1408 meters and a minimum water level of 1350 meters. The gross storage capacity is 232 million cubic meters, while the effective storage capacity 175 million cubic meters.

The spilling system consists of a free flow tunnel spillway on the left bank bypassing 3,400 cubic meters per second, a crest way over the dam discharging 1,400 cubic meters per second, two sluiceways through dam discharging 1,600 cubic meters per second and two Howell Bunger values for emergency.

The powerhouse with three 78,000 kilowatts units is underground located at the left abutmetn of Techi gorge to avoid construction interruption and possible hazard

TABLE 1 DAMS AND RESERVOIRS IN REPUBLIC OF CHINA

| | Name | River | Tributary | Type *1 | Dam Height (m) | Length (m) | Volume (cu.m.) |
|---|---|---|---|---|---|---|---|
| A. Existing | | | | | | | |
| 1. | Chukeng | Tanshui River | Hsintien Creek | CG | 3.6 | 91.8 | |
| 2. | Nuannuan | do. | Keelung Creek | CG | 26.4 | 127.0 | 22,500 |
| 3. | Kweishan | do. | Hsintien Creek | CG | 21.5 | 124.9 | 23,350 |
| 4. | Ayu | do. | Tonghou Creek | CG | 19.5 | 55.0 | 13,920 |
| 5. | Lahao | do. | Nanshih Creek | CG | 27.5 | 62.0 | 30,160 |
| 6. | Shihmen | do. | Tahan Creek | E & R | 133.0 | 360.0 | 6,890,000 |
| 7. | Shihmen Afterbay Weir | do. | do. | CG | 10 | 371.5 | 119,600 |
| | | | | E & R | 29 | 150.0 | 123,700 |
| 8. | Chingtsao Lake | Koya River | | E | 17.0 | 149.0 | 66,000 |
| 9. | Hsiho | Chungkang River | Omei Creek | CG | 5.0 | 90.0 | 5,000 |
| 10. | Chientan | do | Nankang Creek | CG | 10.5 | 34.0 | 2,000 |
| 11. | Tapu | do. | Omei Creek | CG | 21.4 | 98.8 | 27,500 |
| 12. | Patzekang | Houlung River | Laotienliao Creek | E | 12.0 | 44.0 | 72,600 |
| 13. | Tienlun | Tachia River | | CG | 54.0 | 92.0 | 88,000 |
| 14. | Kukuan | do. | | CA | 85.1 | 149.0 | 77,770 |
| 15. | Wuchieh | Choshui River | | CG | 57.6 | 91.0 | |
| 16. | Sun-Moon Lake | do. | Shuisheh Creek | E | 30.0 | 364.0 | 1,590,000 |
| 17. | do | do. | Tousheh Creek | E | 19.1 | 164.0 | |
| 18. | Chongkuei | do. | Chongkuei Creek | CCG | 26.2 | 63.2 | |
| 19. | Wanta | do. | Wanta Creek | CG | 13.0 | 46.4 | 5,740 |
| 20. | Wusheh | do. | Wusheh Creek | CCG | 114.0 | 225.7 | 328,800 |
| 21. | Peishankung | Wu River | Nankang Creek | CG | 9.0 | 119.0 | 3,000 |
| 22. | Luliao | Pachang River | Touchien Creek | E | 30.0 | 270.0 | 168,300 |
| 23. | Hungmao Pei | do. | | E | 31.0 | 340.0 | |
| | | | | | 11.0 | 320.0 | 398,700 |
| | | | | | 19.0 | 17.0 | |
| 24. | Chienshan Pei | Chishui River | Kueichung Creek | E | 30.0 | 256.0 | |
| 25. | Teyuan Pei | do. | Wentsopu Creek | R | 7.0 | 633.0 | 16,000 |
| 26. | Paiho | do. | Paishui Creek | E | 42.5 | 210.0 | 950,000 |
| 27. | | do. | do. | E | 8.5 | 130.0 | |
| 28. | Wushantou | Tsengwen River | Kuantien Creek | E | 56.0 | 1,273.0 | 4,002,000 |
| 29. | Hutou Pei | Yenshui River | | E | 15.3 | 470.0 | |
| 30. | Yenshui Pei | do. | Chiehtung Creek | E | 17.0 | 91.0 | 41,470 |
| 31. | Akungtien | Akungtien River | | E | 31.0 | 200.0 | 816,000 |
| 32. | do | do | | E | 6.0 | 2,180.0 | |
| 33. | Tulung | Kaoping River | | CG | 6.2 | 104.0 | |
| 34. | Lungluan Tan | Paoli River | | E | 8.0 | 1,967.0 | |
| 35. | Liwu | Liwu River | | CG | 25.0 | 125.0 | |
| 36. | Chingshui No. 1 | Hualien River | Mukua Creek | CG | 5.5 | 34.0 | 1,100 |
| 37. | Tungmen | Hualien River | Mukua Creek | CG | 36.3 | 88.0 | |
| 38. | Lungchien | do. | Lung Creek | CG | 29.5 | 83.0 | 15,540 |
| 39. | Houlung | Houlung River | Laotienliao Creek | E | 35.5 | 187.0 | 450,000 |
| 40. | Chingshan | Tachia River | | CG | 45.0 | 100.0 | 57,000 |
| 41. | Tsengwen | Tsengwen River | | E & R | 133.0 | 400.0 | 9,296,100 |
| 42. | Techi | Tachia River | | CA | 180.0 | 290.0 | 456,000 |
| 43. | Chingtan | Tanshui River | Hsintien Creek | CG | 24.0 | 114.0 | 19,034 |
| 44. | Chihtan | " | " | CG | 29.0 | 117.0 | 41,485 |
| 45. | Ihsing | " | Tahan Creek | CG | 31.5 | 100.0 | |
| 46. | Paling | " | " | BG | 38.0 | 80.4 | |
| 47. | Shihkang | Tachia River | | CG | 27.0 | 352.0 | |
| B. Under Construction | | | | | | | |
| 1. | Jonghua | Tanshui River | Tahan Creek | CA | 82.0 | 160.0 | 62,000 |
| 2. | Yuanshan | " | " | CG | 12.0 | 310.0 | 30,000 |
| C. Under Design | | | | | | | |
| 1. | Feitsui | Tanshui River | Hsintien Creek | CA | 120.0 | 500.0 | 840,000 |
| 2. | Minghu | Shuili River | | CG | 57.5 | 169.5 | 190,000 |
| 3. | Kuyuan | Liwu River | | CG | 52.0 | 104.0 | 99,300 |
| 4. | Mingtan | Shuili River | | CG | 63.0 | 316.0 | 508,000 |

ABBREVIATIONS:  *1 Type of Dam
    CA - Concrete arch dam
    CCG - Concrete curved gravity dam
    CG - Concrete gravity dam
    E - Earthfill dam
    E&R - Earth and Rock fill dam
    BG - Buttress gravity dam

*2 Purpose
    FC - Flood control
    I - Irrigation
    P - Power
    WS - Water supply
    PS - Pumped Storage
    O - Other

Jan. 30, 1980

| Dam | Drainage Area above Dam (sq.km) | Normal Water Level (m) | Reservoir Area At NWL (sq.km) | Reservoir Gross Capacity (cu.m) | Reservoir Effective Capacity (cu.m) | Purpose *2 | Year Completed | Owner *3 |
|---|---|---|---|---|---|---|---|---|
| A1. | 651.8 | 51.0 | | 240,000 | 240,000 | P | 1908 | TPC |
| 2. | 6.5 | 71.8 | 0.11 | 580,000 | | WS | 1927 | KLW |
| 3. | 312.7 | 111.0 | | | 250,000 | P | 1941 | TPC |
| 4. | 72.8 | 216.8 | | 133,000 | 105,000 | P | 1951 | do. |
| 5. | 207.0 | 221.2 | | 304,000 | 300,000 | P | 1951 | do. |
| 6. | 763.4 | 245.0 | 8.15 | 316,000,000 | 251,000,000 | I, P, FC&WS | 1964 | SRAE |
| 7. | 3.0 | 137.0 | 0.6 | | 2,200,000 | I | 1964 | do |
| 8. | 30.0 | 36.0 | 0.25 | 1,100,000 | 845,500 | I | 1956 | HCIA |
| 9. | 12.0 | 55.5 | 0.25 | 600,000 | 600,000 | I | 1951 | CNIA |
| 10. | 35.0 | 19.0 | 0.16 | 560,000 | 560,000 | I | 1957 | do. |
| 11. | 100.0 | 69.0 | 1.20 | 9,400,000 | 8,160,000 | I | 1960 | do. |
| 12. | 1.5 | 54.0 | 0.03 | | 70,000 | I | 1956 | MLIA |
| 13. | 796.6 | 747.8 | | 824,400 | 753,600 | P | 1952 | TPC |
| 14. | 707.7 | 950.0 | 0.62 | 17,100,000 | 9,039,000 | P | 1961 | do. |
| 15. | 501.0 | 764.0 | 2.55 | 55,000 | | Diversion | 1937 | do. |
| 16. | 501.3 | 748.5 | 8.00 | 167,800,000 | 147,870,000 | P | 1937 | do. |
| 18. | 3.8 | 400.3 | 0.02 | 110,300 | 51,600 | P | 1937 | do. |
| 19. | 162.1 | 1,177.3 | | | | P | 1943 | do. |
| 20. | 219.0 | 1,005.0 | 3.50 | 150,000,000 | 127,000,000 | P | 1959 | do. |
| 21. | 333.7 | 377.2 | | | | P | 1921 | do. |
| 22. | 7.5 | 72.5 | 0.55 | 3,780,000 | 3,570,000 | WS | 1939 | TSC |
| 23. | 31.0 | 73.0 | 0.70 | 6,150,000 | 5,500,000 | WS | 1944 | CYW |
| 24. | | 82.0 | | 8,110,000 | 4,080,000 | WS | 1938 | TSC |
| 25. | 24.7 | 14.0 | 1.92 | 3,410,000 | 2,970,000 | I | 1956 | CNIA |
| 26. | 26.6 | 109.0 | 1.97 | 21,600,000 | 19,400,000 | I, FC & WS | 1965 | PWCB |
| 28. | 496.0 | 60.6 | 13.00 | | 148,000,000 | I | 1930 | CNIA |
| 29. | | | 0.36 | | 1,360,000 | I | 1921 | CNIA |
| 30. | 5.8 | 30.5 | 0.22 | 833,800 | 753,800 | I | 1955 | do. |
| 31. | 31.9 | 38.0 | 4.11 | 45,000,000 | 35,000,000 | I, FC&WS | 1952 | PWCB |
| 33. | 839.3 | 269.7 | | | | P | 1956 | TPC |
| 34. | | | 1.75 | 3,790,000 | 3,630,000 | I | 1958 | PTIA |
| 35. | 510.0 | 167.8 | | 355,000 | 340,000 | P | 1944 | TPC |
| 36. | 72.1 | 696.5 | | | | P | 1941 | do. |
| 37. | | 310.0 | | | | P | | do. |
| 38. | 45.0 | 1,276.6 | | | 200,000 | P | 1959 | do. |
| 39. | 61.1 | 61.0 | 1.62 | 17,700,000 | 16,500,000 | I & WS | 1970 | MLIA |
| 40. | 517.3 | 1,245.0 | 0.10 | 725,000 | 590,000 | P | 1970 | TPC |
| 41. | 481.0 | 225.0 | 17.14 | 707,530,000 | 598,530,000 | I, P & WS | 1973 | TRAB |
| 42. | 514.0 | 1,408.0 | 4.50 | 232,000,000 | 175,000,000 | P, I & FC | 1974 | TPC |
| 43. | 696.8 | 22.2 | | | 700,000 | WS | 1975 | TWD |
| 44. | 645.7 | 44.7 | 0.83 | | 4,200,000 | WS | 1977 | TWD |
| 45. | 616.0 | | | | | O | 1973 | SRAB |
| 46. | 502.8 | | | | | O | 1977 | SRAB |
| 47. | 1,061.0 | 267.1 | | | 2,700,000 | WS & I | 1977 | PWCB |
| B1. | 561.6 | 413.0 | | | 8,600,000 | P & O | 1982 | SRAB |
| 2. | | 51.5 | | | 1,260,000 | WS | 1980 | TWSC |
| C1. | 303.0 | 170.0 | 10.24 | 406,000,000 | 359,000,000 | WS & P | 1986 | TFRDC |
| 2. | 37.0 | 448.0 | 0.56 | 9,200,000 | 7,400,000 | PS | 1984 | TPC |
| 3. | 152.0 | 690.5 | | | 350,000 | P | 1984 | TPC |
| 4. | 54.8 | 373.0 | 0.64 | 14,400,000 | 12,000,000 | PS | 1988 | TPC |

*3 Owner

| | | | |
|---|---|---|---|
| CNIA | - Chunan Irrigation Association | PTIA | - Pingtung Irrigation Association |
| CYW | - Chiayi Waterworks | PWCB | - Provincial Water Conservancy Bureau |
| HCIA | - Hsinchu Irrigation Association | SRAB | - Shihmen Reservoir Administration Bureau |
| KLW | - Keelung Waterworks | TPC | - Taiwan Power Company |
| MLTA | - Miaoli Irrigation Association | TRAB | - Tsengwen Reservoir Administration Bureau |
| TWD | - Taipei Water Department | TSC | - Taiwan Sugar Corporation |
| TWSC | - Taiwan Water Supply Company | TFRDC | - Taipei Feitsui Reservoir Development Commission |

Upstream View, Tachi Dam

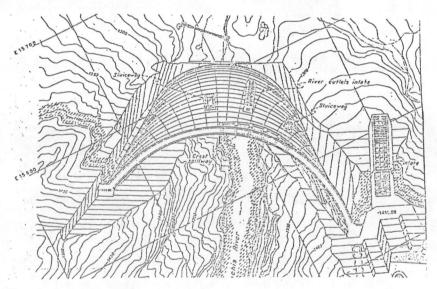

Fig. 3. Techi Dam, Plan

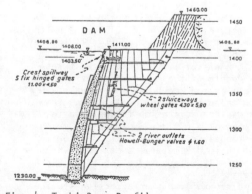

Fig. 4. Techi Dam, Profile

due to rock falling.

The construction work, lasted four years and ten months, was started on December 8, 1969. The whole project was completed on September, 1974.

2. Kukuan Dam

The Kukuan Dam, located in a narrow gorge at mid-stream of the Tachia River in central Taiwan, is primarily a project to store water for generating electrical energy. The construction of the project was started in September 1957 and completed in December 1961.

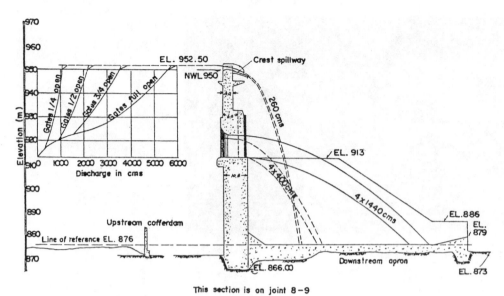

Fig. 5. Kukuan Dam, Vertical Section on Spillway

The Kukuan Dam is of a concrete arch structure of vertical cylinder type with an upstream face radius of 92 meters. Having a total concrete volume of 77,770 cubic meters, the dam is 85.1 meters in height and has a thickness varying from 10 meters at the base to four meters at the crest. Four submerged spillway openings, each 6.6 meters wide and 9.3 meters high controlled by a vertical lift gate, are provided in the main body of the dam to pass a design flood of 5,760 cubic meters per second under the maximum water level of 952 meters. In addition, a crest spillway, 49 meters long, is also provided in the central portion of the crest of the dam to handle a surplus flow the about 260 cubic meters per second.

The normal water level of the reservoir is at elevation 950 meters with the back water of 3.5 kilometers long while the minimum water level for power generation is at elevation 923 meters. The effective storage between these two elevations is about 9,039,000 cubic meters now.

The water from the reservoir is diverted through a two-compartmented two-level intake, a concrete-lined pressure tunnel of 6.2 meters in diameter and 4,856 meters in length, and two embedded vertical steel penstocks to the underground powerstation to feed four 45,000 kilowatts turbine generating units. After serving the generation, the flow discharges back by a tailrace tunnel into the Tachia River.

3. Wusheh Dam

The Wusheh Dam is located on the Wusheh Creek, a tributary of the Choshui River in central Taiwan. It controls a drainage area of 219 square kilometers. The primary purpose is generation of power and furnish water during the dry season to the Sun-Moon Lake downstream.

The Wusheh Dam is of a concrete curved gravity type with a radius of 243.8 meters, a maximum height of 114 meters, a total width of 77 meters at the base and a concrete volume of 328,800 cubic meters. The upstream face of the dam is vertical while the downstream face has a slope of 1 on 0.7. The crest length of the dam is 225.7 meters including an overflow spillway at the middle of dam which provides two openings each controlled by two 13.7 meters wide and 6.25 meters high tainter gates capable to pass a flood of 850 cubic meters per second. Two river outlet works of 1.2 meters square are constructed through the dam for releasing the dead storage of the reservoir to the Sun-Moon Lake in the dry season.

A tunnel spillway, constructed through the saddle of the right abutment with the inlet located about 320 meters upstream from the dam, and the outlet downstream of the Wanta powerstation is designed to carry a flood of 1,670 cubic meters per second. The tunnel spillway has varying diameter from 14 to 8.3 meters, a maximum hydraulic drop of 112 meters and a length of 328.4 meters.

The reservoir has an effective storage capacity of 127 million cubic meters at

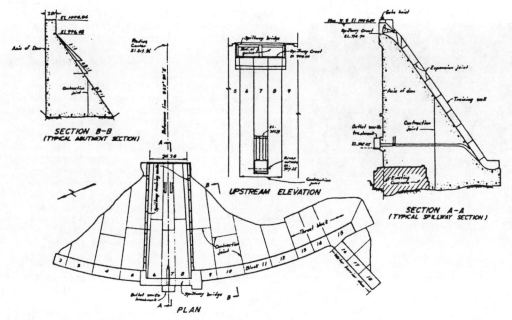

Fig. 6. Wusheh Dam, Plan, Elevation and Section

the normal water level of 1,005 meters and a maximum drawdown of 58 meters for power generation.

The regulated flow from the reservoir is conducted through an intake tunnel and a steel penstock to the Wanta powerstation where two Francis turbine and generator sets having a total capacity of 20,700 kilowatts have been installed. An additional 15,200 kilowatts unit has been installed at the same powerstation supplied with water from the Wanta Dam. After generation, all the tailwater discharges back into the Choshui River.

The Wuchieh Dam, located about 17.6 kilometers downstream from the Wusheh Dam, is the diversion structure of the Sun-Moon Lake. The flow from the Choshui River is diverted to the Sun-Moon Lake at this point by means of a waterway with a total length of 15,000 meters and a capacity of 40.3 cubic meters per second. Thus the Wusheh storage is as important to the Sun-Moon Lake powerstations as the Sun-Moon Lake storage. The seasonal regulation of the Wusheh Reservoir is also beneficial to irrigation of the farmland downstream.

4. Chingshan Dam

The Chingshan Dam, located on the midstream of the Tachia River in Central Taiwan, is a diversion structure to divert flow for generating electrical power between the existing Techi Dam and the existing Kukuan Reservoir. The construction of the project was started in July 1964 and completed in December 1970 with intital installation of two 90,000 kilowatts turbine generating sets.

The Chingshan Dam is of a concrete gravity dam. The dam has a crest length of about 100 meters and a structural height of about 45 meters. Three gated spillways, each provided with a 10.7 meters wide by 20.7 meters high radial gate, are capable to pass the design flood of 6,500 cubic meters per second under the maximum water level at elevation 1247 meters.

The normal water level is at elevation 1245 meters and the minimum water level is at elevation 1236 meters. Live storage between these two elevations, after allowance for sediment deposition, is estimated to be about 590,000 cubic meters.

The flow diverting from the intake to a pressure head tunnel of 6.6 meters in diameter and about 5,436.5 meters long, will be conveyed through the inclined penstock to the underground powerplant to feed two 90,000 kilowatts units in the initial stage. An additional penstock and two turbine generation units, identical in size, was installed after the completion of the upstream Techi Dam. After serving generating, the tailwater then be released to the Kukuan Reservoir through a tailtunnel of 6.9 meters in diameter and 1,602 meters in length.

Downstream View, Chingshan Dam

ern Taiwan, is a key structure of the Shihmen reservoir project. The main functions of the project are for irrigation, power generation, flood control and public water supply with incidental benefit of sediment control and recreation.

The Shihmen Dam is of the earth and rockfill type with an impervious center core. The center core is made of clay cap material; the filters, placed between the core and the rock shells, are made of processed river gravel and sand; and the shells are made of river cobble gravel and terrace cobble gravel. The upstream and downstream faces of dam are protected by boulders. The width of dam is 11.2 meters at the top and 520 meters at the bottom with an average upstream slope 1 on 2.5 and downstream slope 1 on 2. The height of the dam is 133 meters, the crest length of the dam is 360 meters with the crest elevation from 252.1 to 253 meters. The total volume of the dam is about seven million cubic meters including 715,000 cubic meters of clay core.

The saddle chute spillway is located to the east of the dam on the rightbank ridge. Its crest of 100 meters in length is con-

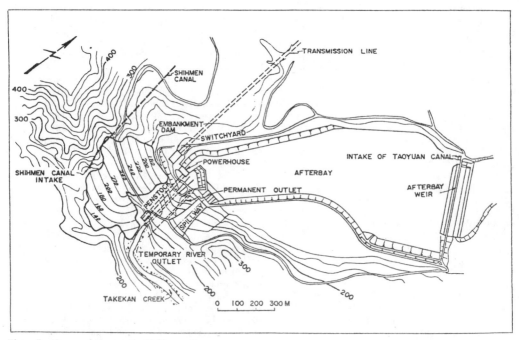

Fig. 7. General Layout, Shihmen Dam

## 5. Shihmen Dam

The Shihmen Dam located on the Tahan Creek, a tributary of the Tanshui River in north- trolled by six tainter gates, each 14 meters wide by 10.6 meters high, capable of discharging a maximum probable flood of 11,400 cubic meters per second.

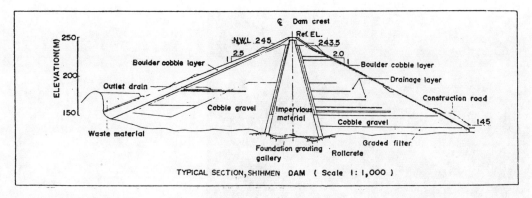

Fig. 8. Typical Section, Shihmen Dam

The reservoir has a gross capacity of 309.12 million cubic meters with the normal water level at elevation 245 meters and reservoir drawdown of 50 meters. The surface area of the reservoir at the normal water level is 8.15 square kilometers and its backwater extends 16.5 kelometers upstream.

The flow from reservoir after serving the two 45,000 kilowatts generating units in the power plant is released to the afterbay, located 1.5 kilometers downstream from the embankment dam. The afterbay consisting of an overflow ogee section and a rockfill portion has a capacity of 2.2 million cubic meters for regulating the water released from peaking generation to supply downstream irrigation uses. The irrigated areas of Shihmen Canal and Taoyuan Canal are 58,000 hectares.

6. Tsengwen Dam

The Tsengwen Dam is located on the midstream of the Tsengwen River in Southwestern Taiwan. It is a multipurpose development designed to supplement the irrigation water supply, to generate electrical power, to augment the supply of water for public consumption, and to protect the project area from floods. The construction was started in October 1967 and completed in October 1973 with a construction period of six years.

The Tsengwen Dam is an earth and rockfill structure, with a maximum height of about 133 meters and a crest length of 400 meters. The crest of the dam is set at elevation 235 meters which provides a freeboard of 2.7 meters above the maximum water level, and 12 meters above the normal water level. A roadway on the crest provides access to the river outlet and power intake structures.

The dam embankment consists principally of materials produced in excavating the spillway channel and structures, supplemented by minimum quantities of selected materials. The volume of the embankment is 9,296,100 cubic meters. The finer materials is compacted to form the impervious central zone, and the coarser materials placed downstream to provide a pervious rolled rockfill zone and downstream riprap. The finer materials are covered by a compacted silty clay membrane to assure watertightness. Sand, gravel, and boulders from the riverbed is placed at the upstream portion to constitute the pervious zone and upstream riprap. The embankment foundation of the impervious membrane and the central zone is excavated to the bedrock, and a grout curtain protects against excessive seepage underneath the dam.

The open channel spillway with gate-controlled crest, located on the right abutment, discharges into a concrete chute with flip bucket. The spillway sill is set at elevation 210 meters to be controlled by three 15x20.5 meters tainter gates. An emergency spillway is provided by utilization diversion tunnel No. 2 for emergency operation.

The reservoir basin is a valley with an average width of about two kilometers. The stream has cut deeply into the Miocene and the Pliocene foundations, which are chiefly sandstone, mudstone and shale. The reservoir has a length of 18 kilometers and a catchment of 481 square kilometers with the normal water level at elevation 225 meters. The gross storage is 707,530,000 cubic meters and the effective storage is 598,530,000 cubic meters with a reservoir drawdown of 60 meters.

Flow released from the Tsengwen Reservoir for irrigation and water supply pass through the powerplant to be equipped with one

Upstream View, Tsengwen Dam

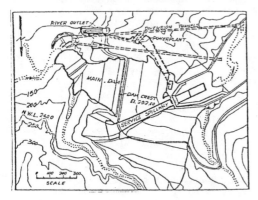

Fig. 9. General Layout, Tsengwen Dam

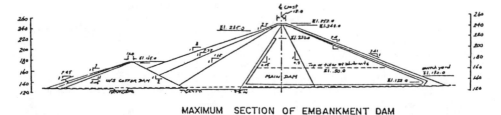

Fig. 10. Typical Section, Tsengwen Dam

## DAM UNDER CONSTRUCTION

Jonghua Dam

The Jonghua Dam site is in the U-shape gorge of Tahan Creek, about 700 meters away from the confluence of Tahan Creek and Piya-Chi, and 27 kilometers away from the Shihmen Dam.

This will be a double curvature and variable thickness concrete arch dam, stretching to both abutments along elevation 410 meters, 160 meters long, 82 meters high, dam radius is 80 meters, the lowest foundation 340 meters. The concrete volume of dam is 70,000 cubic meters. This dam project started on December 1, 1978 and is

50,000 kilowatts turbine generating unit, and then be regulated by the afterbay, six kilometers downstream from the dam. There, the flow is coveyed through the existing Tungkou tunnel to the existing Wushantou reservoir. The combined storage of the Tsengwen and Wushantou reservoirs is released for irrigation and water supply.

scheduled to be completed on May 31, 1982.

The effective storage capacity of the reservoir created by the dam is 7,610,000 cubic meters. The dam, an essential item of work under the Second Phase of Basin Soil and Water Conservation plan for Shihmen Reservoir, was originally designed to intercept the downward movement of silt from the upstream so as to minimize sedimentation in the Shihmen Reservoir. A

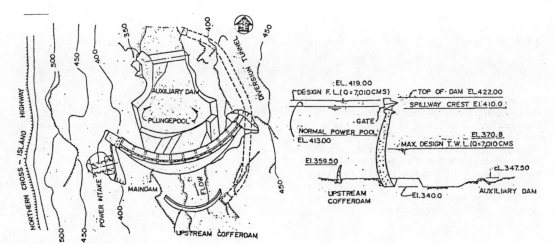

Fig. 11. Jonghua Dam, Plan

Fig. 12. Jonghua Dam, Profile

subsequent study of the project led to the conclusion that it would be adventageous and beneficial to utilize the water head to be derived from construction of the Jonghua Dam and the steep slope of the downstream meandering river channel from the damsite for power generation. Therefore, a power plant at Ihsing is added as a part of the Jonghua Dam Project.

Spillways and sluiceways are provided for release of flood water. Ten overflow spillways are provided with tainter gate to discharge 7,010 cubic meters per second which is a design flood discharge of 100 years' probability. A sluiceway will be installed at the left abutment to discharge 49 cubic meters per second at the normal power elevation of 413 meters. In order to dissipate the energy caused by water flowing over the spillways or through the sluiceways, a plunge pool is to be provided at the downstream of the main dam. Meanwhile, an auxiliary dam is to be constructed about 80 meters away from the downstream of the main dam. The auxiliary dam is a single curvature and variable thickness concrete arch dam and a total crest length of 48 meters, and can be used for overflow. At the downstream of the auxiliary dam will be also provided with plunge pool.

DAM UNDER DESIGN

Feitsui Dam

The Feitsui Dam site is on Peishih Creek, only about 30 kilometers away from Taipei City, safety is the most important factor of consideration.

This will be a double curvature and variable thickness concrete three-centered arch dam, 122.5 meters high and 510 meters crest length. The total storage capacity of the reservoir created by the dam is $406 \times 10^6$ cubic meters. This dam project is scheduled to be finished on June, 1986. The main purpose of the project is to provide the requirement of the water supply of the Greater Taipei Metropolitan Area.

Spillways and sluiceways are provided for release of design flood 9,870 cubic meters per second. Seven overflow spillways with overall width 78 meters are provided between Blocks 12 and 19 in the middle part of the dam crest. Below the spillways, eight sluiceways, each measuring 2.8 meters wide and 4.8 meters high, are to be provided. At the downstream eight fixed-wheel gates are to be installed as control devices for sluiceways. The maximum discharge capacity is 6,670 cubic meters per second for spillways, and 3,200 cubic meters per second for sluiceways. In order to dissipate the energy caused by water flowing over the spillways or through the sluiceways, an auxiliary dam is to be built downstream of the main dam. The middle part of the auxiliary dam shall be of the arch type with a crest elevation of 74 meters and a total crest length of 85 meters, and can be used for overflow. The other parts of the auxiliary dam shall be of the gravity type with a crest elevation of 87.5 meters.

Power intakes which located at elevations of 140 meters and 108 meters will release water of 54.2 cubic meters per second through penstock and powerhouse and generate 35,000 kilowatts peaking power. A river outlet located under sluiceway at No.

15 block of the dam will discharge water of 34 cubic meters per second to satisfy the needs of the various downstream water users.

CONCLUSION

Due to the young geology and narrow ravines, good reservoir sites are rare. Large storm floods add quite an extend to the cost of river diversion and spillway system. Sedimentation is another problem which shortens the life of reservoir and make the economic justification difficult. Good dam sites either have been developed or are being built. Not many sites are left for further development.

*Symposium on Problems and Practice of Dam Engineering / Bangkok / 1-15 December 1980*

# Dam engineering in Indonesia

SOERJONO
*State Electricity Corp., Jakarta, Indonesia*

## 1 INTRODUCTION

The Indonesian Archipelago, which is situated in the tropical zone, does not benefit much of its climatic condition from the hydrological point of view. With the existence of a dry season and a wet season, the rainfall depth is only concentrated in certain months of the year (November, December, January and February), thereby varying from region to region and from island to island.

Such a situation has caused people to look for means of storing up as much as possible water for use in the dry season when rain is scarce if not non-existent.
One answer to this problem is the construction of dams capable of storing up water in large quantities.
The availability of enough water throughout the year (quantitatively as well as qualitatively) will be a guarantee for a decent life on the face of Mother Earth.

In addition water as a renewable energy resource (i.e. for hydropower generation) forms an important factor in the diversification of energy, minimizing the dependability on one type of energy only i.e. natural oil.

## 2 GENERAL FEATURE OF DAM ENGINEERING IN INDONESIA

In history of dam development in Indonesia two distinct periods can be recognized :
 a. The period before 1942.
 b. The period after 1950.
Between 1942 - 1950 no dam development of any significance had been carried out because of the outbreak of World War II.

Dams constructed in the period before 1942 were of a single purpose, their only function being the formation of water reservoirs for the purpose of irrigation (single-purpose dam).
This was due to the climatic conditions in Indonesia, showing a very great difference in water depth between the dry season and and the wet/rainy season.
In addition the development of sugar cane plantations in the beginning of the 20th century required the existence of an irrigation system for the supply of enough water during the dry season.

Therefore the function of the "Waduk" (water reservoir, artificial lake) was far more significant so that the whole construction was called after the "waduk" rather than after dam itself (e.g. the Waduk Pacal dam height = 30 m, the Waduk Malahayu dam height = 28 m, etc.), see Figure 1a and 1b.
The dam itself was usualy of the earth fill type, using manual labour for its foundation excavation and filling (intensive labour force). All these dams are on the island of Java (see Table 1).

Water has also been utilized for hydropower generation in the period before 1942. Almost all the power plants were "run-off river" plants. There plants have been implemented without storing water is reservoirs, but by diversion weirs and having daily pondage only.

Concerning the construction of diversion weirs, we have the Pamarayan Weir in West-Java for irrigation built in the 17th century, another old weir, the Lehgkong Diversion Weirs in East-Java was built in 1857 for flood and irrigation control, in particular for the sugar cane plantation.
 The Lengkong weir was no longer exist but it has been replaced by a modern one. A part of the old weir was preserved for historical purposes.

Observation on pore pressure, settlement and sedimentation have been started ever since.
Some technical details will be presented below.

## JATILUHUR DAM

The largest among the rock fill type dams is the Jatiluhur Dam with a height of up to 100 m and a reservoir capacity of 3 billion m3, the construction of which was completed in 1967.
The Jatiluhur Dam has been erected on fairly homogeneous and solid agglomerate rocks, enabling it to be constructed with inclined core type whereby the part of alluvial deposit has been consolidated. The existence of a clay layer in certain places causes additional counterweight to be required at the downstream side of the dam, and the treatment by consolidation grouting at the approach tunnel side in order to prevent settlement from occuring.
Curtain grouting beneath the foundation (beneath the incline core) as carried out according to the Soletanche method has proved rather successfully.
The "morning glory" spillway structure, which becomes one body with the power station, is a real masterpiece of its designer.

## SELOREJO DAM

The Selorejo Dam with a height of 49 m and a reservoir capacity of 30 million m3 erected on an alluvial deposit in the form of sand & gravel ( thickness $\pm$ 20 m ), requires a special treatment for the formation of a water - tight layer (curtain grouting) at its foundation (See Figure - 3a).
The method of grouting used is the step grouting method using a mixture of cement-milk and bentonite in addtion to the cut-off of the earth-fill to a depth of 5.0. m at the curtain grouting. And at the left abutment foundation, of which the geological structure varies from agglomerate to tuff where the seepage occuring is rather extensive has been treated by earth blanket treatment in the upstream part and sand gravel counterweight in the downstream part.
Since cement-bentonite grouting can not reach the order of $10^{-5}$ cm/sec at the tuff layer part, additional chemical grouting has been applied to the said layer ($\pm$ 250 l/m), which has proved rather successfully.

## WLINGI DAM

To the Wlingi Dam ( zone earth-fill type), which is founded on alluvial deposit of 15 m thickness foundation treatment has been applied by way of cement grouting and chemical grouting in addition to making a cut-off of the earth-fill to a depth of 15 m (See Figure 2b).
Chemical grouting is necessary because of the occurrence of eruption material which, although being porous, is difficult for the cement to penetrate.
A big problem of seepage occurs at the left abutment part, which is of a limestone structure. Although grouting treatment has been applied (cement-chemical grouting) an increase in seepage can be observed from 10 m$^3$/minute to 60 m$^3$/minute), at the time the reservoir being impounded reaches the high water level (EL. 163.50 m).
By way of earth blanket treatment at the upstream part of the left abutment the seepage can be reduced to normal quantity ( 7 m$^3$/minute).
A detailed report concerning the foundation problem in the Wlingi Project has been submitted to the committee of this symposium.
The same problem can be observed at the Klampis Dam on Madura Island (concrete gravity type) experiencing an extensive seepage at its abutment after being impounded for some period.
Although the abutment has been covered with an earth blanket, as a result of the rather high water pressure seepage cannot be withheld.
The treatment to be applied is still under investigation.

## KARANGKATES & LAHOR DAM

Two other dams of the rock fill type, e.g. the Karangkates Dam and the Lahor Dam do not meet with too much foundation trouble (agglomerate and basalt layer at the river bottom of the Karangkates Dam).
However, since clay material to be used for the core does not meet the requirement (too fine, high water content, and low specific gravity), it should be mixed with weathered limestone the latter to be stocked up first in the form of a sandwich layer at the ratio of 70 % : 30 % in volume.
Whereas for rock material limestone rock is used for the Karangkates Dam, and a mixture of limestone and andesitic breccia for the Lahor Dam.
The Lahor reservoir harness the Karangkates reservoir by means of a connecting tunnel to be utilized for the Karangkates Power Station.

## RIAM KANAN DAM

With excellent fill material (sandy clay

All dams constructed before 1942 were only on the island of Java.

After 1950 large dams of up to 100 m height have been constructed with their multipurpose functions such as the Jatiluhur dam with 3 billon m3 and Karangkates dam with 300 million m3 storage capacity.

The construction of these dams were not only directed to the supply of enough water for irrigation purposes, but also for power generation, flood control, sediment control, etc, making an optimum use of the available potential.

Also after 1950, large dams were constructed outside of the island of Java.

As a result of the ever increasing population growth, which is concentrated on the island of Java (population density 850 - 1,000 persons/km2, according to the 1971 census), and the large extent of land reclamation for agricultural and horticultural purposes, the development undertaken is not only directed to the increase of food supply in greater quantities, but also to the improvement of the living standard of the population by way of the following :
  a. The establishment of a flood control system.
  b. The development of industries, using inexpensive energy.

The construction of dams is therefore not only directed to the supply of enough water for irrigation purposes (in particular during the dry season), but also to be utilized for power generation, flood control, sediment control, etc., making an optimum use of the available potential.

## 3 IMPLEMENTATION METHOD

In the period before 1942 the design of dams (generally for irrigation) were carried out by the Department of Public works, for the greater part through its Branch Offices subordinating the irrespective irrigation area, and occasionally through Central Office, whereas the implementation was supervised by the local Branch Office of Public Works. The annual budget was based on the volume of work be done in the respective fiscal year, and payments were effected by the Local Government Treasury Office.

Since 1950 the implementation of large multipurpose dams by the Government is done by the so called Project Organizations subordinated to the Directorate General of Water Resources Development and the State Electricity Corporation (PLN) the both agencies were subordinated to the Ministry of Public works.

In 1978 the P.L.N. was separated from the Ministry of Public Works and attached to the Ministry of Mining & Energy.
However, these two central agencies have ever since maintained a good cooperation in the implementation of the Projects.

The role played by foreign consultants and contractors in the construction implementation of large dams has been ever since a considerable one, due to lack of experienced local consultants and contractors recently been also involved in the construction implementation, either in joint operation with foreign firms or in a combined venture with foreign firms or in a combined venture among themselves.

In order to train the local staff in the design and construction of large dams in Indonesia, an on-force-account system, a "learning while doing" system, a process on the transfer of knowledge and technology has been adopted in a river basin development in East-Java, called the Brantas River Basin Development Project with the assistance of a foreign consultant. This Brantas Project is the first basin-wide development in Indonesia, started in the year 1962.

Transfer of kowledge and technology imposes a long and painstaking process.

In order to obtain the most effective result, each phase of the process of training , i.e. seeing or observing , learning, trying, and doing has to be seriously attended and experienced step by step, supported by a conductive management system.

Transfer of knowledge and technology certainly depends on the attitude of both, provider and recipient.

But by having a common objective, e.g. the construction of a dam or a river improvement project, each party has the opportunity to assess his progress, whereby the neat step in the process of training could be determined .

In essence, it is a huge experimental "laboratory", model scale 1 by 1, in which the participants venture themselves with a real size structure involing real problems.

This policy proved to be very effective in developing expertise, experienced engineers and skilled workers in a relatively short period.

With a gradually lesser role played by the foreign consultant, four multipurpose high dams, one big diversion dam, river improvement works, has been built and some other dam constructions and river improvement works are still being executed.
To illustrate : Total manpower in foreign consultant and construction guidance engineers required for the first project started in 1962 was more the 150,000 man-days

and at the peak of the construction 150 foreign engineers were working at the project site.
On the other hand, for the next fourth multipurpose dam which was started in 1974 and completed in 1978, about 10 foreign engineers only were put in, namely 12,000 man-days in total.
Further more, as a result of exchange of technology and mutual studies which the consultant have been always doing with the Brantas Project Engineers especially is relation to the project planning and design aspects, they have now reached a level as to be able to prepare "bankable" feasibility reports.

"Alumni" from the Brantas Project are now working on many big and important projects throughout the country, besides having important assignments.

It is hoped that in the near future the local staff will have full command on the konwhow of dam engineering and construction, to be further developed by themselves in accordance with the natural conditions prevailing in Indonesia.

As it has been stated before, that during the period before 1942 all the dams were built on the island of Java only. Because of the young geological condition it is not suitable for concrete dams, in addition to the unadequate availability of cement material at that moment.

The design of intake and flush-out structure was generally situated at the dam bottom ( see figure 1 ).
Therefore the dam foundation was always excavated down to the layer of hard soil having a fairly high bearing capacity and at the same time solving the problem of seepage in the foundation.

4. TECHNICAL PARTICULARS

4.1 Due to the successive transfer of power in Indonesia up to the outbreak of World War II, files containing documents on dam development in Indonesia in the period before 1942, were for the greater part being destroyed.
However, from several reports still surviving, the following may be concluded :
As stated in point 1, dams constructed during the period before 1942 were all situated on the island of Java.
Because of the rather young geological condition it is not suitable for the construction of concrete dams, in addition to the inavailability of cement material in an adequate quantity.
In addition the design of intake and flush-out structure was generally situated at the dam bottom (see figure 1).
Therefore the dam foundation was always being down to the layer of hard soil having a fairly high bearing-capacity and at the same time solving the problem of seepage in the foundation.

However, impermeability of the foundation was not always satisfactory.
In this case cut-off walls to a depth of 8 to 18 m was applied in addition to the earth blanket, spread on the upstream part to give more safety to the seepage potential.

Cement grouting techniques was still being known on a very limited basis before 1942 (as have been done to the foundation of Waduk Pacal in 1931). However, as stated earlier, by way of cement grouting the problem of seepage through the foundation could be overcome.

In some cases for this structure either concrete slab as a membrane on the upstream part of the dam, or a reinforcement in the form of a concrete wall was being used ( see figure 1a ).
The use of natural earth material in the design of dam body as an impervious part was being developed.
It is presumed that the leakage occuring after a certain time was caused by this type of structure meeting with damage (i.e. cracked by earthquakes, settlement, etc.), for concrete slabe & concrete wall.

Because of no experience of bulding dams on limestone foundations, in areas having a limestone geological structures, no dams had ever been constructed in that period (i.e. on the island of Madura, and in areas in the southern part of Java).
Almost all of dams, which were constructed earlier, did not use instruments for observation purpose. Only a small number of dams were equipped by benchmarks for the observation of dam settlement and standpipes to observe seepage line through a dam body.

Dam instrumentations required by the dam body (earth pressure, cross arm), were not apllied yet until the construction of the Jatiluhur Dam round about 1960.

4.2 Starting with the construction of Waduk Darma for irrigation (rock fill type), and the Ngebel Dam for hydropower generation (concrete buttress type), the construction of dams in the period after 1950 has been based upon modern techniques.
Modern and advance techniques has been applied in the construction of the Jatiluhur dam in West Java, the Karangkates dam and the Selorejo dam in East Java, and Riam Kanan dam on the island of Kalimantan.

with dry density of $\pm$ 2.7 ton/m3) and relative little foundation trouble, the Riam Kanan Dam (South Kalimantan) is the only dam constructed and completed on the island of Kalimantan of the perfect homogenous earth-fill type.

SEMPOR DAM

The Sempor Dam, also of the rock-fill type, as a result of overtopping of the cofferdam has undergone a modification from a rock-fill type dam with concrete slab membrane to a rock-fill type dam with center core. The foundation, consisting of andesitic, is of a terraced structure showing an inclination toward downstream.
To solve this problem curtain grouting is to be applied as well as providing a blanket in the upstream part and a relief well in the downstream part.

WONOGIRI DAM

A rock-fill dam still under consideration is the Wonogiri Dam (multi-purpose dam with a height of 40 m and a reservoir capacity of $\pm$ 560 million m$^3$) on a foundation consisting of tuff breccia formation and volcanic breccia groups.
The rock formation is not too solid but has the right bearing capacity for a fill-type dam. (See Figure 3b).
Cement curtain grouting is intended to improve the water tightness of the foundation with regard to unforseeable leakage, although the permeability of the foundation has already reached the order of $10^{-5}$ to $10^{-6}$ cm/sec.
As is the case with the Karangkates Dam, the need for core material at the Wonogiri Dam is met with the application of a mixture of clay (too fine material) and sand material from the river bed at the ratio of 40 % : 60 % by way of the "sandwich stockpile" system.
The incline core in the sub-dam is intended to minimize the mixture of material and the risk against foundation settlement for incline core not so serious since the sub-dam is only a temporary structure.

4.2. Damage to existing dams (in particular from the period before 1942) is usually brought about by foundation settlement as well as by cracks in the concrete structure, which is left unrepaired.
Whereas damage occuring in dams constructed during the period after 1950 is not of a significant nature as yet, with the exception leakage, making it necessary to be emptied for the necessary treatment.
The only dam failure occured at the coffer dam of Sempor.
This coffer dam forms part of the rock-fill type main dam with concrete slab membrane on the upstream part, which was overtopped as a result of the diversion tunnel not being capable of discharging then the coming flood.
In a later modification another and larger diversion tunnel was added.

4.3 Dam instrumentation required by the dam body (earth pressure, cross-arm) was not being applied until the construction of the Jatiluhur Dam.
It can be said that almost all of the dams, which were constructed earlier, did not use instruments for observation purposes. Only a small number of dams were only equipped by bench-marks for the observation of the dam settlement and standpipes to observe the seepage line through the dam body.
Observation has also been started on pore pressure, settlement and sedimentation in newly constructed dams (i.e. Jatiluhur, Selorejo, Karangkates, etc.).
By observing continuously the pore pressure meters, settlement by cross arm set in the dam body as well as the sedimentation through measuring the cross section of the reservoirs by using echo sounder (depth recorder).
The preliminary results of this sediment observation show an increase in sedimentation as compared with the figures in the original design.
In this more detailed investigation and observation will be required, in particular in the catchment areas.

4.4 Review of Dam Engineering Application on Present Dam Planning and Construction

A number of dams presently under construction concerns not only the fill-type, but the concrete type as well (concrete gravity or concrete arch type dams), (see table 2).
The Asahan Project which comprise a development scheme, in which a number of dams are simultaneously being constructed, has introduced a new era of dam development with the construction of arch type dams and underground power stations, thus enabling the utilization of hydropower for the production of much more energy. (See Figure - 4).
With the price of natural oil ever on the increase the role of water as renewable energy will be more and more significant.
As an outcome of this situation more and more dams are likely to be constructed with more advanced technology and more improved construction safety.

TABLE - 1   EXISTING DAMS

| No. | Name | Year of Completion | Type | Height (m) | Embankment Volume ($10^3$ m$^3$) | Purpose |
|---|---|---|---|---|---|---|
| (1) | (2) | (3) | (4) | (5) | (6) | (7) |
| | **EAST JAVA** | | | | | |
| 1. | Dawuhan | - | Earth fill with concrete membrane | 12.0 | - | I |
| 2. | Prijetan | 1916 | Earthfill with reinforced concrete core | 20.5 | 144 | I |
| 3. | Selorejo | 1973 | Earthfill | 49.0 | 2,000 | F, I, P |
| 4. | Karangkates | 1974 | Rockfill with vertical canter core | 100.0 | 6,150 | F, P |
| 5. | Wlingi | 1977 | Zoned type earthfill | 28.0 | 630 | I, P, S |
| 6. | Lahor | 1978 | Rockfill | 74.0 | 1,690 | F, P |
| | **CENTRAL JAVA** | | | | | |
| 1. | Nglangon | 1915 | Homogeneous earthfill | 15.5 | 74 | I |
| 2. | Tempuran | 1916 | ditto | 17.8 | 125 | I |
| 3. | Kedung Uling | 1917 | Earthfill | 10.0 | - | I |
| 4. | Plumbon | 1928 | Zoned type earthfill | 18.1 | 23 | I |
| 5. | Gunung Rowo | 1925 | Homogeneous earthfill | 20.5 | 85 | I |
| 6. | Delingan | 1923 | Zoned type earthfill | 27.0 | 300 | I |
| 7. | Pacal | 1933 | Rockfill | 38.0 | 90 | I |
| 8. | Gombong | 1933 | Hydraulic fill | 38.0 | 312 | I |
| 9. | Penjalin | 1934 | Homogeneous earthfill | 24.3 | 396 | I |
| 10. | Malahayu | 1940 | Earthfill with center core | 13.5 | 210 | I |
| 11. | Krisak | 1947 | Earthfill | 13.5 | - | I |
| 12. | Gebyar | 1955 | Earthfill with center core | 14.5 | - | I |
| 13. | Cacaban | 1959 | Homogeneous earthfill | 38.0 | - | I |
| 14. | Ngancar | 1966 | Rockfill, inclined core | 14.3 | - | I |
| 15. | Nawangan | 1975 | Earthfill | 26.0 | - | I |
| 16. | Parangjoho | 1979 | Homogeneous earthfill | 19.9 | - | I |
| 17. | Sempor | 1975 | Rockfill | 58.0 | - | F, I, P |
| | **WEST JAVA** | | | | | |
| 1. | Palajangan | 1924 | Homogeneous earthfill | 19.0 | 78 | I |
| 2. | Situpatok | 1927 | Earthfill with concrete membrane | 27.3 | 217 | I |

TABLE - 1 (CONTINUED)

| (1) | (2) | (3) | (4) | (5) | (6) | (7) |
|---|---|---|---|---|---|---|
| 3. | Cipanunjang | 1930 | Homogeneous earthfill | 32.0 | 182 | I |
| 4. | Cipancuh | -- | ditto | 13.7 | - | I |
| 5. | Darma | -- | Earthfill with concrete membrane | -- | - | I |
| 6. | Jatiluhur | 1967 | Rockfill with inclined core | 100.0 | 9,100 | F, I, P |
| | KALIMANTAN | | | | | |
| 1. | Riam Kanan | 1972 | Earthfill | 57.0 | 670 | P |
| | SOUTH EAST SULAWESI | | | | | |
| 1. | Larona | -- | Rockfill | 32.0 | 750 | P |

LEGEND :

F = Flood Control
I = Irrigation
P = Power Generation
S = Sediment control

TABLE - 2  FUTURE DAMS

| No. | Name | Stage | Type | Height (m) | Embankment Volume ($10^3$ $m^3$) | Purpose |
|---|---|---|---|---|---|---|
| (1) | (2) | (3) | (4) | (5) | (6) | (7) |
| | EAST JAVA | | | | | |
| 1. | Sengguruh | UC | Rockfill | 31.0 | 399 | P |
| 2. | Widas | UC | Earthfill | 35.6 | 700 | I, P |
| 3. | Kesamben | FS | Rockfill | 31.0 | 70 | P |
| 4. | Bendo | FS | ditto | 80.5 | 2,110 | P, E |
| 5. | Badegan | FS | ditto | 60.5 | 7,750 | P, F |
| | CENTRAL JAVA | | | | | |
| 1. | Garung | UC | Concrete gravity | 37.0 | - | P |
| 2. | Wonogiri | UC | Rockfill | 42.0 | 870 | F, I, P |
| 3. | Maung | FS | Earth & Rockfill | 174.0 | 15,000 | I, P |
| 4. | Mrica | DD | ditto | 95.0 | 5,620 | P, F |
| 5. | Jipang | FS | Earthfill | 27.5 | 4,200 | P, F |
| 6. | Kedungombo | DD | Rockfill | 66.0 | 7,100 | P, F |
| 7. | Wadaslintang | DD | ditto | 121.0 | 7,180 | P, I, F |
| 8. | Glapan | FS | ditto | 28.0 | 2,008 | |
| 9. | Nglanji | FS | Earthfill | 40.0 | 2,897 | I, F |
| 10. | Ngrambat | FS | ditto | 41.0 | 1,283 | P, I, F |

TABLE - 2 (CONTINUED)

| (1) | (2) | (3) | (4) | (5) | (6) | (7) |
|---|---|---|---|---|---|---|
| 11. | Banjarejo | FS | Earth & Rockfill | 29.0 | 1,534 | I, F |
| 12. | Kedungwaru | FS | Rockfill | 22.5 | 497 | I |
| 13. | Bandungharjo | FS | ditto | 25.0 | 150 | I |
| | **WEST JAVA** | | | | | |
| 1. | Saguling | DD | Rockfill with center core | 100.0 | 2,959 | P |
| 2. | Jatigede | DD | Rockfill | 115.0 | 6,700 | P, F |
| 3. | Matenggong | FS | ditto | 125.0 | 5,600 | P, F |
| 4. | Binangun | FS | ditto | 30.0 | 2,200 | F |
| | **NORTH SUMATRA** | | | | | |
| 1. | Asahan | UC | Concrete gravity | 31.0 | 19 | P |
| 2. | Sigura-gura | UC | ditto | 48.5 | 38 | P |
| 3. | Tangga | UC | Concrete arch | 78.0 | 52 | P |
| | **SOUTH SUMATRA** | | | | | |
| 1. | Batutegi | DD | Rockfill | 12.0 | 760 | P |
| 2. | Segara midi | DD | ditto | 23.0 | - | I |
| 3. | Way Rarem | UC | Rockfill with center core | 31.0 | 1,100 | I |
| | **KALIMANTAN** | | | | | |
| 1. | Riam Kiwa | FS | Earthfill | 55.0 | 1,000 | P |
| 2. | | | | | | |
| | **B A L I** | | | | | |
| 1. | Ayung | - | -- | - | - | P |
| | **SOUTH SULAWESI** | | | | | |
| 1. | Sadang | DD | Concrete gravity | 15.0 | 20 | P |
| | **NORTH SULAWESI** | | | | | |
| 1. | Tanggari I | FS | -- | - | - | P |
| 2. | Tanggari II | FS | -- | - | - | P |
| 3. | Sawangan | FS | -- | - | - | P |
| | **IRIAN JAYA** | | | | | |
| 1. | Sentani | FS | -- | - | - | P |

LEGEND :

FS = Feasibility Study  
DD = Detail design  
UC = Under construction  

P = Power generation  
I = Irrigation  
F = Flood control  
S = Sediment control

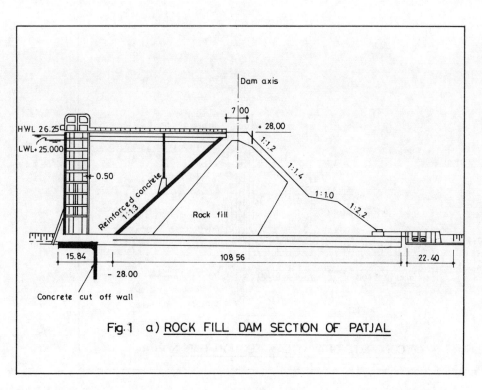

Fig. 1 a) ROCK FILL DAM SECTION OF PATJAL

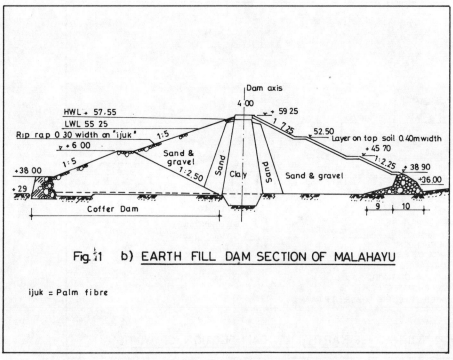

Fig. 1 b) EARTH FILL DAM SECTION OF MALAHAYU

ijuk = Palm fibre

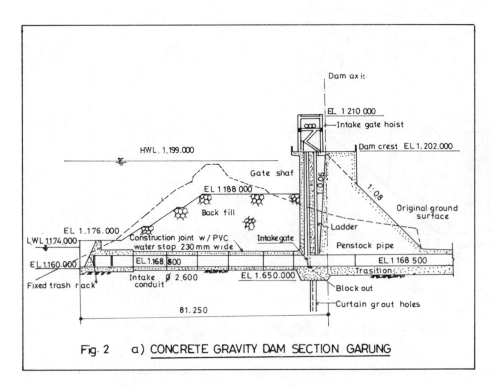

Fig. 2  a) CONCRETE GRAVITY DAM SECTION GARUNG

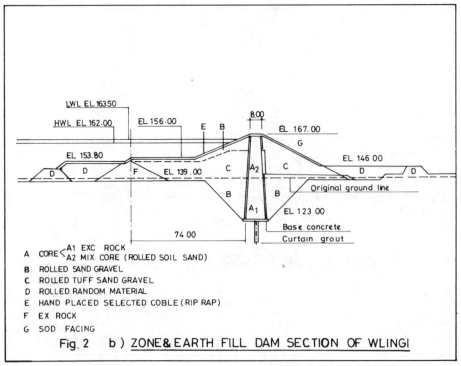

A CORE ⟨ A1 EXC ROCK
       A2 MIX CORE (ROLLED SOIL SAND)
B  ROLLED SAND GRAVEL
C  ROLLED TUFF SAND GRAVEL
D  ROLLED RANDOM MATERIAL
E  HAND PLACED SELECTED COBLE (RIP RAP)
F  EX ROCK
G  SOD FACING

Fig. 2  b) ZONE & EARTH FILL DAM SECTION OF WLINGI

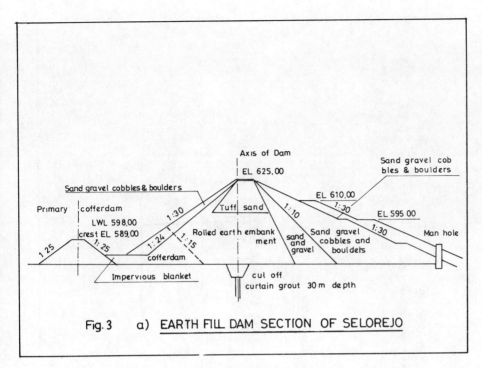

Fig. 3    a) EARTH FILL DAM SECTION OF SELOREJO

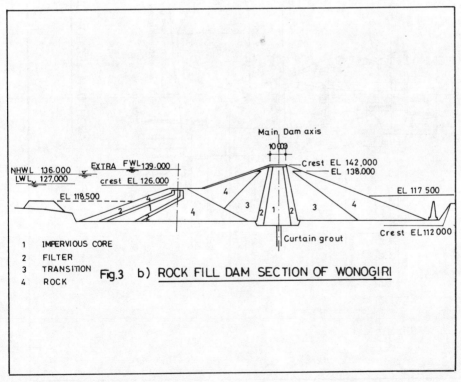

1  IMPERVIOUS CORE
2  FILTER
3  TRANSITION
4  ROCK

Fig. 3    b) ROCK FILL DAM SECTION OF WONOGIRI

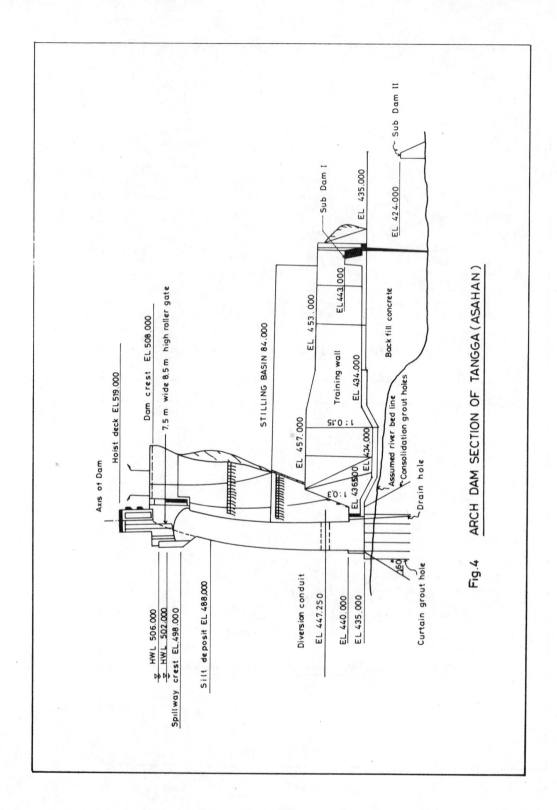

Fig.4  ARCH DAM SECTION OF TANGGA (ASAHAN)

# Dam engineering in Singapore and west Malaysia, 1967-1979

HOOI KAH HUNG
*Binnie dan Rakan Malaysia, Kuala Lumpur*

## INTRODUCTION

In this brief country report I have chosen 5 dams designed and constructed in the Malaysia/Singapore region during the last 13 years. In the time available I have presented the basic features of each of these dams and highlighting some of the data presented with a view for comments from the floor.

Certain aspects are intentionally left out to avoid time consuming discussions but some basic techniques of dam building are included to generate interest in any forgoing discussion. Thus, design methods and soil mechanics aspects are omitted except for an indication of the instrumentation adopted at each of the sites. In all the cases time was an important consideration in the final selection and adoption of the specified materials to be used. The dams were constructed by contractors on the basis of tenders and no serious claims have arisen out of the contracts through a proper understanding of the soils encountered during construction.

The techniques of dam building have been noted to be sometimes based more on improvisation than sophisticated or imported machinery. The economics and method of excavation and filling in earth dams will determine the type of plant to be used. The local method of excavation by face shovel and transporting by locally assembled lorry tippers seem to be ideal in our area and will be used for a long time yet. In fact the availability and abundance of such equipment enables the rate of construction be increased manifold or regulated when the need arises.

## SELETAR DAM, SINGAPORE 1967-1969

| | |
|---|---|
| Height of dam | 95 feet. |
| Volume of fill | 1.1 million yd$^3$. |
| Excavation | 20-30 feet of alluvium. |
| Foundations | 50-100 feet of insitu decomposed granites of soft silty clays. |
| Core cut-off | No special provision. Rolled clay core selected from slope-wash materials or colluvium. |
| Dam shoulders | Coarser materials beneath the colluvium. |
| Borrow area | 0-15 feet of colluvium and/or insitu decomposed granites. |
| Control of fill | Core cutter and sand displacement methods. Filling in specified thicknesses 6-12 in. to 95% of maximum standard proctor compaction dry |

| | | | |
|---|---|---|---|
| Control of fill continued | densities. Moisture contents controlled to P+8% to allow flexibility during construction. | Core cut-off | Slurry trench to depths up to 75 feet after establishment of initial platform. Rolled clay core selected from slopewash materials or colluvium. |
| Instrumentation | Hydraulic piezometers individual mercury gauge, USBR type settlement gauges, total pressure cells, vibrating wire, surface deflection points. | Dam shoulders | Coarser materials beneath the colluvium. |
| | | Borrow area | 0-25 feet of colluvium and/or insitu decomposed granites. |
| Plant aspects during construction | | Control of fill | Core cutter and sand displacement methods, filling in specified thicknesses 6 - 9 in. to 95% of maximum standard proctor compaction dry densities or 100% of proctor effort. Moisture contents controlled to P+8% to allow maximum flexibility during construction. |
| Excavation | Excavation by draglines, dumping by lorry tippers/dump trucks, spreading by dozers using LGP tracks. | | |
| Filling | Excavation by face shovels transporting by lorry tippers/dump trucks, spreading by dozers and compacting by a 75 ton towed rubber tyed rollers. Supplementary excavation and transporting by scrapers was noted to be just as suitable depending on the availability of supporting dozer plant. Rate of filling limited to approximately 10 feet per month particularly within the corefill. | | |
| | | Instrumentation | Hydraulic piezometers with transduser unit, USBR type settlement gauges, total pressure cells (hydraulic), inclinometers, surface deflection points. |
| | | Plant aspects during construction | |
| UPPER PEIRCE DAM, SINGAPORE 1971-1974 | | Excavation | Excavation by hydraulic sluice and pumping was a feature using a well established system of mining excavation, supplemented by dragline excavation necessary. |
| Height of dam | 100 feet. | | |
| Volume of fill | 1.5 million yd$^3$. | | |
| Excavation | 5-10 feet of alluvium excavated by hydraulic means. | | |
| | | Filling | Excavation by face shovels, transporting by lorry tippers, spreading by dozers and compacting by |
| Foundations | 10-20 feet of alluvium and 10-60 feet of insitu decomposed granites. | | |

| | |
|---|---|
| Filling continued | rubber tyred rollers. Other methods not attempted in view of minimum cost factor. Rate of filling limited to approximately 10 feet per month. |

## DURIAN TUNGGAL DAM, MALACCA, MALAYSIA 1973-1976

| | |
|---|---|
| Height of dam | 93 feet. |
| Volume of fill | ½ million yd$^3$. |
| Excavation | 5-20 feet of alluvium. |
| Foundations | Decomposed MICA schist. |
| Core cut-off | No special provisions except for slush grouting in isolated areas of the core trench. Rolled clay selected from top layers of slopewash which includes subsidiary elements of limonites. |
| Dam shoulders | Decomposed MICA schist. |
| Borrow areas | 0-25 feet of decomposed MICA schist including top layers of slopewash materials. |
| Control of fill | Core cutter and sand displacement methods. Filling in specified thicknesses 6 - 9 in. to 95% of maximum standard proctor compaction or 100% of proctor effort. Moisture content controlled to P+8% to allow maximum flexibility during construction. |
| Instrumentation | Hydraulic piezometers with common pressure gauge, standpipe piezometers, surface deflection points. |

### Plant aspects during construction

| | |
|---|---|
| Excavation | Excavation by draglines, dumping by lorry tippers, spreading by dozers using LGP tracks. |
| Filling | Excavation by face shovels, transporting by lorry tippers, spreading by dozers and compacting by a combination of vibration roller and/or rubber tyred roller. The latter is preferred within the core area. Rate of filling limited to approximately 12 feet per month. |

## SUNGEI LEBAM DAM, JOHORE, MALAYSIA 1974-1978

| | |
|---|---|
| Height of dam | 45 feet to be raised to 55 feet in future. |
| Volume of fill | 300,000 yd$^3$. |
| Excavation | 5-10 feet of alluvium. |
| Foundations | 20 - 40 feet alluvium and 20 - 80 feet of insitu decomposed granites. |
| Core cut-off | Interlocking steel sheet piles up to 60 feet lengths after establishment of initial platform. Rolled clay core selected from slopewash materials or colluvium. |
| Dam shoulders | Coarser materials beneath the colluvium. |
| Borrow areas | 0 - 15 feet of colluvium and/or insitu decomposed granites. |

| | | | |
|---|---|---|---|
| Control of fill | Core cutter method filling in specified thicknesses 6 - 9 in. to 95% of maximum standard proctor compaction dry densities or 100% of proctor effort. Moisture contents controlled to P+8% to allow maximum flexibility during construction. | Dam shoulders | Coarser materials beneath the slopewash materials. Close monitoring of the pore pressures during construction enabled an appreciable increase on the spacing of the sand blankets. |
| | | Borrow areas | 0-40 feet of colluvium and/or decomposed granites. |
| Instrumentation | Standpipe piezometers. | Control of fill | Core cutter and sand displacement methods. Filling in specified thicknesses 6 - 12 in. to 95% of maximum standard proctor compaction dry densities. Moisture contents controlled to P+8% to allow maximum flexibility during construction. |

Plant aspects during construction

| | |
|---|---|
| Excavation | Excavation by dragline, dumping by lorry tippers and spreading of dozers with LGP tracks. |
| Filling | Excavation by face shovels, transporting by lorry tippers, spreading by dozers and compacting by a self propelled rubber tyred roller. |

## SUNGEI LANGAT DAM, KUALA LUMPUR 1976 - 1979

| | | | |
|---|---|---|---|
| Height | 196 feet. | Instrumentation | Hydraulic piezometers individual mercury gauges USBR type settlement gauges. Total pressure cells hydraulic, inclinometers, standpipe piezometers abutments, surface deflection points. |
| Volume of fill | 3.2 million yd$^3$. | | |
| Excavation | 5 - 45 feet of alluvium. | | |

Plant aspects during construction

| | | | |
|---|---|---|---|
| Foundations: Core | Insitu decomposed granites of varying depths. | Excavation | Excavation by backactors, dumping by lorry tippers/dumpers spreading by dozers. |
| Shoulders | Alluvium consisting of boulders and gravels in a matrix of clayey silts and sands. | Filling | Excavation by face shovels, transporting by lorry tippers/dump trucks, spreading by dozers and compacting by a dynapac vibrating rollers or self propelled rubber tyred rollers 35 tons minimum. |
| Core cut-off | Curtain grouting maximum of 5 rows. Rolled clay core selected from slopewash materials or colluvium. | | |

# Dam engineering in Korea

S.K.KIM
*Dongguk University, Seoul, Korea*

W.T.KIM
*Industrial Sites & Water Resources Development Corp., Seoul, Korea*

## 1 INTRODUCTION

Small dams for irrigation purpose have been constructed since old times in Korea. As this nation is growing up to a developing country, a large amount of water is needed to meet the demand for industry as well as agriculture, and thus construction of large multi-purpose dams becomes inevitable.

In the 1970's, two multi-purpose high dams, Soyanggang Dam and Andong Dam were completed; they have drawn much interest of dam engineers by the highest one in Korea for the former and the use of highly weathered granite for the latter.

Described herein is the geotechnical behavior of two dams from the start of construction to the present. Prior to the description for the dams the integrated development of major river basins and some problems of dam construction in the basins are presented.

## 2 PROBLEMS OF DAM CONSTRUCTION IN MAJOR RIVER BASINS

There are the four major rivers in the southern part of the Korean Peninsula, three of which flow into Yellow Sea and the other Southern Sea. Big cities, towns, villages are concentrated along the rivers, and they have been experienced flood and drought occasionally.

The Korean government established the integrated development plan for the four major river basins. Dam construction for the purpose of flood control, power generation, and water supply for industry and agriculture is based on this plan.

Fig. 1 shows the four basins and locations of major dams in Korea. The problems for dam construction in each basin are described in some detail below.

### 2.1 Han River Basin

Commercial and industrial cities such as Seoul and Incheon are located downstream of Han River, and therefore water demand in this area has been rapidly increasing as the population increases. The dense population has also been caused pollution of the stream. The solution of such problems is to construct multi-purpose dams for reduction of pollution, flood control, power generation as well as water supply for irrigation. Many large dams had already constructed for these purposes, and Soyanggang multi-purpose dam recently constructed is one of earth and earth-rock dams in Korea. Another multi-purpose concrete dam, Chungju Dam is under construction as a line of the integrated plan of Han River Basin development.

### 2.2 Nakdong River Basin

The second largest city, Busan, and the third one, Daegu, are belong to this basin. Water shortage and gentle geography in this area make difficult to find out proper dam site for the construction of large dams.

Andong multi-purpose dam constructed at upstream of Nakdong River in 1976 is unique by the use of highly weathered granite instead of the usual clay core.

On the other hand, it is necessary to construct a barrage at the estuary of Nakdong River because of the intrusion of salt water during dry seasons. By doing this it is expected that the problem of unsufficient water supply for Busan and its vicinity may be solved.

### 2.3 Geum River Basin

A large part of this basin covers the plain located in the western part of the Korean

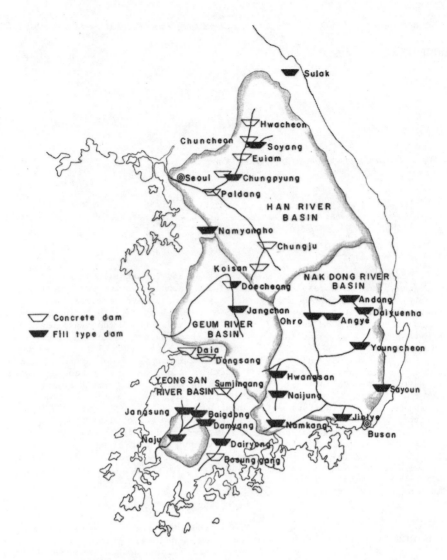

Fig. 1 Concrete and fill type dams in Korea

Peninsula, and thus water is needed mainly for irrigation. During the dry season irrigation by means of pumping is impossible since tidal difference is so high and salt water intrudes into inland. Therefore, the construction of tide embankments is necessary for land expansion as well as water supply.

Major cities located in this basin are Daejeon and Cheongju. Daecheong multi-purpose dam located between those two cities, currently under construction, will solve those pending problems encountered in the Geum River Basin. This dam is characterized by a combination concrete and earth-rock structure.

2.4 Yeongsan River Basin

The length of Yeongsan River is rather short, and thus damage due to drought and flood had been serious every year along the riverside. The water shortage at this basin was solved to some extent by the recent construction of Jangseong Dam and Damyang Dam.

The intrusion problem of salt water exists also in this basin, and thus the construction of the estuary barrage will be inevitable to prevent from intrusion of salt water at the downstream area and also to meet growing demand for water of various purpose.

## 3 GEOTECHNICAL BEHAVIOR OF TWO MAJOR ROCK-FILL DAM

Geotechnical behavior of two major dams, 123 meter-high Soyanggang Dam and 83 meter-high Andong Dam are described.

### 3.1 Soyanggang Dam

#### 3.1.1 General description

Soyanggang multi-purpose dam was planned as a part of the integrated development plan for Han River Basin to control natural river run-off by the construction of a zone-fill dam across a narrow valley in the Soyang River. The dam is located 12 km northeast of Chuncheon, Gangweon Province.

With the installed capacity of 200,000 KW this dam generates a substantial amount of electric power, supplies municipal, industrial, and irrigation water, and also significantly reduces flood damages at downstream of Han River, Seoul and its vicinity. The construction of this dam was started in 1967 and completed in 1973.

#### 3.1.2 Dam section and materials used

Typical section of the dam, which is zoned into five different grade of soil and rock is shown in Fig. 2. The core material of the dam is classified into SC in terms of Unified Soil Classification System, and the inner portion and the outer portion of shell are fine grain rock and coarse grain rock respectively. Sand mixed with fine gravel was used as a filter material.

It is noted from the dam section that the outer shell and core of the dam are connected with a gravel layer horizontally in order to drain easily in sudden drawdown case.

### 3.2 Andong Dam

#### 3.2.1 General description

Andong Dam was constructed as a part of the integrated development plan for four major river basins at the upstream of the Nakdong River, about 4 km northest of Andong city, Gyeongsan Bukdo Province.

With an installed capacity of 90,000 KW, this dam produces hydro-electric energy, controls chronical floods at downstream area, and supplies waters for municipal, industrial and agricultural uses to big

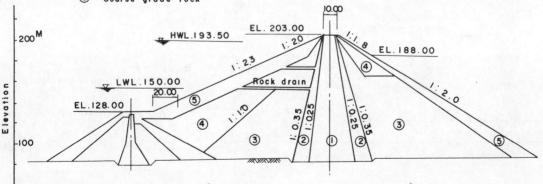

Fig. 2 Maximum cross section of Soyanggang Dam

cities such as Daegu and Busan, rural areas, and also to industrial complexes.

Equipped with water pumping facilities for the first time in Korea, the dam started to constructed in 1971 and completed in 1976.

### 3.2.2 Dam section and materials used

Maximum cross section of Andong Dam, which is classified into five different materials in grading, is shown in Fig. 3. The core material of the dam was originally designed to use clay, but it was changed to use highly weathered granite which was abundant near the dam site. Materials used for the inner portion of the shell was moderately weathered granite, and that for the outer part was rock.

### 3.3 Field measurements

In order to have an overall picture of the dam performance several kinds of instruments were installed for Soyanggang Dam and Andong Dam, and field measurements had been made since the start of construction.

### 3.3.1 Surface movement

Markers were installed on the dam surface to measure the horizontal and vertical displacement for both dams. For Soyanggang Dam one place on the downstream (EL. 167.00) was selected for such measurement, while for Andong Dam the measurements were taken at two places on downstream and one place on upstream. The measured results are shown in Table 1. As shown in the table, Soyanggang Dam which is 1.5 times as high as Andong Dam settled 6.0 times more and moved double downstream.

Table 1. Maximum surface displacement of Soyanggang Dam and Andong Dam

| Dam | Vert. displ. | | Hori. displ. | |
| --- | --- | --- | --- | --- |
| | During Const. | After Const. | During Const. | After Const. |
| Soyanggang* | 57 cm | 9 cm | 17 cm | Negligible |
| Andong** | 15 cm | 10 cm | 8 cm | Negligible |

\* Measured at 75% of dam height, as of June 1972
\*\* Measured at 55% of dam height

### 3.3.2 Pore pressure measurements

Piezometers were installed in shell zone as well as in core zone. For Soyanggang Dam excess pore water pressure was increased during construction with increasing dam height. The maximum ratio of developed pore water pressure to total pressure reached 72 percent at EL. 162.00 (70% H). However,

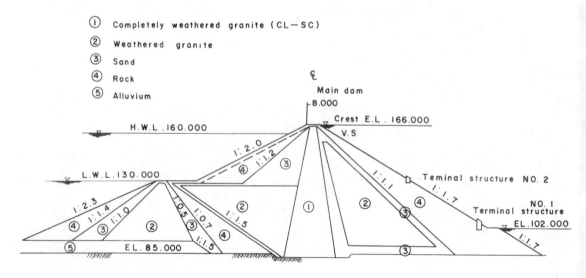

Fig. 3 Maximum cross section of Andong Dam (Sta. 48+4,000)

the developed pore pressure were dissipated rather rapidly through filter zone.

As Andong Dam was impounding water before finishing of the construction, the measured values of pore pressure were that developed due to seepage of water flowing from upstream. It was found from the measurements that head loss from upstream shell to core was negligible because of pervious shell zone.

### 3.3.3 Settlement of core zone

Cross arm type settlement gauges were installed to measure the settlement of core zones. Maximum settlements occurred at the central part of the core height. It is noted that the measured settlements are that of core depth laid below the gauge level and the settlements are due to surcharge filled over the gauge. Measured values of maximum settlements are given in Table 2.

Table 2 Maximum settlements of core zone

| Dam | Max. Settlement | |
|---|---|---|
| | During Const. | After Const. |
| Soyanggang* | 149 cm | - |
| Andong** | 37 cm | 16 cm |

\* Measured at 52% of dam height
\*\* Measured at 60% of dam height, as of Aug. 1980

### 3.3.4 Relative displacement

As a dam section is divided into several zones such as core, the inner portion of shell, and the outer portion of shell, it is anticipated that core zone may settle unevenly even if the materials are subjected to the same surcharge.

It is of interest to note from the measurements that the outer portion of the shell composed of fine grain rock settled more than core zone for both dams. In filling, shell zones were specified to compact over 60 percent of maximum relative density.

Settlements measured at EL. 167.00 (75% height) for Soyanggang Dam are shown in Table 3, and shown in Table 4 are settlements at EL. 102.00 (20% height) for Andong Dam.

### 3.3.5 Dam crest settlement

Markers were installed along the axis of the Andong Dam crest with the interval of 100 meters. Since 80 cm of crest settlements were estimated, this amount of height was

Table 3 Relative settlement for Soyanggang Dam in cm, measured at 75% of dam height at EL. 167.00 as of June 1972

| Core | | Sand & gravel | | Fine rock | | Coarse rock | |
|---|---|---|---|---|---|---|---|
| Dur. Cons. | Aft. Cons. | Dur. Cons. | Aft. Cons. | Dur. Cons. | Aft. Cons. | Dur. Cons. | Aft. Cons. |
| 55 | 27 | 7 | | 62 | 6 | 44 | 5 |

Table 4 Relative settlement for Andong Dam in cm, measured at 20% of dam height at EL. 102.00 as of Aug. 1980

| Core | | Sand | | Gravel | |
|---|---|---|---|---|---|
| During Const. | After Const. | During Const. | After Const. | During Const. | After Const. |
| 28 | 5 | 33 | 10 | 53 | 8 |

raised in advance over the original design height for Andong Dam. The measured settlement of dam crest as of August 1980, four years after construction was only 13 cm at the highest depth of the dam. The settlements decreased towards both sides of the dam since the height descreased. The ratio of settlement to different dam height was almost the same, 0.17 to 0.18 percent.

## 4 CONCLUSION

The integrated development plan for four major river basins was presented and some problems of dam construction in the basins were discussed briefly. Such pending problems would be solved by constructing tide barriers at estuaries and more dams at appropriate dam sites.

Two major rockfill dams constructed in recent years were 123 meter-high Soyanggang Dam and 83 meter-high Andong Dam. Instruments were installed to know the overall behavior of the dams, however, those of the former had not worked well because of some troubles of the instruments after the completion of dam construction.

For Soyanggang Dam the high excess pore water pressure was developed during construction, however, filters worked so well that the pressure dissipated rather rapidly from the dam core. The core zone settled 149 cm during construction of the dam at the central part of the zone.

For Andong Dam shell zones settled more than core zone during construction and this tendency kept the same after construction.

The core zone made of highly weathered granite instead of clay material did not show any geotechnical problems judging from the result obtained by the field measurements. The dam crest settled only 13 cm at the maximum section of the dam, the value of which is much low than estimated one in design.

5 REFERENCES

Industrial Sites and Water Resources Development Corp. 1973, Technical Report for Soyanggang Dam (Korean), Seoul.
Industrial Sites and Water Resources Development Corp. 1977, Technical Report for Andong Dam (Korean), Seoul.